AF334484

# Frontiers in Neurobiology 3

Series Editors:

**Anthony J. Turner (Leeds)**
**F. Anne Stephenson (London)**

FRONTIERS IN NEUROBIOLOGY 3

# Amino Acid Neurotransmission

Edited by
**Anthony J. Turner**
**F. Anne Stephenson**

PORTLAND PRESS
London and Miami

Published by Portland Press Ltd, 59 Portland Place, London W1N 3AJ, U.K.

In North America orders should be sent to Ashgate Publishing Co., Old Post Road, Brookfield, VT 05036-9704, U.S.A.

© 1998 Portland Press Ltd, London

**ISBN 1 85578 080 1    ISSN 1353-6508**

**British Library Cataloguing in Publication Data**
A catalogue record for this book is available from the British Library



Typeset by Portland Press Ltd
Printed in Great Britain by Cambridge University Press, Cambridge, U.K.

# Contents

# Preface

The amino acids L-glutamate, γ-aminobutyric acid (GABA) and glycine are all simple molecules, yet they play an essential role in the mammalian central nervous system in that they are the major mediators of excitatory (L-glutamate) and inhibitory (GABA and glycine) neurotransmission. The realization that these molecules have such an important role came from the pioneering work of Eugene Roberts in California in the 1950s for GABA, and that of David Curtis and colleagues in Australia for the excitatory amino acids and glycine in the late 1950s and early 1960s. There followed a period during which the role of these three amino acids as activators of fast-acting ion channels was substantiated, largely from electrophysiological studies in the late 1960s. Concurrently, it was established that these molecules did indeed satisfy the criteria established for a molecule to function as a neurotransmitter.

In the 1970s, the emphasis changed and it became the turn of the biochemist or, alternatively, of the emerging field of molecular pharmacology to study amino acid transmission by focusing on the receptors using the `new technique' of radioligand binding to study, as was thought, *the* glutamate receptor, *the* GABA receptor and *the* glycine receptor. At the same time, pharmacological evidence was accumulating to suggest the existence of receptor subtypes: the delineation of the $GABA_A$ and $GABA_B$ receptors following the description by Norman Bowery and colleagues of a bicuculline-insensitive but baclofen-sensitive GABA receptor in 1981, and the division of glutamate receptors into $N$-methyl-D-aspartate (NMDA) and non-NMDA receptors following the work by Jeff Watkins in Bristol. Further, the discovery by Joel Bockaert and colleagues of a novel glutamate receptor in 1985 which activated phosphatidylinositol turnover led to the term metabotropic receptor; the ligand-gated ion channels were now termed ionotropic receptors. $GABA_B$ receptors were also identified as metabotropic, whereas to date glycine receptors have only been found to function as ligand-gated ion channels. In the early 1980s, the first of these amino acid receptors was purified. The group of Heinrich Betz reported the isolation of the glycine receptor by strychnine affinity chromatography in 1982. Erwin Sigel and colleagues in Eric Barnard's laboratory in London exploited the fact that the anxiolytic drugs, the benzodiazepines, act via the $GABA_A$ receptor, and purified this protein to apparent homogeneity by benzodiazepine affinity chromatography in 1983. The molecular properties of glutamate receptors remained elusive. It was also during this period that the single-channel properties of the $GABA_A$ and glycine receptors began to be analysed following the advent of the patch-clamp technique first described by Erwin Neher and Bert Sakmann. Again, GABA$_A$ and glycine receptors were side by side when the first genes encoding these subunits were cloned in 1987 by the groups of Heinrich Betz (glycine) and Eric Barnard and Peter Seeburg (GABA). There followed a period of frenetic activity during which multiple genes encoding these subunits were identified, thus revealing an unexpected potential heterogeneity of receptors. It has been a major challenge to elucidate the actual subunit combinations that co-assemble to form functional receptors. Indeed, *in vivo* receptor subunit complements have still not been delineated. However, this potential heterogeneity renewed interest in these proteins from the pharmaceutical companies, several of which began searching for receptor-subtype-selective compounds. This is a particular goal of GABA$_A$ receptor pharmacologists, in order to find a non-sedative anxiolytic drug.

As a result of the advances made in the GABA$_A$ and glycine receptor field, it was widely predicted (but who would admit it now?) that the excitatory amino acid receptors would share many of the features common to these two proteins (fast-acting receptors) and to the G-protein-coupled superfamily (metabotropic receptors). However, progress was slow. It was not until the pioneering work of Michael Hollmann in 1989, at that time working in Stephen Heinemann's laboratory in San Diego, and of Shigetada Nakanishi in 1991 in Japan that the first genes encoding the ionotropic and metabotropic glutamate receptors were identified by expression cloning. Surprisingly these first two genes, *GluR1* and *mGluR1*, encoded relatively high molecular-mass proteins that shared little amino acid sequence identity with other members of the ligand-gated ion channel and G-protein-coupled receptor superfamilies, respectively. As for the GABA$_A$ and glycine receptors, multiple genes encoding homologues of both families were quickly identified. A novel discovery made by Peter Seeburg's group, and described for the first time in the brain, was the mechanism of RNA editing of certain ionotropic glutamate receptor genes.

Attention returned to amino acid neurotransmission as seen from the presynaptic site. Progress in understanding mechanisms and molecular structures of the GABA, glycine and glutamate transporters, and the mechanism of inactivation of these neurotransmitters, was due predominantly to the work of Baruch Kanner in Jerusalem. In the absence of appropriate high-affinity probes for the transporters, Kanner developed a laborious reconstitution assay to measure transporter activity in detergent solution. The availability of this method led to the purification and cloning of, firstly, the GABA transporter. Subsequently the other amino acid transporters were cloned, and again heterogeneity was revealed, although to a lesser extent than for the receptor molecules. Finally, and most recently, the molecular components involved in the release mechanisms of these neurotransmitters have been identified.

Thus this volume aims to cover the most recent developments in all the areas outlined above in order to provide an integrated view, comprising both pre- and post-synaptic events, of amino acid neurotransmission in the mammalian nervous system. Important areas covered include transmitter release mechanisms and properties of the transporter molecules. There are individual chapters devoted to the different receptors, including pharmacological, physiological and molecular aspects, as well as a chapter on the regulation of receptor activity by phosphorylation, a key modulatory mechanism. Finally, the roles of amino acid neurotransmission in long-term potentiation and in neurodegeneration are described.

Finally, we would like to thank all the authors for their excellent contributions to this volume.

**F. Anne Stephenson**, *London*
**Anthony J. Turner**, *Leeds*
1997

# Contributors


**Z.I. Bashir**  Department of Anatomy, School of Medical Sciences, University of Bristol, University Walk, Bristol BS8 1TD, U.K.

**C.-M. Becker**  Institut für Biochemie, Universität Erlangen-Nürnberg, Fahrstrasse 17, D-91054 Erlangen, Germany

**R. Bruton**  Department of Pharmacology, Medical School, University of Birmingham, Edgbaston, Birmingham B15 2TT, U.K.

**J.M. Henley**  Department of Anatomy, School of Medical Sciences, University of Bristol, University Walk, Bristol BS8 1TD, U.K.

**B.I. Kanner**  Department of Biochemistry, Hadassah Medical School, The Hebrew University, P.O. Box 12272, Jerusalem 91120, Israel

**D. Langosch**  Institut für Neurobiologie, Universität Heidelberg, Im Neuenheimer Feld 364, D-69120 Heidelberg, Germany

**D.G. Nicholls**  Department of Biochemistry, University of Dundee, Dundee DD1 4HN, U.K.

**A.M. Palmer**  Department of Neurochemistry, Cerebrus Ltd, Silwood Park, Buckhurst Road, Ascot, U.K.

**R.S. Petralla**  Section on Neurotransmitter Receptor Biology, Laboratory of Neurochemistry, NIDCD, NIH, Bethesda, MD 20892, U.S.A.

**T. Priestley**  Neuroscience Research Centre, Merck Sharp and Dohme Research Laboratories, Terlings Park, Eastwick Road, Harlow, Essex CM20 2QR, U.K.

**L.A. Raymond**  The Kinsmen Laboratory of Neurological Research, Department of Psychiatry, University of British Columbia, 2255 Wesbrook Mall, Vancouver, British Columbia, Canada V6T 1Z3

**S.A. Richmond**  Department of Pharmacology, Medical School, University of Birmingham, Edgbaston, Birmingham B15 2TT, U.K.

**J. Sánchez-Prieto**  Department of Biochemistry, Veterinary Faculty, Complutense University, Madrid, Spain

**T.G. Smart**  The School of Pharmacy, Department of Pharmacology, 29–39 Brunswick Square, London WC1N 1AX, U.K.

**F. Anne Stephenson**  School of Pharmacy, University of London, 29–39 Brunswick Square, London WC1N 1AX, U.K.

**R.J. Wenthold**  Section on Neurotransmitter Receptor Biology, Laboratory of Neurochemistry, NIDCD, NIH, Bethesda, MD 20892, U.S.A.

**P.J. Whiting**  Neuroscience Research Centre, Merck Sharp and Dohme Research Laboratories, Terlings Park, Eastwick Road, Harlow, Essex CM20 2QR, U.K.

# Abbreviations

| | |
|---|---|
| **AA** | arachidonic acid |
| **1S,3R-ACPD** | 1S,3R-aminocyclopentane dicarboxylic acid |
| **AIDS** | acquired immune deficiency syndrome |
| **AMPA** | $\alpha$-amino-3-hydroxy-5-methylisoxazolepropionate |
| **L-AP4** | L-2-amino-4-phosphonobutyrate |
| **AP5** | 2-amino-5-phosphonopentanoate |
| **AP7** | 2-amino-7-phosphonopentanoate |
| **4-AP** | 4-aminopyridine |
| **$[Ca^{2+}]_i$** | cytoplasmic free $Ca^{2+}$ concentration |
| **CAMK** | $Ca^{2+}$/calmodulin-dependent protein kinase |
| **L-CCG-I** | (2S,3S,4S) $\alpha$-(carboxycyclopropyl)glycine |
| **3C4HPG** | 3-carboxy-4-hydroxyphenylglycine |
| **4C3HPG** | 4-carboxy-3-hydroxyphenylglycine |
| **CNS** | central nervous system |
| **4CPG** | 4-carboxyphenylglycine |
| **CSF** | cerebrospinal fluid |
| **EAA** | excitatory amino acid |
| **$[EAA]_i$** | interstitial concentration of EAAs |
| **$E_{Cl}$** | $Cl^-$ equilibrium potential |
| **$E_m$** | membrane potential |
| **(m)EPSC** | current underlying a (miniature) EPSP |
| **EPSP** | excitatory post-synaptic potential |
| **GABA** | $\gamma$-aminobutyric acid |
| **$\gamma2_L$** | long isoform of GABA receptor $\gamma2$ subunit |
| **$\gamma2_S$** | short isoform of GABA receptor $\gamma2$ subunit |
| **GluR** | glutamate receptor |
| **$GlyR_N$** | neonatal glycine receptor isoform |
| **$GlyR_A$** | adult glycine receptor isoform |
| **HIV** | human immunodeficiency virus |
| **3HPG** | 3-hydroxyphenylglycine |
| **5-HT** | 5-hydroxytryptamine (serotonin) |
| **(s)IPSP** | (spontaneous) inhibitory post-synaptic potential |
| **KBP** | kainate binding protein |
| **LTD** | long-term depression |
| **LTP** | long-term potentiation |

| | |
|---|---|
| **MARCKS** | myristoylated alanine-rich C-kinase substrate |
| **MCPG** | $\alpha$-methyl-4-carboxyphenylglycine |
| **mGluR** | metabotropic glutamate receptor |
| **MPP$^+$** | 1-methyl-4-phenylpyridine |
| **MTS** | methanethiosulphonate |
| **NMDA** | $N$-methyl-D-aspartate |
| **PKA** | cAMP-dependent protein kinase |
| **PKC** | protein kinase C |
| **PMA** | phorbol 12-myristate 13-acetate |
| **$P_o$** | probability of channel opening |
| **$P_r$** | probability of release |
| **ROS** | reactive oxygen species |
| **SCC** | Schaffer collateral commissural |
| **sIPSC** | current underlying a sIPSP |
| **STP** | short-term potentiation |
| **TBI** | traumatic brain injury |

# Neurotransmitter release mechanisms

**David G. Nicholls*‡ and Jose Sánchez-Prieto†**

*Department of Biochemistry, University of Dundee, Dundee DD1 4HN, Scotland, U.K. and †Department of Biochemistry, Veterinary Faculty, Complutense University, Madrid, Spain

## Introduction

An understanding of the mechanisms by which the release of glutamate is regulated is central to much of current central nervous system (CNS) research. Firstly, this amino acid is responsible for the great majority of excitatory transmission in the mammalian CNS. Secondly, the patterns of glutamate release associated with the induction and expression of models of memory and learning, such as hippocampal long-term potentiation (LTP) in the hippocampal CA3–CA1 synapse and the mossy-fibre–CA3 synapse, remain controversial. Thirdly, it is of considerable therapeutic importance to establish the processes responsible for the massive excitotoxic release of the transmitter during ischaemia.

In this chapter we shall review the current state of knowledge of the factors that control the release of transmitter glutamate, drawing upon studies at the level of the isolated nerve terminal (or synaptosome) and the isolated cultured neuron, as well as electrophysiological studies with intact brain slices. This synthesis is particularly important since neurochemical and electrophysiological studies have tended to advance in parallel but separately, although the two are interdependent: the neurochemical investigations should employ physiologically relevant conditions, while the electrophysiological approach must encompass the biochemical mechanisms shown to be active in the simpler preparations.

## Dynamics of glutamatergic transmission

Much of our understanding of quantal transmission comes from studies on the neuromuscular junction, where an action potential causes the activation of large arrays of $Ca^{2+}$ channels and the virtually synchronous release of a few hundred synaptic vesicles [1–3]. At the much smaller CNS synapse the number of quanta released per action potential is much lower, and indeed a typical CNS glutamatergic

‡*To whom correspondence should be addressed.*

varicosity may show a mean quantal release of considerably less than unity [4], as estimated from the kinetics of blockade of transmission by the use-dependent *N*-methyl-D-aspartate (NMDA)-receptor inhibitor MK-801. This study suggested that two populations of synapses could be distinguished at the hippocampal CA3–CA1 synapse with release probabilities of 0.03 and 0.3 respectively, implying that the majority of action potentials will pass through a varicosity without releasing even a single synaptic vesicle.

The exocytosis of a single synaptic vesicle can elevate the concentration of the amino acid in the synaptic cleft to around 1 mM [5], thus activating the receptors in the synaptic cleft. The rate-limiting factor in the termination of transmission is not known for certain [6], but a likely candidate is $\alpha$-amino-3-hydroxy-5-methylisoxazolepropionate (AMPA)/kainate receptor desensitization [7], rather than diffusion of glutamate away from the synaptic cleft or re-uptake of glutamate into surrounding neurons and glia [8]. Many glutamatergic terminals are capable of firing at high frequency for prolonged periods of time [9,10].

The synaptosome is the simplest preparation which retains all the pathways for the synthesis, storage, release and re-uptake of transmitter glutamate, together with a range of presynaptic receptors capable of modulating the release of the transmitter. The disadvantages of the preparation are that large populations of terminals are required for individual experiments, so that, however closely defined the brain area from which they are prepared, the terminals will be heterogeneous in terms of their transmitter and precise locus of origin. It is fortunate that glutamatergic terminals appear to greatly outweigh those releasing other transmitters when synaptosomes from the cerebral cortex, hippocampus or cerebellum are prepared.

Glutamate exocytosis from synaptosomes can be monitored continuously by including glutamate dehydrogenase and $NADP^+$ in the incubation medium, the released glutamate being coupled to the generation of fluorescent NADPH [11,12]. KCl-evoked glutamate exocytosis is evoked optimally by 30 mM KCl and shows biphasic kinetics in response to prolonged depolarization: 1 nmol/mg (or 3% of the total glutamate in the preparation) is released within 1 s, whereas a further 4 nmol/mg (12%) is released much more slowly with a half-time of about 60 s [12]. Tetanus toxin and botulinum toxin serotypes A and B preferentially inhibit the slow phase of release, with little effect on the fast phase [13]. These *Clostridial* toxins specifically cleave synaptobrevin (tetanus toxin and botulinum toxin B [14]) and SNAP-25 (botulinum toxin A [15]), but are ineffective when the purified proteins are assembled with syntaxin [16]. This suggests that the fast phase of glutamate exocytosis observed with KCl-depolarized synaptosomes is consistent with the release of vesicles already docked at the active zones via an assembled SNARE complex, whereas the toxin-sensitive slow phase of release may be rate-limited by the transport of synaptic vesicles to the presynaptic docking site.

Suboptimal depolarization of a synaptosomal population by 10 mM KCl would be expected merely to slow the kinetics of exocytosis without affecting the ultimate extent of vesicle release. Instead, what is observed is a partial exocytosis which occurs with the same half-time as that following optimal depolarization [12]. The preparation behaves as though a specific population of vesicles becomes committed to release by this initial depolarization. If the depolarization is now increased by the addition of a further 20 mM KCl, the remaining vesicles are released [12,17]. While this might reflect a heterogeneity in the voltage-dependencies of the presynaptic $Ca^{2+}$ channels or a range of intra-synaptosomal $K^+$ concentrations leading to unequal depolarization in response to a given external $K^+$ concentration, a similar phenomenon can be seen at the level of the single varicosity when exocytosis from glutamatergic pyramidal neurons is monitored by the destaining of synaptic vesicles labelled by endocytotic incorporation of the fluorescent indicator FM1-43 [18].

With continued depolarization by 30 mM KCl, no further exocytosis can be detected after the slow phase has been completed (i.e. by about 3 min). It thus appears that the exocytotic pool has been exhausted and that this protocol only allows a single cycle of exocytosis to occur. No depletion of synaptic vesicle number has been reported in electron micrographs of KCl-exposed synaptosomes, and this is consistent with the rapid endocytotic retrieval of the exocytosed synaptic vesicles following KCl addition which is responsible for the incorporation of FM1-43 into neuronal varicosities following brief KCl depolarization [18]. If amino acid transmitter release from synaptosomes superfused with medium is monitored, repeated challenges with 30 mM KCl leads to repetitive exocytosis – the basis of the standard superfusion assay for monitoring the modulatory effects of pharmaco-logical agents. It is evident, therefore, that the exocytotic machinery must in some way 'reset' itself in the superfusion protocol but not in the case of prolonged KCl depolarization. Possible factors in the resetting might be a decrease in the internal cytoplasmic free $Ca^{2+}$ concentration ($[Ca^{2+}]_i$) (or a cessation of $Ca^{2+}$ entry), or a repolarization of the plasma membrane. In the synaptosomal preparation the second factor is important: temporary repolarization allows a second wave of KCl-evoked depolarization to be observed, whereas removal of external $Ca^{2+}$ followed by its re-addition to synaptosomes exposed to continued depolarization does not (H.T. McMahon and D.G. Nicholls, unpublished work). The nature of this voltage switch is unclear.

Both the fast and slow phases of KCl-evoked glutamate exocytosis are ATP-dependent [12,19], and any decrease in the presynaptic ATP/ADP ratio leads to a rapid inhibition of the extent of KCl-evoked exocytosis [20]. Similarly the addition of EGTA to chelate external $Ca^{2+}$ leads to an immediate cessation of exocytosis during either the fast or the slow phase [12].

The limitations of the KCl-depolarization protocol led many earlier workers to the view that the synaptosomal preparation was in some way damaged and unable to respond to subtle presynaptic regulatory mechanisms. The technique which has helped to revise this view has been the adoption of a second means of evoking synaptosomal excitation: $K^+$-A-type channel inhibition [17]. A-type $K^+$

channels are activated very rapidly in response to slight depolarizations and appear to serve as negative-feedback mechanisms preventing statistical fluctuations in membrane potential from reaching the magnitude at which spontaneous action potentials would fire, while not preventing authentic action potentials. Such channels appear to exist in nerve terminals sensitive to $\alpha$-dendrotoxin or to low concentrations of 4-aminopyridine (4-AP) [17,21–25]. It is notable that both dendrotoxin [26] and 4-AP [27] *in vivo* cause epileptiform seizures and cell death.

Addition of dendrotoxin or 4-AP to synaptosomes results in a $Ca^{2+}$-dependent exocytosis of glutamate from the same pool as that evoked by KCl [24]. As will be discussed below, the same $Ca^{2+}$ channels appear to be involved in both cases. 4-AP-stimulated glutamate release is strongly inhibited by tetrodotoxin [17], indicating that spontaneous action potentials are occurring in the isolated terminals. In addition, 4-AP-evoked glutamate release is, as will be discussed below, under the potent control of presynaptic mechanisms which control the intensity of presynaptic action potentials. As will be discussed in the remainder of this chapter, far from being a degenerate system the synaptosome is capable of showing a subtle interaction of facilitatory and inhibitory presynaptic regulation.

## $Ca^{2+}$ channels associated with glutamate exocytosis

The ability to obtain sustained glutamate exocytosis in response to a clamped KCl depolarization implies that the $Ca^{2+}$ channels coupled to glutamate exocytosis either must not undergo voltage inactivation, or else must possess a substantial non-inactivating component, allowing them to flicker open and closed with undiminished frequency even after several minutes of depolarization, since any rapidly inactivating channel would not contribute to the sustained release.

The $Ca^{2+}$ entry that triggers the release of transmitter glutamate can be monitored in synaptosomes either as the entry of $^{45}Ca^{2+}$ or as the elevation in $[Ca^{2+}]_i$, most usually by fura-2 [28]. These two methods monitor quite different parameters: the isotope follows the total entry of $Ca^{2+}$, which in synaptosomes may be sufficient to increase the total concentration by millimolar amounts, whereas fura-2 measures the volume average $[Ca^{2+}]_i$, which typically does not rise above 0.5 $\mu$M. The enormous discrepancy is due to the avid binding of $Ca^{2+}$ to intracellular anions, including proteins, phospholipids and nucleotides such as ATP.

The fura-2 signal in synaptosomes following a clamped KCl depolarization is biphasic, with an initial spike which then recovers to a plateau which is sustained for as long as the depolarization is prolonged [17,29]. The rate of change in the fura-2 signal at any time point is a reflection of the balance between $Ca^{2+}$ entry into the cytoplasm and $Ca^{2+}$ extrusion from the cytoplasm, largely across the plasma membrane in response to the $Ca^{2+}$-ATPase. Thus the recovery from the initial spike reflects the inactivation of a $Ca^{2+}$ channel following which $Ca^{2+}$

extrusion from the cytoplasm, predominantly by the $Ca^{2+}$-ATPase [12], can lower $[Ca^{2+}]_i$.

The plateau is attained at a value of $[Ca^{2+}]_i$ at which the activity of the $Ca^{2+}$ extrusion pathways matches the sustained rate of $Ca^{2+}$ entry through the opening of non-inactivating $Ca^{2+}$ channels, and typically occurs at an indicated bulk $[Ca^{2+}]_i$ of about 0.4 μM [17]. The $Ca^{2+}$-ATPase is strongly activated by an increase in $[Ca^{2+}]_i$; this is due not only to an increase in the concentration of its substrate (and a decrease in the thermodynamic gradient of the ion), but also to activation by $Ca^{2+}$/calmodulin, leading to a very sharp increase in activity as a function of $[Ca^{2+}]_i$. Any increase in $[Ca^{2+}]_i$ above the value at which uptake through the $Ca^{2+}$ channels balances the rate of efflux through the $Ca^{2+}$-ATPase leads to a decrease in $[Ca^{2+}]_i$, while conversely any decrease in $[Ca^{2+}]_i$ below this balance point leads to a net increase in $[Ca^{2+}]_i$ as influx exceeds efflux.

It should be noted that a high power relationship between the rate at which $Ca^{2+}$ is pumped out of the cytoplasm and $[Ca^{2+}]_i$ means that the steady-state elevation in $[Ca^{2+}]_i$ is not a linear measure of the rate of $Ca^{2+}$ entry: a cube relationship would mean that an 8-fold increase in the rate of $Ca^{2+}$ uptake into the terminal would only result in a doubling of the steady-state $[Ca^{2+}]_i$

The entry of $^{45}Ca$ into KCl-depolarized synaptosomes is strongly biphasic: a very rapid uptake declining rapidly to a slow sustained uptake [29]. That the rapid phase is due to a class of $Ca^{2+}$ channels undergoing rapid voltage inactivation is confirmed by pre-depolarization of synaptosomes (KCl depolarization in the absence of $Ca^{2+}$ followed by readdition of the cation). Under these conditions much of the rapid phase of $Ca^{2+}$ uptake is abolished [29] and the spike in the fura-2 response disappears [12]. Interestingly the net $Ca^{2+}$-dependent release of glutamate is unchanged by this procedure, allowing us to state that the $Ca^{2+}$ channel(s) in presynaptic nerve terminals coupled to the release of transmitter glutamate is coupled to a class of $Ca^{2+}$ channels that remain flickering open and closed throughout a prolonged depolarization.

No spike of $Ca^{2+}$ entry is seen when synaptosomes are depolarized by 4-AP [17]; indeed, the entry of $^{45}Ca^{2+}$ only corresponds to the non-inactivating slow phase of KCl-evoked entry. Thus the puzzling massive $^{45}Ca$ uptake in the initial 1 s of KCl depolarization is not activated by more closely physiological action potentials.

The release-coupled $Ca^{2+}$ channel in cerebrocortical synaptosomes has a high voltage threshold. Using $^{86}Rb$ to monitor the $K^+$ gradient across the membrane, it appears that the plasma membrane potential has to be decreased from −20 to −30 mV in order to stimulate release maximally [11]. When $Ca^{2+}$-dependent glutamate release from cerebellar granule cells is monitored, little release is seen with 30 mM KCl, and 50–70 mM is required for optimal release [30]. This KCl requirement is substantially higher than that required to activate nifedipine-sensitive L-channels in the same cells, and also eliminates any involvement of the low-threshold T-channels.

The pharmacology of the presynaptic glutamate-coupled $Ca^{2+}$ channels is a focus of considerable current research. Since the presynaptic terminal is too small for direct electrophysiology, and since any contribution of presynaptic channels to whole-cell currents measured by somatic patch–clamping is attenuated, indirect techniques have to be adopted, based on neurochemical approaches.

L-Channel agonists and antagonists are without effect on KCl-evoked glutamate release from synaptosomes [31–33]; this is consistent with their localization predominantly on cell soma and the axon hillock [34]. There is, however, one circumstance where L-channels appear to control glutamate release: when cerebellar granule cells are pre-depolarized in the absence of $Ca^{2+}$ and then high concentrations (typically 5 mM) of $Ca^{2+}$ are added back to the incubation [35,36], the majority of the $Ca^{2+}$-dependent glutamate release is found to be sensitive to L-channel antagonists. Glutamate release under these conditions can be regulated by adenosine, which activates an inhibitory receptor coupled via a pertussis-toxin-sensitive G-protein [35,36]. The mechanism underlying this apparent change of channel type (which is not seen in isolated nerve terminals) is currently unclear.

The funnel-web spider *Agelenopsis aperta* contains a multiplicity of $Ca^{2+}$-channel antagonists and one, termed Aga-GI, is effective against high-KCl-evoked glutamate release from both synaptosomes and granule cells [30,33]. Aga-GI blocks the very-high-threshold $Ca^{2+}$ channel which is coupled to glutamate release in granule cells depolarized conventionally by elevated KCl [30,36]. The responsive channels appear to be concentrated in neurite regions of the granule cells rather than on the cell body, and the concentration of KCl required to release glutamate correlates well with that required to activate the Aga-GI-sensitive channel [30]. N-channels do not appear to be coupled to glutamate release in this preparation [37].

The classic N-channel inhibitor is ω-conotoxin-GVIA, from *Conus geographicus*. ω-CgTx-GVIA is highly effective in inhibiting $^{45}Ca$ entry into chick brain synaptosomes [38–40] but is much less effective in the rat CNS, where KCl depolarization allows some inhibition of catecholamine release to be seen [41] but little inhibition of glutamate release [33].

Glutamate exocytosis from KCl-depolarized synaptosomes is thus predominantly linked to $Ca^{2+}$ entry through channels which do not fall into the classic categories of L, T or N. A partially characterized polyamine fraction from the venom of the funnel-web spider *Agelenopsis aperta*, termed FTx, inhibits the non-L non-N post-synaptic $Ca^{2+}$ channel on dendritic spines of cerebellar Purkinje cells [42]. These channels were termed P (for Purkinje). There has been an unfortunate tendency to expand the class of P-channels to encompass all high-threshold channels insensitive to L-channel inhibitors or to CgTx-GVIA, despite increasing evidence that this is a distinctly heterogeneous group. One toxin which is described as a P-channel inhibitor is Aga-IVA, a peptide toxin from the same spider [43]. Aga-IVA inhibits somatic and dendritic P-channels as defined by FTx, but has somewhat variable effects on presynaptic $Ca^{2+}$ channels [44–46]. The release of glutamate from cerebrocortical synaptosomes has been reported to be inhibited by Aga-IVA only over a narrow range of partial depolarization, with no effect on release evoked by 30–50 mM KCl [44].

It is difficult to quantify the effectiveness of these presynaptic $Ca^{2+}$-channel antagonists on the basis of their effect on the elevation of the plateau phase of the fura-2 response, since the response is non-linear with respect to the rate of $Ca^{2+}$ cycling, as discussed above. However, Aga-GI can lower the sustained plateau elevation of fura-2 following either KCl or 4-AP by some 50% at saturating concentrations. It is likely that this represents a greater than 50% inhibition of $Ca^{2+}$ entry, and complete inhibition of glutamate exocytosis is observed [33].

Since the exocytosis of glutamate commences within considerably less than 1 ms following a physiological action potential [47], and since a single terminal is capable of firing several times per second, the $Ca^{2+}$ signal within a terminal must be initiated and terminated with great rapidity. This essentially eliminates on two counts the possibility that bulk elevations in $[Ca^{2+}]_i$ are required. Firstly, the $Ca^{2+}$ trigger that initiates exocytosis must be located in the immediate vicinity of the voltage-activated $Ca^{2+}$ channel, particularly since the extensive binding of $Ca^{2+}$ to relatively immobile cytoplasmic components greatly slows the diffusion of free $Ca^{2+}$ within the cytoplasm. Equally important is a means for rapidly terminating exocytosis and resetting the process in time for the next action potential. This demands a fast rate of dissociation of $Ca^{2+}$ from the exocytotic trigger, which would be assisted by a low-affinity interaction between $Ca^{2+}$ and the trigger. A low-affinity interaction would be possible if the $Ca^{2+}$ trigger were to be located in a region of high localized $Ca^{2+}$ concentration in the immediate vicinity of the $Ca^{2+}$ channel itself. There is convincing evidence that such local 'hot-spots' exist. At the presynaptic terminal of the (glutamatergic) squid giant synapse, these hot-spots can be detected by injecting the terminal with the luminescent $Ca^{2+}$ indicator protein aequorin that has been genetically modified to decrease its affinity for $Ca^{2+}$ so that it responds to concentrations of the order of 100 μM [48], or by using fluorescent indicators [49]. In these studies, action potentials evoked highly localized arrays of elevated $Ca^{2+}$ immediately under the presynaptic plasma membrane [48].

Additional evidence for localized coupling in squid giant axons has involved injecting $Ca^{2+}$-channel chelators of various affinities and rate constants for binding. It was found that a fast-acting low-affinity chelator, a derivative of BAPTA [bis-($o$-aminophenoxy)ethane-$N,N,N',N'$-tetra-acetic acid], was more effective in inhibiting transmission than a much-higher-affinity, but considerably slower, chelator such as EGTA [50]. It was concluded that a high localized concentration triggered release, since this could be blocked by a low-affinity but rapidly acting chelator.

A uniquely large vertebrate glutamatergic terminal is found in the goldfish retina. Bipolar cells have a large (5 μm diameter) terminal which tonically releases glutamate from small synaptic vesicles. The isolated terminals can be patch–

clamped in whole-cell mode and the capacitance increase in response to exocytotic fusion of the vesicles with the plasma membrane determined, adapting techniques first exploited for secretory granules in mast and chromaffin cells [51,52]. Exocytosis can be initiated by a sudden elevation in $[Ca^{2+}]_i$ following photolysis of the $Ca^{2+}$ chelator DM-nitrophen. It was found that exocytosis (which was detectable within 1–2 ms) required optimal peak $Ca^{2+}$ concentrations of up to $100\mu$ M.

With this exception, the sizes of vertebrate CNS terminals are generally too close to the wavelength of light to permit any resolution in $Ca^{2+}$ imaging experiments. It is, however, possible to obtain indirect evidence for a high localized $[Ca^{2+}]_i$ by comparing the effectiveness of depolarization (which should create a local hot-spot immediately under the channel) with that of ionomycin (which should generate a uniform elevation in $[Ca^{2+}]_i$) [53]. When ionomycin was added to give exactly the same bulk average $Ca^{2+}$ signal as that following KCl or 4-AP, no release of glutamate occurred. Glutamate release was only observed when sufficient ionomycin was added to flood the terminal with $Ca^{2+}$. This is consistent with the need for a high local concentration of $Ca^{2+}$ in the immediate vicinity of the $Ca^{2+}$ channel to provide the trigger for amino acid release.

CNS terminals, with a typical quantal release of less than one per action potential, differ considerably from those at the neuromuscular junction or at the squid giant axon, where a single action potential may induce the exocytosis of more than 100 synaptic vesicles and where there is good evidence for an extended array of $Ca^{2+}$ channels which fire synchronously to form a two-dimensional 'sheet' of elevated $Ca^{2+}$ immediately under the membrane [48]. Such a mechanism is ideally suited to the neuromuscular junction, where an enormous muscle fibre must be rapidly and unequivocally depolarized in order to trigger its contraction, and where there is no scope for a partial contraction of a single fibre. In the case of the relatively tiny CNS glutamatergic terminal, in contrast, the dimensions would not appear to permit the generation of an inwardly directed wave of high $[Ca^{2+}]_i$. In any case, it is not apparent how such a mechanism, designed for all-or-none release at the neuromuscular junction, could be used to allow precise control over the probability of exocytosis.

A more likely mechanism in the CNS is that a release assembly consists of a single $Ca^{2+}$ channel, or a small array of closely associated channels with the associated docking site and protein machinery, operating essentially autonomously from other $Ca^{2+}$ channels in the vicinity. Investigations with presynaptic patching of the uniquely large cholinergic calyx-type terminal formed during development of the chick ciliary ganglion [54] are consistent with exocytosis in response to activation of a single $Ca^{2+}$ channel.

Control over the probability of exocytosis at a single $Ca^{2+}$-channel/exocytosis site could then be exerted by controlling the activity of the $Ca^{2+}$ channel either directly, perhaps via interaction with a receptor-controlled G-protein, or indirectly, by controlling the duration or intensity of the presynaptic action potential. Since $Ca^{2+}$-channel opening may passively follow the action potential, prolonging the potential would increase the probability of exocytosis occurring at this site. As we shall see, studies with synaptosomes provide evidence for both of these types of presynaptic regulation. However, we have not been able to obtain evidence for a third possible means of presynaptic regulation, in which the intra-terminal $Ca^{2+}$–secretion coupling would be directly modulated, perhaps by protein phosphorylation. Such direct regulation is well established for chromaffin cells [55–58], where protein kinase C (PKC) activation potentiates the exocytotic response to a given $Ca^{2+}$ elevation in permeabilized cells where there is no longer any possibility of channel modulation. This distinction emphasizes the importance of avoiding unsupported extrapolation between different exocytotic systems.

Some terminals can co-release glutamate together with a peptide neurotransmitter such as dynorphin [59] or substance P [60]. The same experiments which give evidence for localized $Ca^{2+}$ coupling for glutamate release suggest that a relatively small but delocalized elevation in bulk $[Ca^{2+}]_i$ is required to release neuropeptides [53], since the $Ca^{2+}$ ionophore ionomycin can effectively release neuropeptides while failing to release glutamate. Physiologically this implies that the well established frequency-dependent release of neuropeptides is due to the need for repeated action potentials to elevate the bulk $[Ca^{2+}]_i$.

## Vesicle traffic and the exocytotic mechanism

Synaptosomes contain a number of prominent phosphoproteins. These include the synapsins [61] responsive to $Ca^{2+}$/calmodulin-dependent protein kinases I (CAMKI) and II (CAMKII), and three PKC substrates: MARCKS (myristoylated alanine-rich C-kinase substrate) [62–65], B-50 (also known as GAP-43 and neuromodulin) [66,67] and dynamin (P96; dephosphin) [68,69].

Synapsins IA and IB are alternatively spliced variants that possess sites for CAMKII-dependent phosphorylation and undergo a rapid cycle of phosphorylation/dephosphorylation when synaptosomes are depolarized by 30 mM KCl [61]. The functional significance of this cycle has been intensively investigated. *In vitro*, synapsin I readily associates both with synaptic vesicles and with the cytoskeleton [70,71]. Phosphorylation by CAMKII decreases the affinity of this association. In addition, deep-etch freeze-fracture electron microscopy of presynaptic terminals reveals the presence of synapsin cross-links between synaptic vesicles and the cytoskeleton [72]. In the light of these findings it has been proposed that synapsin I phosphorylation plays a key role in releasing synaptic vesicles from their cytoskeletal attachment, making them available for transport to the active zones and eventual exocytosis [61]. Some evidence supports this view; thus injection of exogenous CAMKII into the squid giant synapse enhances transmission, while injection of dephosphosynapsin I is inhibitory [73].

Freeze–thawed synaptosomes allowed to incorporate CAMKII release more glutamate in response to subsequent depolarization [74], but the total extent of release is very low, suggesting that the treatment has compromised the integrity of the preparation. One approach which does not support the obligatory requirement for a calmodulin-mediated event during the KCl-evoked release of glutamate from the synaptosomal preparation is the finding that replacing $Ca^{2+}$ in the medium by $Ba^{2+}$ allows normal KCl-evoked exocytosis to be observed, even though $Ba^{2+}$, which does not activate calmodulin, does not allow a cycle of synapsin I phosphorylation and dephosphorylation to occur [75,76]. Two more lines of evidence which argue against the phosphorylation-dependent vesicle release hypothesis are, first, the observation that many terminals contain synapsin II (which lacks the CAMKII phosphorylation site) and no synapsin I, and, finally, that synapsin I knock-out mice show a remarkable absence of any dramatic neurological phenotype [77].

P96 (dephosphin) is a presynaptic PKC substrate that is dephosphorylated during KCl depolarization of synaptosomes [68,69]. It has recently become apparent [69] that the protein is identical to the microtubule-associated protein dynamin, which is altered in the temperature-sensitive endocytosis mutation *Shibere* in *Drosophila* [78]. Dephosphorylation of P96 is catalysed by calcineurin, and the phosphatase in turn can be inhibited by cyclosporin; however, inhibition of dephosphorylation has no detectable effect on KCl-evoked glutamate exocytosis from synaptosomes [79]. The functional significance of the changed phosphorylation state of dephosphin is thus not clear.

## Presynaptic receptors and glutamate exocytosis

Physiologically relevant presynaptic modulation could control the probability of vesicle exocytosis in response to an action potential. Additionally, the duration of the period during which glutamate is elevated in the synaptic cleft could be modulated by controlling the rate of re-uptake into the terminal and surrounding glia. Some of the investigations into the presynaptic mechanisms surrounding glutamate exocytosis can address these questions, and these will form the topic of the remainder of this chapter.

## PKC and the facilitation of glutamate exocytosis

Several isoforms of PKC have been detected in presynaptic nerve terminals [80,81]. PKC activation by phorbol esters has been generally observed to increase transmitter release from synaptosomes, brain slices and neuronal cultures (reviewed in [82]). From first principles there are two classes of mechanism that could be responsible for enhanced transmitter release: a modification of the properties of the presynaptic ion channels, resulting directly or indirectly in enhanced entry of $Ca^{2+}$ into the terminal, and facilitation of an intra-synaptosomal locus.

We were able to throw some light upon the mechanism of PKC potentiation when we observed [83] that the sensitivity of synaptosomal glutamate release to phorbol esters was far greater when release was evoked with 4-AP than with KCl. Thus 4-AP-evoked glutamate exocytosis can be modulated from maximal (when PKC is activated by phorbol esters) to essentially zero when PKC is fully inhibited by the PKC inhibitor Ro 31-8220 [64]. In total contrast, KCl-evoked glutamate release is entirely insensitive to phorbol ester or Ro 31-8220. There is one exception to this last statement, and that is when leaked endogenous adenosine is present and exerts a tonic inhibition of glutamate exocytosis; the ability of PKC-mediated pathways to suppress such tonic effects will be discussed below.

PKC activity in the intact synaptosome can be followed by monitoring the phosphorylation state of the characteristic 13 kDa peptide generated by *Staphlococcal* V8 protease treatment of the *in situ* PKC substrate MARCKS. Thus the failure of phorbol esters to modulate release in the absence of 4-AP is not due to inactivation of PKC, since MARCKS is still phosphorylated under these conditions. Since KCl causes a clamped depolarization of the plasma membrane and evokes glutamate release which is independent of PKC activity [64], this eliminates a locus for the kinase at the release-coupled $Ca^{2+}$ channel (which, as discussed above, is not inactivated during prolonged KCl depolarization) or downstream of $Ca^{2+}$ entry at an intra-synaptosomal locus. This conclusion differs from that obtained from studies on the exocytosis of large secretory granules from platelets, chromaffin cells and mast cells, where phorbol esters potentiate exocytosis in permeabilized preparations (e.g. [84]), and emphasizes that it is essential to avoid unsupported extrapolation from one exocytotic system to another.

If, by elimination, PKC must be acting upstream of the $Ca^{2+}$ channel, this strongly suggests that the kinase might be modulating the ion channels which are responsible for the spontaneous action potentials initiated by 4-AP. There are two possibilities: activation of the voltage-activated $Na^+$ channels which initiate the action potentials, or inhibition of the 4-AP-resistant $K^+$ channels which help to terminate the action potentials. While there are PKC phosphorylation sites on $Na^+$ channels which decrease peak conductance and slow inactivation [85], a more likely locus is the 4-AP-resistant $K^+$-channel (Fig. 1). Thus PKC-mediated inhibition of delayed rectifier-type $K^+$ channels by phorbol esters can be detected at the Purkinje cell soma [86], while $^{86}Rb$ fluxes across the synaptosomal plasma membrane are inhibited by diacylglycerol [87]. It should be noted that mammalian terminals possess rapidly activating delayed rectifier-type channels, permitting them to propagate short action potentials allowing firing at up to 200 Hz [88].

4-AP has limited pharmacological specificity, and delineation of the PKC-sensitive $K^+$ channel is easier if the high-affinity $K^+_A$-channel inhibitor α-dendrotoxin is substituted for 4-AP. Dendrotoxin evokes the same release of glutamate as does 4-AP and release can be potentiated to the same extent by phorbol esters [82]. This indicates that the putative $K^+$ channel regulated by PKC is dendrotoxin-resistant. The channel is, however, sensitive to 1 mM $Ba^{2+}$ (acting as a $K^+$-channel inhibitor rather than a secretogogue), since $Ba^{2+}$ mimics the effects of phorbol 12,13-dibutyrate in stimulating release, while the phorbol esters no longer potentiate in the presence of $Ba^{2+}$ [82].

**Fig. 1.**       **Schematic pathways involved in the presynaptic regulation of glutamate exocytosis from cerebrocortical nerve terminals**

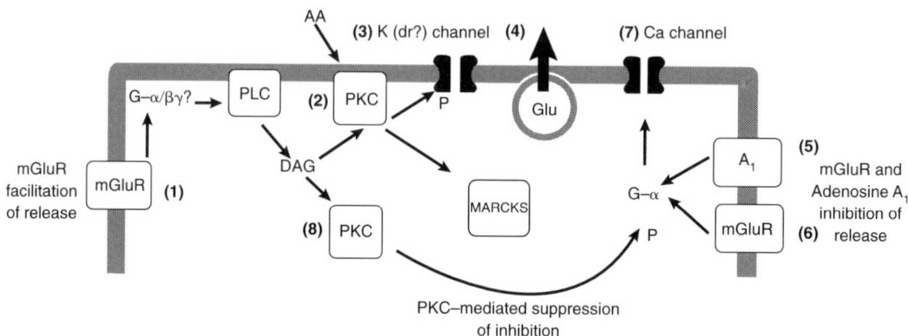


*The facilitation pathway is activated by mGluR agonists acting at a receptor (1) coupled to phospholipase C (PLC) via a G-protein. This interaction is sensitive to pertussis toxin and may be mediated by G-protein βγ subunits. Diacylglycerol (DAG) generated by the phospholipase C acts synergistically with exogenous AA to activate a PKC (2) which is able to phosphorylate MARCKS. Functional studies also indicate that the PKC enhances presynaptic action potentials, most plausibly by inactivating a delayed rectifier (dr) $K^+$ channel (3), and thus increases the release of glutamate (4). Inhibition of glutamate release is produced by activation of an adenosine $A_1$ receptor (5) and, in young rats, by a separate mGluR (6) coupled via a G-protein to the release-coupled $Ca^{2+}$ channel (7). In adult rats a PKC which does not require AA (8) suppresses the inhibitory action of adenosine. In young rats AA is required for this suppression, and the same PKC species that mediates facilitation may be involved.*

The locus for PKC action on glutamate release appears, therefore, to be a dendrotoxin-insensitive, $Ba^{2+}$-sensitive, $K^+$ channel (Fig. 1). It should be noted that action potential prolongation has been shown to be a highly effective way of enhancing transmission at the squid giant synapse [25]. Although action potential prolongation in the synaptosomal preparation cannot be directly observed, it would be predicted that the average depolarization of the synaptosomal preparation would be enhanced and that this would lead to an increased elevation of the bulk $[Ca^{2+}]_i$ reported by fura-2. Both are observed in practice [83].

The concentration of phorbol esters required to enhance 4-AP-evoked glutamate release (100 nM [89]) is rather high in relation to many other PKC-mediated events. However, the concentration required can be decreased by the parallel addition of 1–2 μM arachidonic acid (AA) [90]. This is consistent with the known properties of several isoforms of PKC, which are synergistically activated by diacylglycerol and AA [91]. PKC-$\alpha$, -$\beta$I and -$\epsilon$ are possible candidates, the last being particularly likely in view of its predominantly presynaptic location [92] and the apparent $Ca^{2+}$-independence of the phorbol ester effects. Thus enhanced depolarization can be seen in synaptosomes incubated in the presence of EGTA [83], while MARCKS phosphorylation (see below) is not increased by the depolarization-evoked increase in $[Ca^{2+}]_i$ [64].

## Presynaptic facilitatory receptors

While there have been reports of presynaptic ionotropic receptors that are present on synaptosomes and capable of enhancing glutamate exocytosis (e.g. [93]), the most clearly documented effects are due to the operation of G-protein-coupled receptors, and these will now be discussed in detail.

A major physiological activator of presynaptic PKC is glutamate itself (Fig. 1), acting via a phospholipase $A_2$-coupled metabotropic glutamate receptor (mGluR) [94,95]. In the presence (but not the absence) of AA, the non-selective mGluR agonist 1$S$,3$R$ aminocyclopentane dicarboxylic acid (1$S$,3$R$-ACPD) can mimic the effects of added phorbol esters on cerebrocortical synaptosomes, suggesting that the AA sensitization of PKC to diacylglycerol may be physiologically relevant [94]. That AA does not act prior to PKC (for example on the receptor, G-protein or phospholipase C) is shown by the AA-insensitivity of diacylglycerol production in the presence of 1$S$,3$R$-ACPD [65,96]. Similarly, AA does not act downstream of PKC, as phosphorylation of MARCKS has the same requirement for AA as does the potentiation of release itself [65].

Presynaptic PKC, whether activated by phorbol esters or by mGluR agonists, has no detectable effect on the presynaptic release-coupled $Ca^{2+}$ channel or on the process of $Ca^{2+}$–secretion coupling itself. Thus the clamped depolarization of synaptosomes evoked by high KCl causes an elevation of $[Ca^{2+}]_i$ and exocytosis of glutamate which is totally insensitive to the activation state of PKC, even when this is inhibited by the PKC inhibitor Ro 31-8220 [64,65]. Interestingly, the inhibitory effect on 4-AP-evoked glutamate release, which is caused by the addition of AA alone [97], and which is PKC independent [98], is also not seen during KCl evoked depolarization, consistent with an action as an activator of the 4-AP-insensitive $K^+$ channel(s). The competence of KCl to release glutamate in the presence of Ro 31-8220 also demonstrates that MARCKS phosphorylation, or indeed the phosphorylation of any protein by a Ro 31-8220-sensitive PKC, is not essential for glutamate exocytosis [64,65].

The physiological source and significance of the AA that potentiates presynaptic PKC is currently a matter of speculation. While synaptosomal preparations contain significant amounts of AA, this does not appear to activate the kinase as judged by the need for exogenous fatty acid [64,65,94,95]. It should be emphasized that these concentrations of AA are far below those at which nonspecific damage to the integrity of nerve terminals is observed. An attractive possibility, but one that still remains to be proven, is that the AA originates post-synaptically and that the phospholipase $A_2$ is activated under conditions where the NMDA receptor is active during the establishment of synaptic plasticity. The additional presence of transmitter glutamate in the synaptic cleft would provide the synergism necessary for diacylglycerol production and PKC sensitization, and could serve to reinforce the release of glutamate occurring during the period required for the establishment of synaptic plasticity [99]. A further factor, which will be discussed below, is that PKC activation seems to decouple inhibitory presynaptic receptors, thus further enhancing the facilitation of release [89,100–102].

The proposed sequence of events at the presynaptic membrane (Fig. 1) would thus be: (a) glutamate binds to the presynaptic mGluR, activating a G-protein coupled to phospholipase C; (b) the diacylglycerol generated will only activate PKC in the presence of AA; (c) activated PKC directly or indirectly phosphorylates a presynaptic $K^+$ channel resulting in its inactivation; this prolongs or intensifies physiological action potentials; (d) since the $Ca^{2+}$ channel coupled to glutamate release does not appear to undergo voltage-dependent inactivation, its probability of opening will increase with prolonged depolarization; and (e) the tight linkage between $Ca^{2+}$ entry and glutamate release will result in an increased quantal release of transmitter.

The physiological significance of this positive-feedback autoreceptor can prompt intriguing speculation. Firstly, it should be emphasized that what is being observed must be an extremely widespread phenomenon in the cerebral cortex to produce such a robust control of glutamate release from the isolated synaptosomes; if only a subpopulation of glutamatergic terminals possessed this regulatory pathway, the response would be diluted out by non-responsive terminals. A number of possible sources of AA *in vivo* can be proposed; one would be phospho-lipase $A_2$ activated post-synaptically in response to NMDA receptor activation [103–105], thus providing a mechanistic basis for proposals that AA could function as a retrograde messenger during plastic changes at the synapse. While it is unreasonable that such a mechanism would chronically elevate glutamate release, a short-lasting enhancement during the induction of potentiation could act to ensure that glutamate release remains high until a synapse is securely switched to the potentiated form. An acute action of the facilitatory receptor is supported by the observation that it undergoes rapid desensitization in the continued presence of the agonist [106].

## Other proposed presynaptic facilitatory pathways

High concentrations of forskolin enhance presynaptic facilitatory pathways at hippocampal CA3–CA1 synapses [107]. A presynaptic mechanism in which cAMP directly enhances release has been proposed, and this is supported by electrophysio-logical evidence for presynaptic β-adrenoceptors at this synapse [108].

### Inhibitory presynaptic receptors

Glutamate exocytosis from synaptosomes, primary neuronal cultures and brain slices can be inhibited by a variety of agonists acting on presynaptic receptors; these include $GABA_B$ receptor agonists (where GABA is $\gamma$-aminobutyric acid) [35,109,110], adenosine acting at $A_1$ receptors [100,111,112], muscarinic acetylcholine receptor agonists [113], mGluR agonists [114–120] (at least in young rats [121]), dynorphin [122], neuropeptide Y [123], $5\text{-}HT_2$ receptor agonists (where 5-HT is 5-hydroxytryptamine) [124], dopamine $D_2$ receptor agonists [125] and $\alpha_2$-adrenergic agonists [126]. The conventional action of an inhibitory receptor is either to inhibit $Ca^{2+}$ channels or to activate $K^+$ channels, both of which would decrease release by respectively hyperpolarizing or reducing $Ca^{2+}$ entry. Electrophysiological studies of somatic inhibitory receptors report either or both of these effects.

### Adenosine-mediated inhibition

At the presynaptic nerve terminal the lack of direct electrophysiology has generally implied that the ambiguity between these two channel effects cannot be resolved. However, the ability to stimulate synaptosomes in two distinct ways, i.e by elevated KCl and by 4-AP, has enabled the action of one inhibitory presynaptic receptor, the adenosine $A_1$ receptor, to be localized to an inhibition of $Ca^{2+}$ entry [100] (Fig. 1). Adenosine agonists cause potent inhibition of 4-AP-evoked glutamate release; indeed, endogenously leaked adenosine is sufficient to inhibit release partially and can be removed by adenosine deaminase. In contrast to the PKC-mediated effects discussed above, adenosine agonists and antagonists have no effect on the population average membrane potential in the presence or absence of 4-AP, as determined with a cyanine dye, suggesting that the receptor does not activate a $K^+$ channel [100]. Furthermore, adenosine agonists have almost exactly the same inhibitory effect on KCl-evoked glutamate release. Activation of a $K^+$ channel that is active during prolonged KCl depolarization would, if anything, enhance release by moving the membrane potential closer to the $K^+$ equilibrium potential.

### The inhibitory presynaptic mGluR

There is some confusion as to whether $1S,3R$-ACPD exerts facilitatory or inhibitory presynaptic effects in specific systems. It is evident that, in addition to the facilitatory phospholipase C-coupled mGluR discussed above, terminals can express an inhibitory presynaptic receptor, which may be an L-2-amino-4-phosphonobutyrate (L-AP4)-sensitive mGluR4 [114,115,127–130], and that this can inhibit release [131,132]. Recent developmental studies with cerebrocortical synaptosomes have revealed that the inhibitory mGluR (Fig. 1) is only expressed during a tight developmental window (2 3 weeks *post partum*) [133]. The presence of the inhibitory receptor can be demonstrated when L-AP4 or $1S,3R$-ACPD is added, the facilitatory receptor being inactive due to the absence of AA. The inhibition of glutamate release by L-AP4 is consistent with a pertussis-toxin-sensitive decrease in $[Ca^{2+}]_i$ rather than the decrease in intra-synaptosomal cAMP levels which seems to underlie a secondary mechanism of the L-AP4 receptor coupling [102].

## Interactions between facilitatory and inhibitory presynaptic receptors

The very extent of the facilitatory and inhibitory modulations discussed above suggests that a significant proportion of presynaptic terminals might possess both regulatory pathways. It is therefore of interest to establish how a nerve terminal might respond when exposed at the same time to both facilitatory and inhibitory agonists.

Phorbol esters suppress the inhibitory action of adenosine [100,128]. This can be seen with synaptosomes stimulated by elevated KCl as well as by 4-AP [89], indicating that the PKC-mediated suppression must be due to an action distinct from $K^+$-channel inhibition (which is not seen in this condition, as discussed above). Incubation of cells with phorbol esters has been known to cause receptor desensitization [134], and it is therefore of considerable interest to establish whether such suppression can also be observed during more physiological, receptor-mediated, stimulation of PKC.

1S,3R-ACPD activation of the presynaptic phospholipase C-coupled mGluR can suppress the inhibition of glutamate exocytosis exerted by the adenosine receptor [89] and the inhibitory mGluRs [102] as effectively as can phorbol 12,13-dibutyrate (Fig. 1). The conditions required, however, are different, since the suppression of inhibitory mGluR action in young rats requires 1S,3R-ACPD and added AA [102], whereas 1S,3R-ACPD suppresses the adenosine inhibition in adult rats in the absence of added AA [89]. In addition, inhibition of the endogenous tonic PKC activity in the terminals from young rats does not alter the effects of inhibitory mGluRs [133] but enhances the initial extent of adenosine inhibition of release [89]. It is possible that some of the chronic enhancements of exocytosis reported in the presence of phorbol esters (e.g. [81,135,136]) may relate to such a removal of a tonic inhibition, since PKC-mediated enhancement of KCl-evoked glutamate can be seen in synaptosomal preparations containing leaked adenosine, e.g. in the absence of added adenosine deaminase [89].

## Presynaptic aspects of synaptic plasticity

Enhanced synaptic transmission can be due to an increase in the post-synaptic response to an unchanged release of transmitter, to a presynaptic enhancement in the amount of transmitter released, or to a combination of both factors. Quantal analysis is a powerful technique at the neuromuscular junction to distinguish between increased miniature endplate potential frequency (indicative of enhanced presynaptic release) and amplitude (indicating an enhanced post-synaptic responsiveness). In contrast, quantal analysis in the CNS is extremely difficult [137], and even in the case of intensively investigated phenomena such as hippocampal LTP there is currently no consensus as to the interpretation of the quantal analysis [138].

That the coupling of the facilitatory mGluR to enhanced glutamate release observed with the synaptosomal preparation may have a correlate in synaptic plasticity observed in hippocampal slice preparations is suggested in a number of recent reports. The facilitation of *in situ* PKC activation by AA either with exogenous diacylglycerol [139] or with mGluR-generated diacylglycerol [95] is seen in hippocampal terminals, while in hippocampal slices diacylglycerol and AA added together, but not separately, result in LTP in response to low-frequency stimulation [140,141].

There are some features of the presynaptic effects of the retrograde messenger AA that are attractive in relation to synaptic plasticity. AA would enhance glutamate release only in those nerve terminals that are 'marked' with an additional factor that is provided by presynaptic activity, i.e. an increased diacylglycerol level resulting from presynaptic phospholipase C-coupled mGluR activation. In contrast, in those nerve terminals that do not have a recent experience of firing, i.e. with low diacylglycerol, AA would be inhibitory by itself [97] or would reinforce the inhibition caused by L-AP4 or adenosine $A_1$ agonists [101]. However, the facilitation of glutamate release may not be sufficient to potentiate synaptic transmission if presynaptic inhibition remains active, as suggested by the finding that adenosine $A_1$ agonists are inhibitory and antagonists facilitatory in controlling the expression of LTP in the hippocampus [142].

It is here that a second aspect of the presynaptic action of AA becomes relevant. The activation of PKC suppresses presynaptic inhibition by L-AP4 [102] and adenosine [89] to allow an effective facilitation of glutamate release to occur. Thus the arrival of AA at the nerve terminal would switch the presynaptic control of glutamate release from inhibitory to facilitatory. The transient nature of the suppression of the inhibitory responses suggests that the presynaptic changes upon the arrival of AA are to produce a transient rather than a sustained increase in release that could be needed to reach a threshold of post-synaptic stimulation which is not obtained under normal synaptic transmission. If the source of AA is the activation of post-synaptic NMDA receptors, as discussed above, then the increase in glutamate release provided by the retrograde messenger AA could be necessary to activate post-synaptic glutamate receptors other than NMDA receptors.

Recently, evidence has been obtained that mGluRs are involved in LTP, since the mGluR antagonist $(R,S)$-α-methyl-4-carboxyphenylglycine blocks the induction of LTP at synapses in the CA1 region of the hippocampus [143]. Since post-synaptic mGluRs are localized perisynaptically in the hippocampus [144], the facilitation of glutamate release provided by AA may well be needed to activate these post-synaptic mGluRs. However, the lack of selective antagonists for mGluRs prevents the determination of the contribution of facilitatory phospholipase C-coupled mGluRs to LTP.

Nitric oxide generators have been reported to induce presynaptic inhibition in hippocampal Schaffer collateral synapses [145,146] or LTP [146], depending on whether NO donors are paired with a low- or a high-frequency stimulation. Similar effects are produced by cGMP analogues, suggesting that NO may produce its effects by activating soluble guanylate cyclase. However, the interpretation of such experiments is complicated by the avidity with which NO can bind to the haem group of mitochondrial cytochrome oxidase and inhibit respiration [147]. It is important to eliminate trivial effects on mitochondrial bioenergetics before ascribing physiological roles to such factors.

Long-term depression is an activity-dependent decrease in synaptic strength observed in the CA1 region of the hippocampus after low-frequency stimulation. Long-term depression, as well as LTP, is induced by an increase in the post-synaptic $Ca^{2+}$ concentration. However, during high-frequency stimulation, $Ca^{2+}$ reaches high levels and preferentially activates a protein kinase producing LTP, while the lower levels of $Ca^{2+}$ achieved during low-frequency stimulation preferentially activate protein phosphatases, decreasing the synaptic strength [148]. A recent study has reported that, although long-term depression in the hippocampus is *induced* post-synaptically, it is *expressed* presynaptically as a result of a long-lasting decrease in transmitter release [149], suggesting that it is likely to require the production of a retrograde messenger.

## Glutamate release and excitotoxicity

Glutamate exocytosis shows a strict requirement for ATP [19], and any decrease in the presynaptic ATP/ADP ratio leads to a rapid inhibition of the extent of KCl-evoked exocytosis [20]. The origin of the massive release of glutamate which is observed *in vivo* following brain ischaemia [150–153] would of necessity have to be cytoplasmic (by reversal of the plasma membrane glutamate transporter) once ATP levels have collapsed, but this does not preclude an initial phase of $Ca^{2+}$-dependent exocytosis [154] due to spontaneous neuronal excitability which might be intrinsic to a neuron or a consequence of a prior failure of inhibitory interneurons. Consistent with an exocytotic component is the finding that adenosine $A_1$ agonists (which, as discussed below, inhibit the $Ca^{2+}$ channel coupled to glutamate release) substantially reduce the ischaemic release of glutamate *in vivo* [155–157] (but see [158]), while $A_1$ antagonists prevent the depression of excitatory transmission seen in hypoxia [159] and hypoglycaemia [160].

While the physiological function of the neuronal and glial transporters is to scavenge glutamate from the synaptic cleft, the transporter operates close to thermodynamic equilibrium (for a review, see [161]). Thus any depolarization of the plasma membrane and/or increase in the internal $Na^+$ concentration that persists for more than a few seconds leads to an efflux of cytoplasmic glutamate by reversal of the transporter [162,163]. This release of glutamate is $Ca^{2+}$-independent and is characterized by being independent of ATP, insensitive to *Clostridial* neurotoxins and accompanied by substantial amounts of endogenous aspartate, which is a largely cytoplasmic marker [164].

This $Ca^{2+}$-independent release of glutamate would not be significant during the brief physiological depolarization required to evoke exocytosis. However, there are pathological conditions in which reversal of the plasma membrane transporter makes a major contribution to the release of glutamate. An example is following anoxia or ischaemia, when the neuronal ATP levels have collapsed sufficiently to inhibit exocytosis [12,19,20] and where the lack of ATP for the $Na^+/K^+$-ATPase, which is the major utilizer of ATP in the resting neuron, results in a decrease in the $Na^+$ electrochemical potential and a resultant reversal of the transporter which, instead of scavenging glutamate from the synaptic cleft, now instead releases the cytoplasmic pool of glutamate and aspartate. Much of the massive excitotoxic release of glutamate which occurs during ischaemia *in vivo* can be attributed to this pathway [20].

*Research into synaptosomal exocytosis has been supported by the Medical Research Council, the Wellcome Trust, DGICYT (Spain), and the European Union (Biomed I). We acknowledge the contributions to this work by Harvey McMahon, Talvinder Sihra, Jennifer Pocock, Anne Barrie, Immaculada Herrero, Elena Vásquez, David Budd and Gareth Tibbs.*

## References


1. Barrett, E.F. and Stevens, C.F. (1972) J. Physiol. (London) **227**, 691–708
2. Heuser, J.E., Reese, T.S., Dennis, M.J., Jan, Y., Jan, L. and Evans, L. (1979) J. Cell Biol. **81**, 275–300
3. Fesce, R., Grohvoaz, F., Hurlbut, W.P. and Ceccarelli, B. (1980) J. Cell Biol. **85**, 337–345
4. Hessler, N.A., Shirke, A.M. and Malinow, R. (1993) Nature (London) **366**, 569–572
5. Clements, J.D., Lester, R.A.J., Tong, G., Jahr, C.E. and Westbrook, G.L. (1992) Science **258**, 1498–1501
6. Jonas, P. and Spruston, N. (1994) Curr. Opin. Neurobiol. **4**, 366–372
7. Colquhoun, D., Jonas, P. and Sakmann, B. (1992) J. Physiol. (London) **458**, 261–287
8. Sarantis, M., Ballerini, L., Miller, B., Silver, R.A., Edwards, M. and Attwell, D.A. (1993) Neuron **11**, 541–549
9. Gabbiani, F., Midtgaard, J. and Knöpfel, T. (1994) J. Neurophysiol. **72**, 999–1009
10. Jaeger, D. and Bower, J.M. (1994) Exp. Brain Res. **100**, 200–214
11. Nicholls, D.G. and Sihra, T.S. (1986) Nature (London) **321**, 772–773
12. McMahon, H.T. and Nicholls, D.G. (1991) J. Neurochem. **56**, 86–94
13. McMahon, H.T., Foran, P., Dolly, J.O., Verhage, M., Wiegant, V.M. and Nicholls, D.G. (1992) J. Biol. Chem. **267**, 21338–21343
14. Schiavo, G., Benfenati, F., Poulain, B., Rossetto, O., Polverino-de-Laureto, P., Dasgupta, B.R. and Montecucco, C. (1992) Nature (London) **359**, 832–835
15. Blasi, J., Chapman, E.R., Link, E., Binz, T., Yamasaki, S., De Camilli, P., Südhof, T.C., Niemann, H. and Jahn, R. (1993) Nature (London) **365**, 160–163

16. Hayashi, T., McMahon, H., Yamasaki, S., Binz, T., Hata, Y., Südhof, T.C. and Niemann, H. (1994) EMBO J. **13**, 5051–5061

17. Tibbs, G.R., Barrie, A.P., Van-Mieghem, F., McMahon, H.T. and Nicholls, D.G. (1989) J. Neurochem. **53**, 1693–1699

18. Ryan, T.A., Reuter, H., Wendland, B., Schweizer, F.E., Tsien, R.W. and Smith, S.J. (1993) Neuron **11**, 713–724

19. Sánchez-Prieto, J., Sihra, T.S. and Nicholls, D.G. (1987) J. Neurochem. **49**, 58–64

20. Kauppinen, R.A., McMahon, H.T. and Nicholls, D.G. (1988) Neuroscience **27**, 175–182

21. Tapia, R. and Sitges, M. (1982) Brain Res. **250**, 291–299

22. Dolezal, V. and Tucek, S. (1983) Naunyn-Schmiedeberg's Arch. Pharmacol. **323**, 90–95

23. Agoston, D.V., Hargittai, P. and Nagy, A. (1983) J. Neurochem. **41**, 745–751

24. Tibbs, G.R., Dolly, J.O. and Nicholls, D.G. (1989) J. Neurochem. **52**, 201–206

25. Tibbs, G.R., Nicholls, D.G. and Dolly, J.O. (1989) FEBS Lett. **255**, 159–162

26. Bagetta, G., Nair, S., Nistico, G. and Dolly, J.O. (1994) Neurochem. Int. **24**, 81–90

27. Perreault, P. and Avoli, M. (1992) J. Neurosci. **12**, 104–115

28. Grynkiewicz, G., Poenie, M. and Tsien, R.Y. (1985) J. Biol. Chem. **260**, 3440–3450

29. Nachshen, D.A. (1985) J. Physiol. (London) **361**, 251–268

30. Pocock, J.M., Cousin, M.A. and Nicholls, D.G. (1993) Neuropharmacology **32**, 1185–1194

31. Suszkiw, J.B., O'Leary, M.E., Murawsky, M.M. and Wang, T. (1986) J. Neurosci. **6**, 1349–1357

32. Terrian, D.M., Dorman, R.V. and Gannon, R.L. (1990) Neurosci. Lett. **119**, 211–214

33. Pocock, J.M. and Nicholls, D.G. (1992) Eur. J. Pharmacol. **226**, 343–350

34. Westenbroek, R.E., Ahlijanian, M.K. and Catterall, W.A. (1990) Nature (London) **347**, 281–284

35. Huston, E., Scott, R.H. and Dolphin, A.C. (1990) Neuroscience **38**, 721–730

36. Pocock, J.M., Cousin, M.A., Parkin, J. and Nicholls, D.G. (1995) Neuroscience **67**, 595–607

37. Grignon, S., Seagar, M.J. and Couraud, F. (1993) Neurosci. Lett. **155**, 87–91

38. Cruz, L.J. and Olivera, B.M. (1986) J. Biol. Chem. **261**, 6230–6233

39. Suszkiw, J.B., Murawsky, M.M. and Fortner, R.C. (1987) Biochem. Biophys. Res. Commun. **145**, 1283–1286

40. Lundy, P.M., Stauderman, K.A., Goulet, J.C. and Frew, R. (1989) Neurochem. Int. **14**, 49–54

41. Reynolds, I.J., Wagner, J.A., Snyder, S.H., Thayer, S.A., Olivera, B.M. and Miller, R.J. (1986) Proc. Natl. Acad. Sci. U.S.A. **83**, 8804–8807

42. Llinás, R., Sugimori, M., Lin, J.W. and Cherksey, B. (1989) Proc. Natl. Acad. Sci. U.S.A. **86**, 1689–1693

43. Adams, M.E., Myers, R.A., Imperial, J.S. and Olivera, B.M. (1993) Biochemistry **32**, 12566–12570

44. Turner, T.J., Adams, M.E. and Dunlap, K. (1992) Science **258**, 310–313

45. Uchitel, O.D., Protti, D.A., Sanchez, V., Cherksey, B.D., Sugimori, M. and Llinás, R. (1992) Proc. Natl. Acad. Sci. U.S.A. **89**, 3330–3333

46. Luebke, J.I., Dunlap, K. and Turner, T.J. (1993) Neuron **11**, 895–902

47. Muller, D. and Lynch, G. (1989) Synapse **3**, 67–73

48. Llinás, R., Sugimori, M. and Silver, R.B. (1992) Science **256**, 677–679

49. Smith, S.J., Buchanan, J., Osses, L.R., Charlton, M.P. and Augustine, G.J. (1993) J. Physiol. (London) **472**, 573–593

50. Adler, E.M., Augustine, G.J., Duffy, S.N. and Charlton, M.P. (1991) J. Neurosci. **11**, 1496–1507

51. Von Gersdorff, H. and Matthews, G. (1994) Nature (London) **367**, 735–739

52. Heidelberger, R., Heinemann, C., Neher, E. and Matthews, G. (1994) Nature (London) **371**, 513–515

53. Verhage, M., McMahon, H.T., Ghijsen, W.E.J.M., Boomsma, F., Wiegant, V. and Nicholls, D.G. (1991) Neuron **6**, 517–524

54. Stanley, E.F. (1993) Neuron **11**, 1007–1011

55. Knight, D.E. and Baker, P.F. (1983) FEBS Lett. **160**, 98–100

56. Pocotte, S.L., Frye, R.A., Senter, R.A., TerBush, D.R., Lee, S.A. and Holz, R.W. (1985) Proc. Natl. Acad. Sci. U.S.A. **82**, 930–934

57. Bittner, M.A. and Holz, R.W. (1990) J. Neurochem. **54**, 205–210

58. Vitale, M.L., Del Castillo, A.R. and Trifaro, J.M. (1992) Neuroscience **51**, 463–474

59. Conner-Kerr, T.A., Simmons, D.R., Peterson, G.M. and Terrian, D.M. (1993) J. Neurochem. **61**, 627–636
60. Conti, F., Fabri, M. and Minelli, A. (1992) Brain Res. **599**, 140–143
61. Greengard, P., Valtorta, F., Czernik, A.J. and Benfenati, F. (1993) Science **259**, 780–785
62. Albert, K.A., Walaas, S.I., Wang, J.K.T. and Greengard, P. (1986) Proc. Natl. Acad. Sci. U.S.A. **83**, 2822–2826
63. Blackshear, P.J. (1993) J. Biol. Chem. **268**, 1501–1504
64. Coffey, E.T., Sihra, T.S. and Nicholls, D.G. (1993) J. Biol. Chem. **268**, 21060–21065
65. Coffey, E.T., Herrero, I., Sihra, T.S., Sánchez-Prieto, J. and Nicholls, D.G. (1994) J. Neurochem. **63**, 1303–1310
66. Gispen, W.H. (1985) Brain Res. **328**, 381–385
67. De Graan, P.N.E., Dekker, L.V., Oestreicher, A.B., Van-der-Voorn, L. and Gispen, W.H. (1989) J. Neurochem. **52**, 17–23
68. Robinson, P.J. (1991) Neurosci. Res. Commun. **9**, 167–176
69. Robinson, P.J., Sontag, J.M., Liu, J., Fykse, E.M., Slaughter, C., McMahon, H.T. and Südhof, T.C. (1993) Nature (London) **365**, 163–166
70. Benfenati, F., Valtorta, F., Bahler, M. and Greengard, P. (1989) Cell Biol. Int. Rep. **13**, 1007–1021
71. Benfenati, F., Valtorta, F. and Greengard, P. (1991) Proc. Natl. Acad. Sci. U.S.A. **88**, 575–579
72. Hirokawa, N., Sobue, K., Kanda, K., Harada, A. and Yorifuji, H. (1989) J. Cell Biol. **108**, 111–126
73. Llinás, R., McGuiness, T.L., Leonard, C.S., Sugimori, M. and Greengard, P. (1985) Proc. Natl. Acad. Sci. U.S.A. **82**, 3035–3039
74. Nichols, R.A., Sihra, T.S., Czernik, A.J., Nairn, A.C. and Greengard, P. (1990) Nature (London) **343**, 647–650
75. McMahon, H.T. and Nicholls, D.G. (1993) J. Neurochem. **61**, 110–115
76. Sihra, T.S., Piomelli, D. and Nichols, R.A. (1993) J. Neurochem. **61**, 1220–1230
77. Rosahl, T.W., Geppert, M., Spillane, D., Herz, J., Hammer, R.E., Malenka, R.C. and Südhof, T.C. (1993) Cell **75**, 661–670
78. Koenig, J.H. and Ikeda, K. (1989) J. Neurosci. **9**, 3844–3860
79. Nichols, R.A., Brown, J.M. and Suplick, G.R. (1993) Soc. Neurosci. Abstr. **19**, 373.19
80. Shearman, M.S., Shinomura, T., Oda, T. and Nishizuka, Y. (1991) J. Neurochem. **56**, 1255–1262
81. Terrian, D.M., Ways, D.K., Gannon, R.L. and Zetts, D.A. (1993) Hippocampus **3**, 205–220
82. Nicholls, D.G. and Coffey, E.T. (1994) in Molecular and Cellular Mechanisms of Neurotransmitter Release (Stjarne, L., Greengard, P., Hokfelt, T. and Ottoson, D., eds.), pp. 189–204, Raven Press, New York
83. Barrie, A.P., Nicholls, D.G., Sánchez-Prieto, J. and Sihra, T.S. (1991) J. Neurochem. **57**, 1398–1404
84. Burgoyne, R.D., Morgan, A. and O'Sullivan, A.J. (1989) Cell. Signalling **1**, 323–334
85. Hell, J.W., Appleyard, S.M., Yokoyama, C.T., Warner, C. and Catterall, W.A. (1994) J. Biol. Chem. **269**, 7390–7396
86. Linden, D.J., Smeyne, M., Sun, S.C. and Connor, J.A. (1992) J. Neurosci. **12**, 3601–3608
87. Colby, K.A. and Blaustein, M.P. (1988) J. Neurosci. **8**, 4685–4692
88. Forsythe, I.D. (1994) J. Physiol. (London) **479**, 381–388
89. Budd, D.C. and Nicholls, D.G. (1995) J. Neurochem. **65**, 615–621
90. Herrero, I., Miras-Portugal, M.T. and Sánchez-Prieto, J. (1992) J. Neurochem. **59**, 1574–1577
91. Tanaka, C. and Nishizuka, Y. (1994) Annu. Rev. Neurosci. **17**, 551–567
92. Saito, N., Itouji, A., Totani, Y., Osawa, I., Koide, H., Fujisawa, N., Ogita, K. and Tanaka, C. (1993) Brain Res. **607**, 241–248
93. Barnes, J.M., Dev, K.K. and Henley, J.M. (1994) Br. J. Pharmacol. **113**, 339–341
94. Herrero, I., Miras-Portugal, M.T. and Sánchez-Prieto, J. (1992) Nature (London) **360**, 163–166
95. Vázquez, E., Herrero, I., Miras-Portugal, M.T. and Sánchez-Prieto, J. (1994) Neurosci. Res. Commun. **15**, 187–194
96. Vázquez, E., Herrero, I., Miras-Portugal, M.T. and Sánchez-Prieto, J. (1994) Neurosci. Lett. **174**, 9–13
97. Herrero, I., Miras-Portugal, M.T. and Sánchez-Prieto, J. (1991) J. Neurochem. **57**, 718–721
98. Herrero, I., Miras-Portugal, M.T. and Sánchez-Prieto, J. (1992) FEBS Lett. **296**, 317–319

99. Nicholls, D.G. (1992) Nature (London) **360**, 106–107
100. Barrie, A.P. and Nicholls, D.G. (1993) J. Neurochem. **60**, 1081–1086
101. Vázquez, E., Budd, D., Herrero, I., Nicholls, D.G. and Sánchez-Prieto, J. (1995) Neuropharmacology **34**, 919–927
102. Herrero, I., Vázquez, E., Miras-Portugal, M.T. and Sánchez-Prieto, J. (1996) Eur. J. Neurosci. **8**, 700–709
103. Lazarewicz, J.W., Wroblewski, J.T. and Costa, E. (1990) J. Neurochem. **55**, 1875–1881
104. Dumuis, A., Pin, J.P., Oomagari, K., Sebben, M. and Bockaert, J. (1990) Nature (London) **347**, 182–184
105. Tapia-Arancibia, L., Rage, F., Récasens, M. and Pin, J.P. (1992) Eur. J. Pharmacol. Mol. Pharmacol. **225**, 253–262
106. Herrero, I., Miras-Portugal, M.T. and Sánchez-Prieto, J. (1994) Eur. J. Neurosci. **6**, 115–120
107. Chavez-Noriega, L.E. and Stevens, C.F. (1994) J. Neurosci. **14**, 310–317
108. Gereau, R.W. and Conn, P.J. (1994) J. Neurophysiol. **72**, 1438–1442
109. Pende, M., Lanza, M., Bonanno, G. and Raiteri, M. (1993) Brain Res. **604**, 325–330
110. Tarelius, E., Schoch, J. and Breer, H. (1994) Neurochem. Int. **24**, 349–361
111. Poli, A., Lucchi, R., Zottini, M. and Traversa, U. (1993) Brain Res. **620**, 245–250
112. Alzheimer, C., Sutor, B. and Ten Bruggencate, G. (1993) Neuroscience **57**, 565–575
113. Marchi, M., Bocchieri, P., Garbarino, L. and Raiteri, M. (1989) Neurosci. Lett. **96**, 229–234
114. Gannon, R.L., Baty, L.T. and Terrian, D.M. (1989) Brain Res. **495**, 151–155
115. Clements, J.D. and Forsythe, M.H. (1989) J. Physiol. (London) **410**, 57P
116. Lovinger, D.M. (1991) Neurosci. Lett. **129**, 17–21
117. Rainnie, D.G. and Shinnick-Gallagher, P. (1992) Neurosci. Lett. **139**, 87–91
118. Calabresi, P., Mercuri, N.B. and Bernardi, G. (1992) Neurosci. Lett. **139**, 41–44
119. Goh, J.W. and Musgrave, M.A. (1993) Neuropharmacology **32**, 311–312
120. Glaum, S.R. and Miller, R.J. (1993) J. Neurophysiol. **70**, 2669–2672
121. Baskys, A. and Malenka, R.C. (1991) J. Physiol. (London) **444**, 687–701
122. Weisskopf, M.G., Zalutsky, R.A. and Nicoll, R.A. (1993) Nature (London) **362**, 423–427
123. Bleakman, D., Harrison, N.L., Colmers, W.F. and Miller, R.J. (1992) Br. J. Pharmacol. **107**, 334–340
124. Maura, G., Carbone, R., Guido, M., Pestarino, M. and Raiteri, M. (1991) Eur. J. Pharmacol. **202**, 185–190
125. Maura, G., Giardi, A. and Raiteri, M. (1994) J. Pharmacol Exp. Ther. **269**, 246–255
126. Kamisaki, Y., Hamada, T., Maeda, K., Ishimura, M. and Itoh, T. (1993) J. Neurochem. **60**, 522–526
127. Gannon, R.L. and Terrian, D.M. (1991) Neuroscience **41**, 401–410
128. Thomsen, C., Kristensen, P., Mulvihill, E., Haldeman, B. and Suzdak, P.D. (1992) Eur. J. Pharmacol. Mol. Pharmacol. **227**, 361–362
129. Kristensen, P., Suzdak, P.D. and Thomsen, C. (1993) Neurosci. Lett. **155**, 159–162
130. Prézeau, L., Carrette, J., Helpap, B., Curry, K., Pin, J.P. and Bockaert, J. (1994) Mol. Pharmacol. **45**, 570–577
131. Burke, J.P. and Hablitz, J.J. (1994) J. Neurosci. **14**, 5120–5130
132. Johnson, C.R., Glazewski, S. and Fox, K. (1994) Soc. Neurosci. Abstr. **20**, 279
133. Vázquez, E., Herrero, I., Miras-Portugal, M.T. and Sánchez-Prieto, J. (1995) Neuroscience **68**, 117–124
134. Lai, W.S., Rogers, T.B. and El-Fakahany, E.E. (1990) Biochem. J. **267**, 23–29
135. Terrian, D.M. (1995) J. Neurochem. **64**, 172–180
136. Terrian, D.M. and Ways, D.K. (1995) J. Neurochem. **64**, 181–190
137. Bekkers, J.M. (1994) Curr. Opin. Neurobiol. **4**, 360–365
138. Thomson, A.M. (1992) Trends Neurosci. **15**, 167–168
139. Zhang, L. and Dorman, R.V. (1993) Brain Res. Bull. **32**, 437–441
140. Collins, D.R. and Davies, S.N. (1993) Eur. J. Pharmacol. **240**, 325–326
141. Bramham, C.R., Alkon, D.L. and Lester, D.S. (1994) Neuroscience **60**, 737–743
142. De Mendonça, A. and Ribeiro, J.A. (1994) Neuroscience **62**, 385–390
143. Bashir, Z.I., Bortolotto, Z.A., Davies, C.H., Berretta, N., Irving, A.J., Seal, A.J., Henley, J.M., Jane, D.E., Watkins, J.C. and Collingridge, G.L. (1993) Nature (London) **363**, 347–350

144. Baude, A., Nusser, Z., Roberts, J.D.B., Mulvihill, E., McIlhinney, R.A.J. and Somogyi, P. (1993) Neuron **11**, 771–787
145. Boulton, C.L., Irving, A.J., Southam, E., Potier, B., Garthwaite, J. and Collingridge, G.L. (1994) Eur. J. Neurosci. **6**, 1528–1535
146. Zhuo, M., Kandel, E.R. and Hawkins, R.D. (1994) Neuroreport **5**, 1033–1036
147. Schweizer, M. and Richter, C. (1994) Biochem. Biophys. Res. Commun. **204**, 169–175
148. Bear, M.F. and Malenka, R.C. (1994) Curr. Opin. Neurobiol. **4**, 389–399
149. Bolshakov, V.Y. and Siegelbaum, S.A. (1994) Science **264**, 1148–1152
150. Sandberg, M., Butcher, S.P. and Hagberg, H. (1986) J. Neurochem. **47**, 178–184
151. Ikeda, M., Nakazawa, T., Abe, K., Kaneko, T. and Yamatsu, K. (1989) Neurosci. Lett. **96**, 202–206
152. Graham, S.H., Shiraishi, K., Panter, S.S., Simon, R.P. and Faden, A.I. (1990) Neurosci. Lett. **110**, 124–127
153. Baker, A.J., Zornow, M.H., Scheller, M.S., Yaksh, T.L., Skilling, S.R., Smullin, D.H., Larson, A.A. and Kuczenski, R. (1991) J. Neurochem. **57**, 1370–1378
154. Katayama, Y., Kawamata, T., Tamura, T., Hovda, D.A., Becker, D.P. and Tsubokawa, T. (1991) Brain Res. **558**, 136–144
155. Simpson, R.E., O'Regan, M.H., Perkins, L.M. and Phillis, J.W. (1992) J. Neurochem. **58**, 1683–1690
156. Héron, A., Lasbennes, F. and Seylaz, J. (1993) Brain Res. **608**, 27–32
157. Héron, A., Lekieffre, D., Le Peillet, E., Lasbennes, F., Seylaz, J., Plotkine, M. and Boulu, R.G. (1994) Brain Res. **641**, 217–224
158. Cantor, S.L., Zornow, M.H., Miller, L.P. and Yaksh, T.L. (1992) J. Neurochem. **59**, 1884–1892
159. Katchman, A.N. and Hershkowitz, N. (1993) Neurosci. Lett. **159**, 123–126
160. Zhu, P.J. and Krnjevic, K. (1993) Neurosci. Lett. **155**, 128–131
161. Attwell, D.A., Barbour, B. and Szatkowski, M. (1993) Neuron **11**, 401–407
162. Nicholls, D.G., Sihra, T.S. and Sánchez-Prieto, J. (1987) J. Neurochem. **49**, 50–57
163. Bernath, S. (1992) Prog. Neurobiol. **38**, 57–91
164. McMahon, H.T., Rosenthal, L., Meldolesi, J. and Nicholls, D.G. (1990) J. Neurochem. **55**, 2039–2047

# 2

# Structure, function and regulation of sodium-coupled neurotransmitter transporters

**Baruch I. Kanner**

Department of Biochemistry, Hadassah Medical School, The Hebrew University, P.O. Box 12272, Jerusalem 91120, Israel

## Introduction

Sodium-coupled neurotransmitter transporters, which are located in the plasma membranes of nerve terminals and glial processes, serve to keep the extracellular transmitter levels below those that are neurotoxic. Moreover, they help, in conjunction with diffusion, to terminate neurotransmitter action in synaptic transmission. Such a termination mechanism operates with most transmitters, including $\gamma$-aminobutyric acid (GABA), L-glutamate, glycine, dopamine, 5-hydroxytryptamine (5-HT; serotonin) and noradrenaline (norepinephrine). Another termination mechanism is observed with cholinergic transmission. After dissociation from its receptor, acetylcholine is hydrolysed into choline and acetate. The choline moiety is then recovered by sodium-dependent transport as described above. As the concentrations of the transmitters in the nerve terminals are much higher than in the cleft (typically by four orders of magnitude), energy input is required. The transporters that are located in the plasma membranes of nerve endings and glial cells obtain this energy by coupling the flow of neurotransmitters to that of sodium. The $(Na^+ + K^+)$-ATPase generates an inwardly directed electrochemical sodium gradient which is utilized by the transporters to drive 'uphill' transport of the neurotransmitters (reviewed in Kanner, 1983, 1989; Kanner and Schuldiner, 1987).

Neurotransmitter uptake systems have been investigated in detail by using plasma membranes obtained upon osmotic shock of synaptosomes. It appears that these transporters are coupled not only to sodium, but also to additional ions such as potassium or chloride (Table 1).

These transporters are of considerable medical interest. Since they function to regulate neurotransmitter activity by removing it from the synaptic cleft, specific transporter inhibitors can be potentially used as novel drugs for

**Table 1.          Comparison of GABA and glutamate transporters**

| Property | GABA transporter | Glutamate transporter |
|---|---|---|
| Co-substrates | $Na^+$, $Cl^-$ | $Na^+$, $K^+$, $OH^-$ |
| Electrogenicity | + | + |
| Localization | Neuronal, glial | Neuronal, glial |
| 'Sociology' | Belongs to large family of transporters for all neurotransmitters excluding glutamate | Belongs to separate small family of glutamate transporters |
| Relationship with bacterial transporters | – | glt-P, glutamate transporter; dct-A, dicarboxylic acid transporter |
| Predicted topology | Twelve transmembrane domains; N- and C-termini are cytoplasmic | Between six and ten transmembrane domains; N- and C-termini are cytoplasmic |
| Glycosylation | + | + |
| Possible regulation | Protein kinase C | Protein kinase C; arachidonic acid |

treatment of neurological disease. For instance, attenuation of GABA removal will prolong the effect of this inhibitory transporter, thereby potentiating its action. Thus inhibitors of GABA transport could represent a novel class of anti-epileptic drugs. Well known inhibitors that interfere with the functioning of biogenic amine transporters include anti-depressant drugs and stimulants such as amphetamines and cocaine. The neurotransmitter glutamate, at excessive local concentrations, causes cell death by activating $N$-methyl-D-aspartic acid receptors and subsequent calcium entry. The transmitter has been implicated in neuronal destruction during ischaemia, epilepsy, stroke, amyotropic lateral sclerosis and Huntington's disease. Neuronal and glial glutamate transporters may have a critical role in preventing glutamate from acting as an exitotoxin (Johnston, 1981; McBean and Roberts, 1985).

In the last few years, major advances have been made in the cloning of these neurotransmitter transporters. After the GABA transporter was purified (Radian et al., 1986), the ensuing protein sequence information was used to clone it

(Guastella et al., 1990). Subsequently the expression cloning of a noradrenaline transporter (Pacholczyk et al., 1991) provided evidence that these two proteins represented the first members of a novel superfamily of neurotransmitter transporters. This result led, using PCR and other technologies relying on sequence conservation, to the isolation of a growing list of neurotransmitter transporters (reviewed in Schloss et al., 1992; Uhl, 1992; Amara and Kuhar, 1993). This list includes various subtypes of GABA transporters and those for all the above-mentioned neurotransmitters, except glutamate. All of the members of this superfamily are dependent on sodium and chloride and, by analogy with the GABA transporter (Keynan and Kanner, 1988), are likely to co-transport their transmitter with both sodium and chloride. Interestingly, sodium-dependent glutamate transport is not chloride-dependent, but rather sodium and glutamate are counter-transported with potassium (Table 1; Kanner and Sharon, 1978; Kanner and Bendahan, 1982). More recently, three distinct but highly related glutamate transporters have been cloned (Kanai and Hediger, 1992; Pines et al., 1992; Storck et al., 1992). These represent a distinct family of transporters.

In this chapter, I describe current knowledge of two prototypes of these distinct families, i.e. the GABA transporters and the glutamate transporters.

## Mechanism

The GABA transporter co-transports the neurotransmitter with sodium and chloride in an electrogenic fashion (Kanner, 1983; Keynan and Kanner, 1988; Table 1). The available measurements include tracer fluxes (Keynan and Kanner, 1988) and electrophysiological approaches (Kavanaugh et al., 1992; Mager et al., 1993).

The latter approach has revealed that, at very negative (inside) potentials, the chloride dependency is not absolute (Mager et al., 1993). At the present time it is not clear what the mechanistic interpretation of this result is. One possibility is that, under these conditions, another anion (such as hydroxyl) may take over the role of chloride. In addition, a transient current is observed in the absence of GABA. This transient can be blocked by bulky GABA analogues which can bind to the transporter but are not translocated by it (Mager et al., 1993). It probably reflects a conformational change of the transporter that occurs after the sodium has bound. These measurements also permit determination of the turnover number of the transporter. These estimates of a few cycles per second (Mager et al., 1993) are in agreement with biochemical ones (Radian et al., 1986). It is of interest to note that, although both GABA and 5-HT transporters belong to the same transporter superfamily (Table 1), the latter appears to exhibit quite distinct properties (Mager et al., 1994). It appears that the 5-HT transporter is electroneutral, but that a transporter-associated current can be detected. This is probably related to the observation that, under some conditions, this transporter may act as a channel. This transporter-channel appears to be less sodium selective than the transporter mode which carries 5-HT in a coupled fashion (Mager et al., 1994).

The mechanism of sodium-dependent L-glutamate transport was investigated initially by tracer flux studies employing radioactive glutamate. These studies indicated that the process is electrogenic, with positive charge moving in the direction of the glutamate (Kanner and Sharon, 1978). This observation suggested that it is possible to monitor L-glutamate transport electrically using the whole-cell patch–clamp technique (Brew and Atwell, 1987). In addition to L-glutamate, D- and L-aspartate are transportable substrates with affinities in the lower micromolar range. The system is stereospecific with regard to glutamate, the D-isomer being a poor substrate. Glutamate uptake is driven by an inwardly directed sodium ion gradient, and at the same time potassium moves outwards. The movement of potassium is not a passive movement in response to the charge carried by the transporter. Rather, it is an integral part of the translocation cycle catalysed by the transporter. Its role is described further below. Evidence has also been presented that another ionic species is countertransported (in addition to potassium), namely hydroxyl ions (Bouvier et al., 1992).

The first-order-dependence of the carrier current on internal potassium (Barbour et al., 1988), together with the well known first-order-dependence on external L-glutamate and the sigmoidal dependence on external sodium, suggest a stoichiometry of 3 $Na^+$:1 $K^+$:1 glutamate (Kanner and Sharon, 1978; Barbour et al., 1988). This stoichiometry implies that one positive charge moves inwards per glutamate anion entering the cell. If a hydroxyl anion is countertransported as well (Bouvier et al., 1992), the stoichiometry could be 2 $Na^+$:1 $K^+$:1 gluta mate:1 $OH^-$, and transport would still be electrogenic. A stoichiometry of 2 $Na^+$:1 glutamate is also favoured by direct experimental evidence obtained by kinetic (Stallcup et al., 1979) and thermodynamic (Erecinska et al., 1983) methods.

The study of the ion dependence of partial reactions of the glutamate transporter has revealed that glutamate transport is an ordered process. First, sodium and glutamate are translocated. After their release inside the cell, potassium binds and is translocated outwards, so that a new cycle can be initiated (Kanner and Bendahan, 1982; Pines and Kanner, 1990).


## Reconstitution, purification and localization

Using methodology that enables one to reconstitute many samples simultaneously and rapidly, one of the subtypes of the GABA transporter (Radian et al., 1986) and one of the subtypes of the L-glutamate transporter (Danbolt et al., 1990) have been purified to apparent homogeneity. Both are glycoproteins and both have an apparent molecular mass of 70–80 kDa. The two transporters retain all the properties observed in membrane vesicles. They are distinct not only because of their different functional properties. Antibodies generated against the GABA transporter (Radian et al., 1986) react (as detected by immunoblotting) only with fractions containing GABA transport activity, and not with those containing L-glutamate transport activity (Danbolt et al., 1990). The opposite is true for antibodies generated against the glutamate transporter (Danbolt et al., 1992). More recently, the glycine transporter has also been purified and reconstituted. Interestingly, it appears to be a larger protein than the GABA and glutamate transporters, of about 100 kDa in size (Lopez-Corcuera et al., 1991). The 5-HT transporter has also been purified, but these preparations, containing a band around 70 kDa, have been shown to be active only in the binding of [$^3$H]imipramine and not in 5-HT transport (Graham et al., 1992; Launay et al., 1992). Immuno-cytochemical localization studies of the GABA transporter revealed that, in most brain areas, it is located in the membranes of nerve terminals (Radian et al., 1990), although in some areas, such as the substantia nigra, glial processes were labelled (Table 1).

Using the antibodies raised against the glutamate transporter, the immunocytochemical localization of this transporter was studied at the light- and electron-microscopic levels in the rat central nervous system. In all regions examined (including cerebral cortex, caudato-putamen, corpus callosum, hippocampus, cerebellum and spinal cord) the transporter was found to be located in glial cells rather than in neurons. In particular, fine astrocytic processes were strongly stained. Putative glutamatergic axon terminals appeared non-immuno-reactive (Danbolt et al., 1992). The uptake of glutamate by such terminals (for which there is strong previous evidence), therefore, may be due to a subtype of the glutamate transporter that is different from the glial transporter. Using a monoclonal antibody raised against this transporter, a similar glial localization of the transporter was found (Hees et al., 1992). Another glial transporter, in addition to a neuronal one, has been identified (see below).

## A new superfamily of sodium-dependent neurotransmitter transporters

Partial sequencing of the purified GABA$_A$ transporter allowed the cloning of the GABA transporter GAT-1, the first member of the newly discovered family of Na$^+$-dependent neurotransmitter transporters (Guastella et al., 1990). After expression cloning of the noradrenaline transporter (Pacholczyk et al., 1991), it became clear that it had significant sequence identity with the GABA$_A$ transporter. The use of functional cDNA expression assays and amplification of related sequences by PCR resulted in the cloning of additional transporters belonging to this family, such as the dopamine (Kilty et al., 1991; Shimada et al., 1991; Usdin et al., 1991) and 5-HT (Blakely et al., 1991; Hoffman et al., 1991) transporters, additional GABA transporters (Borden et al., 1992; Clark et al., 1992; Lopez-Corcuera et al., 1992; Liu et al., 1993a), transporters of glycine (Guastella et al.,

1992; Liu et al., 1992b; Smith et al., 1992), proline (Fremeau et al., 1992), taurine (Liu et al., 1992a; Uchida et al., 1992) and betaine (Yamauchi et al., 1992), and two 'orphan' transporters whose substrates are still unknown (Uhl et al., 1992; Liu et al., 1993c). In addition, another family member that was originally thought to be a choline transporter (Mayser et al., 1992) probably is in fact a creatine transporter (Guimbal and Kilimann, 1993). A novel glycine transporter cDNA encoding for a 799-amino-acid protein was been isolated by Liu et al. (1993b). This is significantly larger than most members of the superfamily. If we take into account the fact that part of the mass of these transporters is constituted by sugars, this sequence could encode the 100 kDa glycine transporter that was purified and reconstituted by Lopez-Corcuera et al. (1992).

The deduced amino acid sequences of these proteins revealed 30–65% identity between different members of the family. Based on these differences, the family can be divided into four subgroups: (a) transporters of biogenic amines (noradrenaline, dopamine and 5-HT); (b) various GABA transporters as well as transporters of taurine and creatine; (c) transporters of proline and glycine; and (d) orphan transporters. These proteins share some features of a common secondary structure. Each transporter is composed of 12 hydrophobic putative trans-membrane α-helices. The lack of a signal peptide suggests that both the N- and C-termini face into the cytoplasm. These regions contain putative phosphorylation sites, which may be involved in regulation of the transport process (see below). The second extracellular loop between helices 3 and 4 is the largest, and contains putative glycosylation sites.

Alignment of the deduced amino acid sequences of 13 different members of this superfamily whose substrates are known (subgroups a–c) revealed that some segments within these proteins share a higher degree of identity than others. The most highly conserved regions (>50% identity) are helix 1 together with the extracellular loop connecting it with helix 2, and helix 5 together with a short intracellular loop connecting it with helix 4 and a larger extracellular loop connecting it with helix 6. These domains may be involved in stabilizing a tertiary structure that is essential for the function of all these transporters. Alternatively, they may be related to a common function of these transporters, such as the trans-location of sodium ions. The region stretching from helix 9 onwards is far less conserved than the segment containing helices 1–8. Possibly, this domain contains some residues that are involved in translocating the different substrates. The least conserved segments are the N- and C-termini. As was mentioned above, these areas may be involved in regulation of the transport process. The orphan transporters differ from all other members of the family in three regions. They contain much larger extracellular loops between helices 7 and 8 and helices 11 and 12.

## Molecular cloning and predicted structure of glutamate transporters

Transporters for many neurotransmitters were cloned on the assumption that they are related to the GABA (Guastella et al., 1990) and noradrenaline (Pacholczyk et al., 1991) transporters (reviewed by Schloss et al., 1992; Uhl, 1992; Amara and Kuhar, 1993). This approach was unsuccessful for the glutamate transporter. Three different glutamate transporters have now been cloned using different approaches: GLAST (Storck et al., 1992), GLT-1 (Pines et al., 1992) and EAAC 1 (Kanai and Hediger, 1992). The first two appear to be of glial origin (Danbolt et al., 1992; Storck et al., 1992) and the third of neuronal origin (Kanai and Hediger, 1992). Indeed, the three transporters are not related to the above superfamily (Pines et al., 1992; Storck et al., 1992; Kanai and Hediger, 1992). On the other hand, they are very similar to each other, displaying ~50% identity and ~60% similarity. They also appear to be related to the proton-coupled glutamate transporter of *Escherichia coli* and other bacteria (glt-P; Tolner et al., 1992) and the dicarboxylate transporter of *Rhizobium meliloti* (dct-A; Jiang et al., 1989). In these cases the identities are around 25–30%. Thus these transporters form a distinct family. They each contain between 500 and 600 amino acids. In addition, it has been shown that this family also encodes sodium-dependent transporters which do not use dicarboxylic acids as substrates, but rather use neutral amino acids (Arriza et al., 1993; Shafqat et al., 1993).

GLT-1, which encodes the glutamate transporter that was purified (Danbolt et al., 1990, 1992; Pines et al., 1992), contains 573 amino acids and has a relative molecular mass of 64 kDa, in good agreement with the value of 65 kDa for the purified and deglycosylated transporter (Danbolt et al., 1992). Hydropathy plots are relatively straightforward at the N-terminal side of the protein and the three different groups have predicted the presence of six transmembrane α-helices at very similar positions (Kanai and Hediger, 1992; Pines et al., 1992; Storck et al., 1992). On the other hand, there is much more ambiguity at the C-terminal side, where no (Storck et al., 1992), two (Pines et al., 1992) and four (Kania and Hediger, 1992) α-helices have been predicted. However, all three groups note uncertainty in assigning transmembrane α helices in this part of the protein, taking into account alternative possibilities including membrane-spanning β-sheets (Storck et al., 1992). It is clear that experimental approaches to delineate the topology of these transporters are badly needed.

## Regulation of neurotransmitter transport

It is conceivable that the re-uptake process is subject to physiological regulation. However, our knowledge of this aspect of neurotransmitter function is rudimentary. It has been shown that arachidonic acid, which may be released via phospholipase A$_2$, can inhibit several sodium-coupled uptake systems (Rhoads et al., 1983), including the uptake systems for glycine (Zafra et al., 1990) and glutamate (Barbour et al., 1989). Nieoullon et al. (1983) found that *in vivo* electrical stimulation (for 10 min) of the frontal cortex increased high-affinity glutamate uptake in rat striatum. The increase was due to an increase in affinity. The uptake measurements were done in tissue samples dissected out 20 min after the cessation of stimulation. This increase from the basal level could be inhibited by dopaminergic activity (Kerkerian et al., 1987). The existence of putative phosphorylation sites (Pines et al., 1992) indicates that this glial glutamate transporter may be regulated by protein kinases and phosphatases. The finding that glutamate transport activity ($V_{max}$, but not $K_m$) is increased in cultured glial cells after incubation of the cells with phorbol esters (Casado et al., 1991) suggests that the putative phosphorylation sites are physiologically relevant. We have provided evidence that this stimulation of glutamate transport by phorbol esters is due to a direct phosphorylation of the transporter by protein kinase C. A single serine residue (serine-113) located in the loop connecting putative transmembrane helices 2 and 3 appears to be the major site of this phosphorylation (Casado et al., 1993).

In the case of the GABA transporter, modulation of GABA transport activity by phorbol esters has been reported. However, a point of uncertainty regarding this phenomenon is that, whereas one group reported stimulation by the phorbol ester (Corey et al., 1994), another group found preincubation of oocytes with this compound to be inhibitory (Osawa et al., 1994).

## Structure–function relationships in the superfamily of neurotransmitter transporters

It has been shown previously that parts of N- and C-termini of the GABA$_A$ transporter are not required for function (Mabjeesh and Kanner, 1992). In order to define these domains, a series of deletion mutants of the GABA transporter was studied (Bendahan and Kanner, 1993). Transporters truncated at either end until just a few amino-acids distance from the beginning of helix 1 and the end of helix 12 retained their ability to catalyse sodium- and chloride-dependent GABA transport. These deleted segments did not contain any residues conserved among the different members of the superfamily. In contrast, when the truncated segment included some of these conserved residues, the transporter's activity was severely reduced. However, the functional damage was not due to impaired turnover or impaired targeting of the truncated proteins (Bendahan and Kanner, 1993).

Fragments of the $(Na^+ + Cl^-)$-coupled GABA$_A$ transporter were produced by proteolysis of membrane vesicles and reconstituted preparations from rat brain (Mabjeesh and Kanner, 1993). The former were digested with Pronase; the latter with trypsin. Fragments with different apparent molecular masses were recognized by sequence-directed antibodies raised against this transporter. When

GABA was present in the digestion medium, the generation of these fragments was almost entirely blocked (Mabjeesh and Kanner, 1993). At the same time, the neurotransmitter largely prevented the loss of activity caused by the protease. The effect was specific for GABA; protection was not afforded by other neurotransmitters. It was only observed when the two co-substrates, sodium and chloride, were present on the same side of the membrane as GABA (Mabjeesh and Kanner, 1993). The results indicate that the transporter may exist in two conformations. In the absence of one or more of the substrates, multiple sites located throughout the transporter are accessible to the proteases. In the presence of all three substrates, i.e. conditions favouring the formation of the translocation complex, the conformation is changed such that these sites become inaccessible to protease action.

We have investigated the role of the hydrophilic loops connecting the putative transmembrane α-helices in GAT-1. Deletions of randomly chosen non-conserved single amino acids in the loops connecting helices 7 and 8 or 8 and 9 result in inactivation of transport upon expression in HeLa cells. However, transporters in which these amino acids are replaced by glycine retain significant activity. The expression levels of the inactive mutant transporters were similar to that of the wild-type, but one of them, $\Delta$Val-348, appears to be defectively targeted to the plasma membrane. Our data are compatible with the idea that a minimal length of the loops is required, presumably to enable the transmembrane domains to interact optimally with each other (Kanner et al., 1994).

The substrate translocation performed by the various members of the superfamily is sodium- and usually chloride-dependent. In addition, some of the substrates contain charged groups as well. Therefore charged amino acids in the membrane domain of the transporters may be essential for their normal function. This was tested using the GABA transporter (Pantanowitz et al., 1993). Out of five charged amino acids within its membrane domain, only one, arginine-69 in helix 1, is absolutely essential for activity. It is not merely the positive charge that is important, as the substitution of this residue with other positively charged amino acids does not restore activity. The functional damage is not due to impaired turnover or impaired targeting of the mutated protein. The three other positively charged amino acids and the only negatively charged one are not critical (Pantanowitz et al., 1993). It is possible that arginine-69 may be involved in chloride binding.

The transporters of biogenic amines contain an additional negatively charged residue in helix 1, i.e. aspartate-79 (dopamine transporter numbering). Replacement of aspartate-79 in the dopamine transporter with alanine, glycine or glutamate significantly reduced the transport of dopamine and 1-methyl-4-phenylpyridine (MPP$^+$) (a Parkinsonism-inducing neurotoxin), and the binding of $(-)$-2$\beta$-carbomethoxy-3$\beta$-fluorophenyl)tropane (CFT; a cocaine analogue) without affecting the $B_{max}$. Apparently, aspartate-79 in helix 1 interacts with the amine of dopamine during the transport process. Serine-356 and serine-359 in helix

7 are also involved in dopamine binding and translocation, perhaps by interacting with the hydroxy groups on the catechol (Kitayama et al., 1992).

Studies of other proteins indicate that, in addition to charged amino acids, aromatic amino acids containing $\pi$-electrons are also involved in maintaining the structure and function of these proteins (Sussman and Silman, 1992). Tryptophan residues in the membrane domain of the GABA transporter were mutated into serine or leucine (Kleinberger-Doron and Kanner, 1994). Mutations at positions 68 and 222 (in helix 1 and helix 4 respectively) led to a decrease of over 90% in GABA uptake. On the basis of the sequence alignments of the transporters of the superfamily, it was postulated that tryptophan-222 is involved in the binding of the amino group of GABA.

In the case of the glutamate transporter GLT-1, in addition to the putative protein kinase C consensus sites, the conserved charged amino acids have been mutated and analysed. These include a conserved lysine located in helix 5 and a histidine in helix 6. While the former is not critical, the latter, histidine-326, is essential (Zhang et al., 1994). Furthermore, five conserved negatively charged amino acids are clustered in the C-terminal hydrophobic part of the transporter for which the hydropathy plot is ambiguous. Three of these residues appear to be critical for the function of the transporter, and one of them appears to be involved in substrate discrimination (Pines et al., 1995).

*I thank Mrs. Beryl Levene for expert secretarial assistance. The work from my laboratory was supported by the Bernard Katz Minerva Center for Cell Biophysics, and by grants from the U.S.–Israel Binational Science Foundation, the Basic Research Foundation administered by the Israel Academy of Sciences and Humanities, the National Institutes of Health, and the Bundesministerium fur Forschung und Technologie.*

## References


Amara, S.G. and Kuhar, M.J. (1993) Annu. Rev. Neurosci. **16**, 73–93

Arriza, J.L., Kavanagh, M.P., Fairman, W.A., Wu, Y.-N., Murdoch, G.H., North, R.A. and Amara, S.G. (1993) J. Biol. Chem. **268**, 15329–15332

Barbour, B., Brew, H. and Atwell, D. (1988) Nature (London) **335**, 433–435

Barbour, B., Szatkowski, M., Ingledew, N. and Atwell, D. (1989) Nature (London) **342**, 918–920

Bendahan, A. and Kanner, B.I. (1993) FEBS Lett. **318**, 41–44

Blakely, R.D., Benson, H.E., Fremeau, Jr., R.T., Caron, M.G., Peek, M.M., Prince, H.K. and Bradley, C.C. (1991) Nature (London) **353**, 66–70

Borden, L.A., Smith, K.E., Hartig, P.R., Branchek, T.A. and Weinshank, R.L. (1992) J. Biol. Chem. **267**, 21098–21104

Bouvier, M., Szatkowski, M., Amato, A. and Atwell, D. (1992) Nature (London) **335**, 433–435

Brew, H. and Atwell, D. (1987) Nature (London) **327**, 707–709

Casado, M., Zafra, F., Aragon, C., and Gimenez, C. (1991) J. Neurochem. **57**, 1185–1190

Casado, M., Bendahan, A., Zafra, F., Danbolt, N.C., Aragon, C., Gimenez, C. and Kanner, B.I. (1993) J. Biol. Chem. **268**, 27313–27317

Clark, J.A., Deutch, A.Y., Gallipoli, P.Z. and Amara, S.G. (1992) Neuron **9**, 337–348

Corey, J.L., Davidson, N., Lester, H.A., Brecha, N. and Quick, M.W. (1994) J. Biol. Chem. **269**, 14759–14767

Danbolt, N.C., Pines, G. and Kanner, B.I. (1990) Biochemistry **29**, 6734-6740

Danbolt, N.C., Storm-Mathisen, J. and Kanner, B.I. (1992) Neuroscience **51**, 295–310

Erecinska, M., Wantorsky, D. and Wilson, D.F. (1983) J. Biol. Chem. **258**, 9069–9077

Fremeau, Jr., R.T., Caron, M.G. and Blakely, R.D. (1992) Neuron **8**, 915–926

Graham, D., Esnaud, H. and Langer, S.Z. (1992) Biochem. J. **286**, 801–805

Guastella, J., Nelson, N., Nelson, H., Czyzyk, L., Keynan, S., Miedel, M.C., Davidson, N., Lester, H. and Kanner, B.I. (1990) Science **249**, 1303–1306

Guastella, J., Brecha, N., Weigmann, C. and Lester, H.A. (1992) Proc. Natl. Acad. Sci. U.S.A. **89**, 7189–7193

Guimbal, C. and Kilimann, M.W. (1993) J. Biol. Chem. **268**, 8418–8421

Hees, B., Danbolt, N.C., Kanner, B.I., Haase, W., Heitmann, K. and Koepsell, H. (1992) J. Biol. Chem. **267**, 23275–23281

Hoffman, B.J., Mezey, E. and Brownstein, M.J. (1991) Science **254**, 579–580

Jiang, J., Gu, B., Albright, L.M. and Nixon, B.T. (1989) J. Bacteriol. **171**, 5244–5253

Johnston, G.A.R. (1981) in Glutamate: Transmitter in the Central Nervous System (Roberts, P.J., Storm-Mathisen, J. and Johnston, G.A.R., eds.), pp. 77–87, John Wiley and Sons, Chichester/New York/Brisbane/Toronto

Kanai, Y. and Hediger, M.A. (1992) Nature (London) **360**, 467–471

Kanner, B.I. (1983) Biochim. Biophys. Acta **726**, 293–316

Kanner, B.I. (1989) Curr. Opin. Cell Biol. **1**, 735–738

Kanner, B.I. and Bendahan, A. (1982) Biochemistry **21**, 6327–6330

Kanner, B.I. and Schuldiner, S. (1987) CRC Crit. Rev. Biochem. **22**, 1–39

Kanner, B.I. and Sharon, I. (1978) Biochemistry **17**, 3949–3953

Kanner, B.I., Bendahan, A., Pantanowitz, S. and Su, H. (1994) FEBS Lett. **356**, 192–194

Kavanaugh, M.P., Arriza, J.L., North, R.A. and Amara, S.G. (1992) J. Biol. Chem. **267**, 22007–22009

Kerkerian, L., Dusticier, N. and Nieoullon, A. (1987) J. Neurochem. **48**, 1301–1306

Keynan, S. and Kanner, B.I. (1988) Biochemistry **27**, 12–17

Kilty, J.E., Lorang, D. and Amara, S.G. (1991) Science **254**, 578–579

Kitayama, S., Shimada, S., Xu, H., Markham, L., Donovan, D.M. and Uhl, G.R. (1992) Proc. Natl. Acad. Sci. U.S.A. **89**, 7782–7785

Kleinberger-Doron, N. and Kanner, B.I. (1994) J. Biol. Chem. **269**, 3063–3067

Launay, J.M., Geoffroy, C., Mutel, V., Buckle, M., Cesura, A., Alouf, J.E. and Da-Prada, M. (1992) J. Biol. Chem. **267**, 11344–11351

Liu, Q.R., Lopez-Corcuera, B., Nelson, H., Mandiyan, S. and Nelson, N. (1992a) Proc. Natl. Acad. Sci. U.S.A. **89**, 12145–12149

Liu, Q.R., Nelson, H., Mandiyan, S., Lopez-Corcuera, B. and Nelson, N. (1992b) FEBS Lett. **305**, 110–114

Liu, Q.R., Lopez-Corcuera, B., Mandiyan, S., Nelson, H. and Nelson, N. (1993a) J. Biol. Chem. **268**, 2104–2112

Liu, Q.R., Lopez-Corcuera, B., Mandiyan, S., Nelson, H. and Nelson, N. (1993b) J. Biol. Chem. **268**, 22802–22808

Liu, Q.R., Mandiyan, S., Lopez-Corcuera, B., Nelson, H. and Nelson, N. (1993c) FEBS Lett. **315**, 114–118

Lopez-Corcuera, B., Vazquez, J. and Aragon, C. (1991) J. Biol. Chem. **266**, 24809–24814

Lopez-Corcuera, P., Liu, Q.R., Mandiyan, S., Nelson, H. and Nelson, N. (1992) J. Biol. Chem. **267**, 17491–17493

Mabjeesh, N.J. and Kanner, B.I. (1992) J. Biol. Chem. **267**, 2563–2568

Mabjeesh, N.J. and Kanner, B.I. (1993) Biochemistry **32**, 8540–8546

Mager, S.J., Naeve, J., Quick, M., Guastella, J., Davidson, N. and Lester, H.A. (1993) Neuron **10**, 177–188

Mager, S., Min, C., Henry, D.J., Chavkin, L., Hoffman, B.J., Davidson, N. and Lester, H.A. (1994) Neuron **12**, 845–859

Mayser, W., Schloss, P. and Betz, H. (1992) FEBS Lett. **305**, 31–36

McBean, G.J. and Roberts, P.J. (1985) J. Neurochem. **44**, 247–254

Nieoullon, A., Kerkerian, L. and Dusticier, N. (1983) Neurosci. Lett. **43**, 191–196

Osawa, I., Saito, N., Koga, T. and Tanaka, C. (1994) Neurosci. Res. **19**, 287–293

Pacholczyk, T., Blakely, R.D. and Amara, S.G. (1991) Nature (London) **350**, 350–353

Pantanowitz, S., Bendahan, A. and Kanner, B.I. (1993) J. Biol. Chem. **268**, 3222–3225

Pines, G. and Kanner, B.I. (1990) Biochemistry **29**, 11209–11214

Pines, G., Danbolt, N.C., Bjoras, M., Zhang, Y., Bendahan, A., Eide, L., Koepsell, H., Storm-Mathisen, J., Seeberg, E. and Kanner, B.I. (1992) Nature (London) **360**, 464–467

Pines, G., Zhang, Y. and Kanner, B.I. (1995) J. Biol. Chem. **270**, 17093–17097

Radian, R., Bendahan, A. and Kanner, B.I. (1986) J. Biol. Chem. **261**, 15437–15441

Radian, R., Ottersen, O.L., Storm-Mathisen, J., Castel, M. and Kanner, B.I. (1990) J. Neurosci. **10**, 1319–1330

Rhoads, D.E., Ockner, B.K., Peterson, N.A. and Raghupathy, E. (1983) Biochemistry **22**, 1965–1970

Schloss, P., Mayser, W. and Betz, H. (1992) FEBS Lett. **307**, 76–78

Shafqat, S., Tamarappoo, B.K., Kilberg, M.S., Puranam, R.S., McNamara, J.O., Guadaño-Ferraz, A. and Fremeau, R.T. (1993) J. Biol. Chem. **268**, 15351–15355

Shimada, S., Kitayama, S., Lin, C.L., Patel, A., Nanthakumar, E., Gregor, P., Kuhar, M. and Uhl, G. (1991) Science **254**, 576–578

Smith, K.E., Borden, L.A., Hartig, P.A., Branchek, T. and Weinshank, R.L. (1992) Neuron **8**, 927–935

Stallcup, W.B., Bullock, K. and Baetge, E.E. (1979) J. Neurochem. **32**, 57–65

Storck, T., Schulte, S., Hofmann, K. and Stoffel, W. (1992) Proc. Natl. Acad. Sci. U.S.A. **89**, 10955–10959

Sussman, J.L. and Silman, I. (1992) Curr. Opin. Struct. Biol. **2**, 721–729

Tolner, B., Poolman, B., Wallace, B. and Konings, W. (1992) J. Bacteriol. **174**, 2391–2393

Uchida, S., Kwon, H.M., Yamauchi, A., Preston, A.S., Marumo, F. and Handler, J.S. (1992) Proc. Natl. Acad. Sci. U.S.A. **89**, 8230–8234

Uhl, G.R. (1992) Trends Neurosci. **15**, 265–268

Uhl, G.R., Kitayama, S., Gregor, P., Nanthakumer, E., Persico, A. and Shimada, S. (1992) Mol. Brain Res. **16**, 353–359

Usdin, T.B., Mezey, E., Chen, C., Brownstein, M.J. and Hoffman, B.J. (1991) Proc. Natl. Acad. Sci. U.S.A. **88**, 11168–11171

Yamauchi, A., Uchida, S., Kwon, H.M., Preston, A.S., Robey, R.B., Garcia-Perez, A., Burg, M.B. and Handler, J.S. (1992) J. Biol. Chem. **267**, 649–652

Zafra, F., Alcantara, R., Gomeza, J., Aragon, C. and Gimenez, E. (1990) Biochem. J. **271**, 237–242

Zhang, Y., Pines, G. and Kanner, B.I. (1994) J. Biol. Chem. **269**, 19573–19577

# 3

# Electrophysiology of GABA$_A$ receptors

Trevor G. Smart

The School of Pharmacy, Department of Pharmacology,
29–39 Brunswick Square, London WC1N 1AX, U.K.

## Introduction

It has become increasingly clear over the last 30 years that excitatory and inhibitory synaptic transmission play major roles in the neuronal processing of sensory and motor information in the mammalian central nervous system (CNS). The control of neuronal excitation is, not surprisingly, multi-faceted, and ranges from the regulation of the release of the main excitatory neurotransmitter, glutamate, to the activation of inhibitory receptors or ion channels. The later aspect can be accomplished by activation of voltage-gated ion channel populations that are mostly permeable to $K^+$, driving the cell membrane potential away from the threshold for action potential generation. Another mechanism for the control of excitation is achieved by fast inhibition of neurons following the release of the amino acid $\gamma$-aminobutyric acid (GABA), which then activates membrane-bound proteins commonly referred to as GABA$_A$ receptors [1]. Two other receptors can be activated by GABA: a slower, more prolonged, inhibition results from GABA activating G-protein-coupled GABA$_B$ receptors, invariably leading to an increased membrane conductance to $K^+$ [2]; and a novel receptor, termed GABA$_C$, that is directly gated by GABA permitting an increased $Cl^-$ conductance which is insensitive to bicuculline [3–5]. The significance of GABA$_C$ receptors with regard to synaptic inhibition awaits detailed study. The importance of GABA$_A$ receptors to neural regulation of excitability is reflected by their ubiquitous distribution throughout the CNS. GABA$_A$ receptors appear on the majority of neurons in the CNS and the peripheral autonomic nervous system, and can also be found on various types of glia.

Pharmacological studies of GABA$_A$ receptors have revealed two unifying features that have become almost diagnostic criteria for identifying the presence of this class of receptor, namely their permeability to $Cl^-$ and their sensitivity to inhibition by the phthalide isoquinoline alkaloid bicuculline. Quite how many different types of GABA$_A$ receptor are present in the CNS remains unsolved, but recent molecular biological studies have supported long-standing pharmacological evidence for the likely presence of more than one type of GABA$_A$ receptor. The basis for this receptor heterogeneity may well originate from the structure of the receptors. Cloning and electron microscopic studies have demonstrated that the GABA$_A$ receptor is a hetero-oligomeric protein, probably composed of five polypeptide subunits forming an integral ion channel that is permeable to Cl⁻ [6–8]. The number of subunits synthesized in the CNS is in excess of five, and comparisons of their amino acid sequence identities have indicated that four major discrete families of subunits exist, possessing similar transmembrane architecture and defined as: α1–α6, β1–β4, γ1–γ4 and δ1. With the exception of the δ subunit, the three remaining subunit families are composed of multiple members. The structural diversity of the subunits is further increased by the ability of the β2, β4 and γ2 genes, in some species, to undergo alternative splicing, each yielding two products with and without a splice insert in the amino acid sequence.

The provision of this relatively and unexpectedly large number of GABA$_A$ receptor subunit cDNAs has required the use of cell-based expression systems, such as the *Xenopus laevis* oocyte and numerous secondary cell lines, in order to assemble and express recombinant GABA$_A$ receptors. This has enabled the crucial question, 'how does receptor structure influence receptor function?', to be posed. The aim of this chapter is to review recent evidence regarding the physiological and pharmacological properties of native and recombinant GABA$_A$ receptors, and to examine how the 'structure–function question' is being addressed.


# Physiology of GABA$_A$ receptor ion channels

## Synaptic physiology: inhibitory transmission

The major role of GABA$_A$ receptors in the CNS is to provide fast and reversible inhibition of neurons. This is achieved by GABA being released from interneurons [9–11] and markedly increasing the conductance of post-synaptic membranes to Cl⁻. Electrophysiological recordings from neurons either maintained in dissociated or organotypic slice cultures, or from acutely prepared brain slices, are the preparations of choice for investigating synaptic inhibition. Previously, most studies were based on the intracellular recording technique, but the later use of patch-clamp recording methods has significantly enhanced the resolution of small synaptic currents hitherto unseen. There are at least two main types of inhibitory post-synaptic potential (IPSP) that can be evoked by the release of GABA: the stimulus-evoked IPSP is similar to many other synaptic potentials in being sensitive to tetrodotoxin and external Ca$^{2+}$ [1], whereas the much-smaller-amplitude spontaneous IPSPs (sIPSPs; also known as miniature IPSPs), which constantly bombard neurons in the resting state in the absence of any overt stimulation, are relatively unaffected by tetrodotoxin or external Ca$^{2+}$ [12–15] (Fig. 1). The relationship between the IPSP and the sIPSPs was believed to be strictly quantal [12], but resolution of a third form of GABA synaptic potential, involving

## Fig. 1.        IPSPs in CA1 hippocampal pyramidal neurons

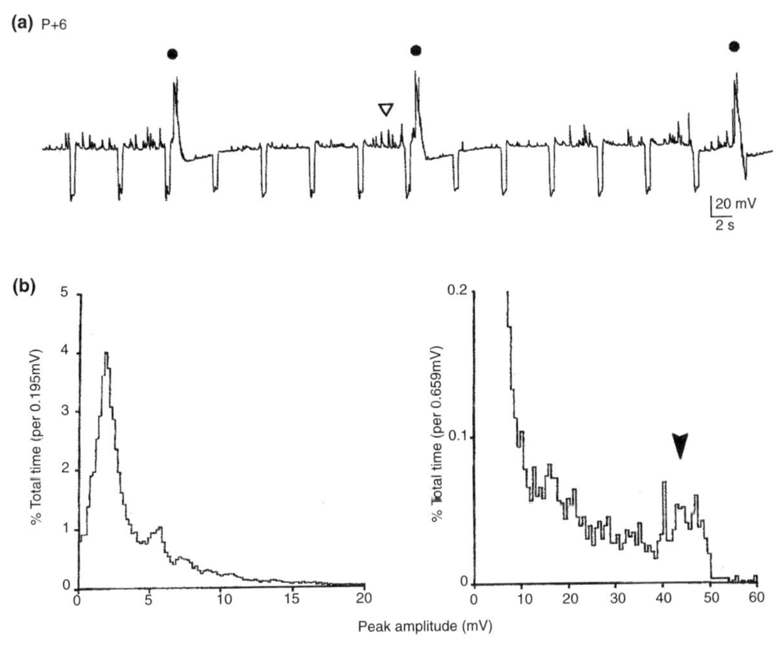



(a) Intracellular recording from a single neonatal (postnatal day 6; P+6) pyramidal neuron, demonstrating small- and medium (arrowhead)-amplitude sIPSPs, in addition to large-amplitude GABA$_A$-receptor-mediated events (●). The downward deflections (electrotonic potentials) monitor the conductance of the membrane. Note the fluctuations and increased conductance during the sIPSPs. (b) sIPSP amplitude histogram showing multimodal behaviour (left-hand plot); the mean amplitude of the large events is between 40 and 50 mV (right-hand plot). Data from X. Xie and T.G. Smart (unpublished work).

'subquantal' or small-mode synaptic conductances elicited by GABA in hippocampal neurons, suggests that a revision of the classic quantal theory may be necessary at some GABAergic synapses [16,17]. The different amplitudes of the small-mode and large-mode sIPSPs were explained on the basis of different frequencies in the release of presynaptic GABA from synaptic boutons [17]. Furthermore, from the peak conductance of the synaptic potentials, small-mode IPSPs might result from the synchronized opening of only three to five GABA-activated ion channels, whereas the large mode IPSPs required 50–200 synchronously activated channels [16]. In comparison, the number of GABA-activated channels underlying a single 'quantal event' during inhibitory synaptic transmission in the dentate gyrus could be approx. 6–30 [12].

Irrespective of the different release mechanisms that might influence the amplitudes of the sIPSPs, the mode of inhibition will primarily be dependent upon an increased flux of $Cl^-$. An increased membrane permeability to $Cl^-$ often results in a shift in the membrane potential ($E_m$), which can be either hyperpolarizing or depolarizing, towards the $Cl^-$ equilibrium potential ($E_{Cl}$). Even if $E_m = E_{Cl}$, activation of the $Cl^-$ channels by GABA will result in inhibition, primarily because this conductance increase provides a 'shunt' for excitatory synaptic currents and currents underlying the action potential [18].

The relatively small numbers of $GABA_A$ receptor ion channels activated during a sIPSP and the inability of positive modulators such as benzodiazepines to enhance the amplitude of the current underlying the sIPSP (sIPSC) [19] (Fig. 2), as well as the small variance of the peak IPSC amplitude, are all in accordance with the $GABA_A$ receptors in the synapse being saturated by presynaptically released GABA [20]. The likely concentration of GABA in the synaptic cleft could be at least as high as 500 μM [21]. This concentration of GABA could reasonably be expected to induce receptor desensitization, but this phenomenon probably does not influence the profile of IPSCs, since the decay time constants of the current are invariably much smaller than the measured time constants associated with desensitization [22] (although recent evidence suggests that some degree of rapid desensitization can occur at the $GABA_A$ receptor, which might interfere with the IPSC profile [23]). A further factor that could shape the profile of the IPSC is the process of GABA uptake. The potential large concentrations of GABA in the synaptic cleft will either diffuse away from the $GABA_A$ receptors after dissociation or be

---

**Fig. 2.**   **Inhibitory synaptic transmission and effects of benzodiazepines in granule neurons of hippocampal brain slices**

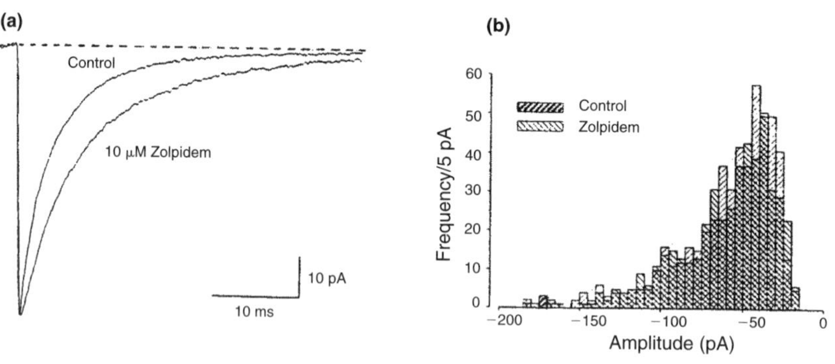


*(a) Averaged sIPSCs are shown in control Krebs solution and after the addition of the benzodiazepine agonist zolpidem (10 μM). Note the prolongation in the duration of the IPSC without an increase in the amplitude. (b) The amplitude histogram collates data from 400–500 sIPSCs before and after exposure to zolpidem. Adapted from [20], with permission.*

transported into neurons or adjacent glia. GABA uptake inhibitors have no discernable effect on the decays of sIPSPs, suggesting that GABA uptake is not usually required to remove synaptically released GABA [23a], but the decays of evoked IPSCs are significantly prolonged by uptake inhibition [23a]. GABA transporters might, therefore, limit the spread of GABA to adjacent synapses, preventing activation of juxtaposed GABA$_A$ and also GABA$_B$ receptors [24].

The modulation of synaptic inhibition can take various forms. A presynaptic enhancement of release will lead to an increased frequency of sIPSCs, and enhancing post-synaptic GABA$_A$ receptors might be expected to lead to an increased sIPSC amplitude and/or a prolonged duration. If the GABA$_A$ receptors present in the synaptic cleft are saturated by GABA, then quite probably an enhancement of synaptic inhibition will only be achieved by prolonging the duration of the sIPSC. Since individual interneurons will form multiple synapses with nearby 'principal neurons' [11], it is conceivable that localized stimulation of the interneuron will result in the release of GABA from many, if not all, of the synaptic boutons; however, the release may not be synchronized, resulting in a compound IPSC failing to attain maximum amplitude. If an agent can promote synchronization, then the compound IPSC could be enhanced in amplitude and duration. This is most likely to occur via presynaptic mechanisms affecting the release of GABA. Two agents that can apparently promote synchronization are 4-aminopyridine [25–27] and Zn$^{2+}$ [28,29] (Fig. 3). Zn$^{2+}$ appears to be more specific for synchronizing release from GABAergic neurons, yielding large-amplitude synaptic potentials with typical GABA$_A$ receptor pharmacology [29–31] (Figs. 3 and 4).

---

**Fig. 3.**     **Spontaneous large-amplitude synaptic potentials corresponding to synchronous release of GABA**

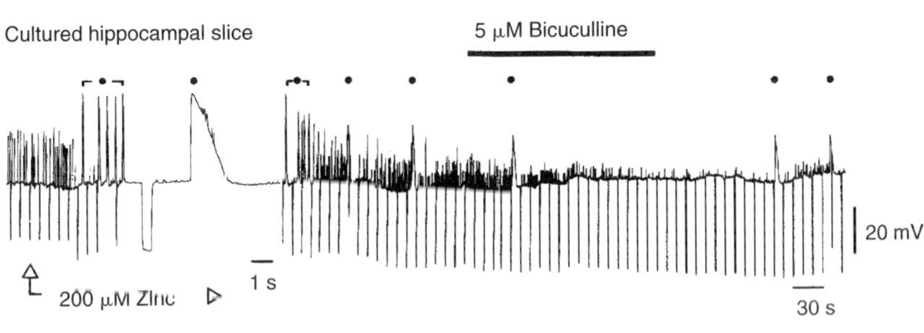


*Intracellular recordings are shown of spontaneous large-amplitude GABA$_A$-receptor-mediated synaptic potentials (●) induced by exposing an organotypically cultured hippocampal brain slice to 200 µM Zn$^{2+}$ at −52 mV. Exposure to 5 µM bicuculline blocks these synaptic events and the smaller-amplitude sIPSPs. Data from U. Gerber, X. Xie, B. Gahwiler and T.G. Smart (unpublished work).*

**Fig. 4.    Comparison of large-amplitude GABA$_A$-receptor-mediated synaptic potentials in adult and neonatal hippocampal brain slices**

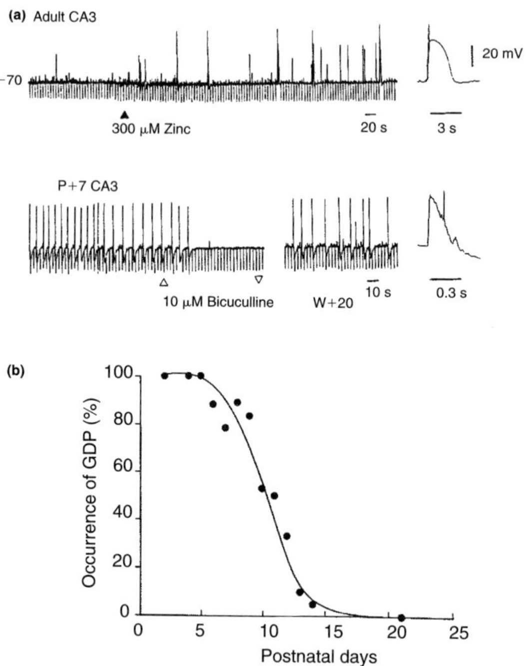


(*a*) *Intracellular recordings (at −70 mV) from adult (upper) and neonatal (postnatal day 7; P+7) (lower) CA3 neurons. Large-amplitude synaptic potentials were induced in the adult slice using 300 μM Zn$^{2+}$. In the neonatal slices, similar events occurred naturally and were inhibited by bicuculline, recovering after 20 min in control Krebs solution (W+20).The right-hand traces illustrate individual synaptic potentials (note the expanded time scales). (**b**) In neonatal hippocampal neurons, the frequency of occurrence of spontaneous large-amplitude events (giant polarizing potentials; GDP) decreases with postnatal development (to 50% after 10 days), becoming difficult to detect after 2 weeks. Data from 168 neurons. (**a**) Reproduced from Nature [Xie, X. and Smart,T.G. (1991) Nature (London) 349, 521–524], copyright 1991, Macmillan Magazines Limited. (**b**) X. Xie and T.G. Smart, unpublished work.*

In the immature hippocampus, an alternative role for GABA has been suggested following the observation of depolarizing synaptic potentials mediated by GABA$_A$ receptors [28,32–34]. Unusually, the large-amplitude depolarizing synaptic potentials occur rhythmically, although they exhibit pharmacological properties consistent with the activation of GABA$_A$ receptors (Fig. 4). The depolarization probably results from an unusual Cl$^-$ gradient across the cell membranes of neonatal neurons, making $E_{Cl}$ reside at a more depolarized level with respect to the resting potential when compared with $E_{Cl}$ in the adult CNS [35]. As the CNS matures, $E_{Cl}$ adopts a level that is hyperpolarized with respect to $E_m$, and the spontaneous, rhythmically occurring, GABA$_A$-receptor-mediated synaptic potentials become hyperpolarizing and eventually, within 2 weeks of postnatal development, they wane and disappear [32] (Fig. 4). The large depolarizing synaptic potentials can initiate action potential firing in hippocampal CA3 pyramidal neurons, suggesting that GABA may be acting as an excitatory transmitter enabling a secondary increase in intracellular $Ca^{2+}$ to occur [36], which could be important for subsequent neuronal development [35]. While the large depolarization is overtly excitatory by initiating action potential firing, the accompanying large increase in $Cl^-$ conductance could still act as a shunt for excitation and thus, at this stage of development, GABA may have a dual role, finely balanced between excitation and inhibition.

## Single GABA-activated ion channels

Patch-clamp recording, in addition to improving the resolution of inhibitory synaptic currents, has also enabled a direct measure of current flow through single GABA-activated ion channels. Most investigators have preferred the use of the more flexible excised patches, usually the outside-out configuration and less frequently the inside-out patch. Single GABA channel currents are analysed according to various criteria that include: channel conductance ($\gamma$), mean open and closed lifetimes ($\tau$) and the probability of channel opening ($P_o$), which is also composed of the number of active channels in the patch ($N$) that are activated by GABA and their frequency of opening ($f$). Single-channel conductance measurements taken from native GABA$_A$ receptors resident on cultured neurons have revealed a preferred main conductance state (range 20–30 pS) that is responsible for up to 90% of the GABA-activated $Cl^-$ current across the membrane, and usually a number of less frequently adopted sub- or supra-conductance states (8–71 pS) [36a–42] (Fig. 5). These conductance levels appear to feature regularly in different preparations, suggesting some consistency in the structure of the ion channel amongst different GABA$_A$ receptors, although the underlying reason for the fluctuation between conductances is unknown (see the next section).

Analyses of single-channel lifetimes or dwell times, in open (conducting) and closed or shut (non-conducting) states, have been used to gain insight into the gating kinetics of single GABA channels. The open and closed time distributions are routinely fitted by two or invariably more exponential functions, suggesting that the channel can adopt multiple open and closed states, and perhaps also desensitized states [39,42–46]. Increasing the concentration of GABA did not alter the time constants describing the open times for GABA-activated channels in cultured spinal cord neurons, but increased the frequency of long openings at the expense of the shorter openings [39]. In comparison, the longer closed time periods, usually associated with closed times between channel activation, were reduced as the GABA concentration increased [39]. To enable resolution of discrete single-

**Fig. 5.**          **Single GABA-activated ion channels in cultured large**
                     **cerebellar neurons**

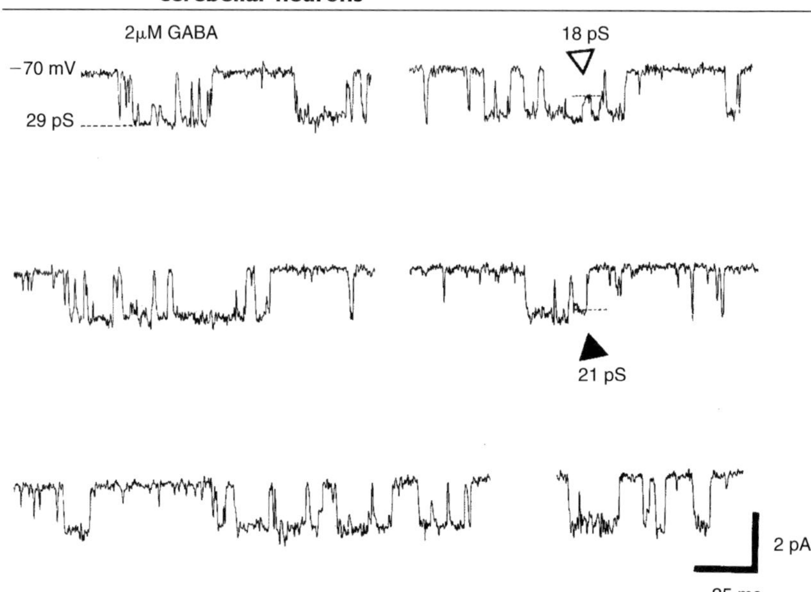

An outside-out patch exposed to 2 $\mu$M GABA produced single-channel openings (downward deflections) to various conductance levels at a holding potential of $-70$ mV. The main-state conductance was 29 pS and accounted for the majority of current carried through the ion channel (>90%). Less frequently visited smaller subconductance states were also observed at 18 pS (open arrowhead) and 21 pS (closed arrowhead). T.G. Smart, unpublished work.

channel events, most studies use low concentrations of GABA (0.5–2 $\mu$M) to elicit single-channel currents. The use of higher concentrations in sympathetic neurons demonstrated that $P_o$ increased with increasing GABA concentration as expected; however, the values of $P_o$ at each GABA concentration (10–2000 $\mu$M) exhibited considerable variation that is not consistent with a single species of $GABA_A$ receptor ion channel [41] (Fig. 6). One explanation for these data might simply be that there is a heterogeneous mix of $GABA_A$ receptors in the patch, with each one possessing different kinetic properties; alternatively, a single $GABA_A$ receptor ion channel could exhibit 'seed changes' in kinetic behaviour with time, perhaps entering different 'gating modes'.

Anion substitution experiments have been used to find out more about the permeability properties of the GABA-activated ion channel. Replacing $Cl^-$ in the external Krebs and pipette electrolyte either with the halides ($Br^-$, $I^-$ or $F^-$) or with larger polyanions resulted in a relative permeability sequence for GABA-activated channels in mouse spinal cord neurons of: $I^- > Br^- > Cl^- > F^- >$ formate >

## Fig. 6.  Variation in $P_o$ in cultured sympathetic neurons

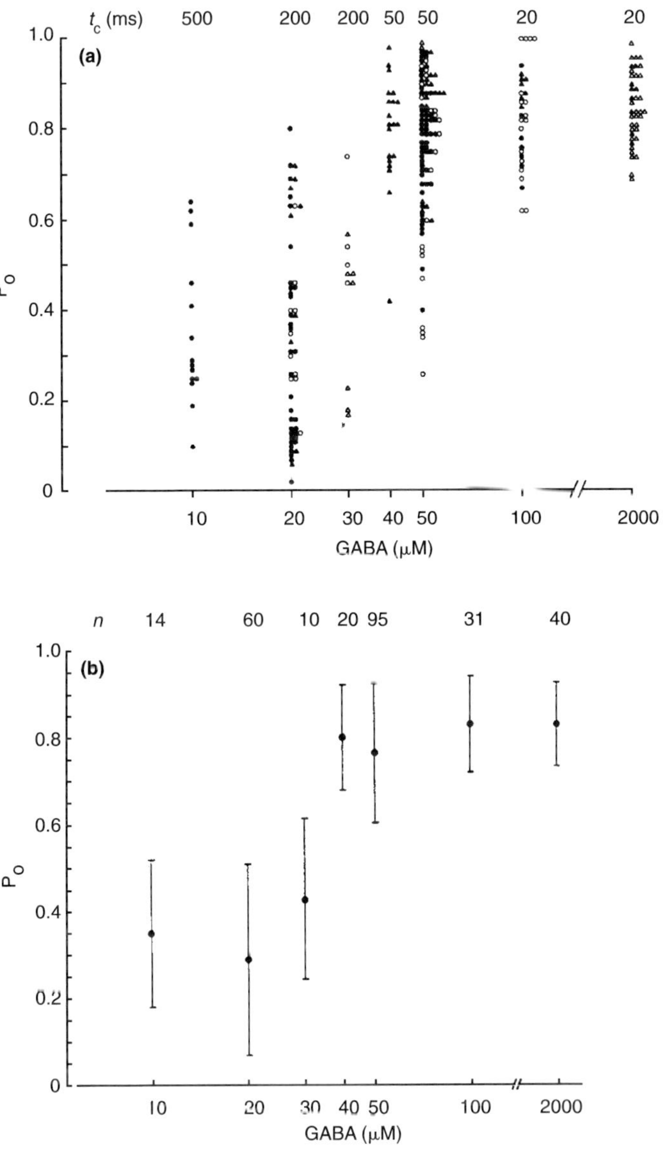



*(a) The probability of the channel opening, $P_o$, is represented as a function of the GABA concentration. Each symbol represents a burst of channel activity defined by the critical gap length ($t_c$, ms). (b) Mean data calculated from (a), with error bars representing S.D.; n is the number of bursts averaged at each concentration of GABA. Note the wide variation in $P_o$. Results reproduced from [41], with permission.*

bicarbonate >acetate >phosphate >proprionate [47]. The diameter of the ion channel was estimated from bulk space filling models of the various inorganic and organic anions to be approx. 5.6 Å [47] (cf. a value for the pore diameter of approx. 3.25 Å [48], obtained by measuring the effect of anions on IPSPs in spinal motor neurons). The value of 5.6 Å is similar to estimates derived from cortical neurons (5.8 Å) [49], and is compatible with an ion channel formed from a pentameric arrangement of polypeptide subunits and with recent electron microscopic data on the quaternary structure of a porcine native $GABA_A$ receptor [8]. The ability of $HCO_3^-$ to permeate through the GABA-activated ion channel raises the interesting possibility of $HCO_3^-$ efflux causing transient changes in external pH [49a].

## Recombinant $GABA_A$ receptor ion channels

Some of the variability associated with the single-channel data collected with native neuronal $GABA_A$ receptors might be partly explained by studying cloned $GABA_A$ receptors. The conductance and gating properties of recombinant $GABA_A$ receptors do vary depending on the polypeptide subunit composition. For hetero-oligomeric $GABA_A$ receptors, the smallest main-state conductances of 15–18 pS are produced from receptors comprising $\alpha1\beta1$ subunits [50–52]. Although it is presently unclear whether the single-channel conductance will vary with different $\alpha$ and/or $\beta$ subunits, lower single-channel conductances of 11 pS are apparent with $\alpha1\beta2$ $GABA_A$ receptors [53]. Moreover, replacing the $\beta2$ subunit with $\gamma2$, giving $\alpha1\gamma2$ subunit receptors, results in an increased channel conductance of 30 pS [53]. Expression of $GABA_A$ receptors containing $\alpha1\beta1\gamma2S$ (29 pS) [51] or $\alpha1\beta2\gamma2S$ (32 pS) [53] subunits (where $\gamma2S$ is the short form of $\gamma2$) also produced channels with conductances higher than those of $\alpha\beta$-subunit-containing receptors.

Recombinant $GABA_A$ receptor ion channels are also characterized by the presence of subconductance states, which usually appear to occur at approx. 50% of the main conductance state. For $\alpha1\beta1$- and $\alpha1\beta1\gamma2S$-subunit-containing receptors, subconductance states occur at 10 and 15 pS respectively [51], and for $\alpha1\gamma2_S$ at 16 pS [53]. Overall, as for some native $GABA_A$ receptors, 90% of the total $Cl^-$ current passing across the cell membrane is carried by the main conductance state for $\alpha1\beta1$- and $\alpha1\beta1\gamma2S$-subunit-containing receptors. Therefore the significance of the subconductance states is yet to be demonstrated. It is worth noting that, whereas there is broad agreement that $GABA_A$ receptors containing $\alpha\beta$ or $\alpha\beta\gamma$ subunits can be readily expressed for functional assays, $\alpha\gamma$-subunit-containing receptors have proved problematic, with some investigators demonstrating robust large GABA-activated currents (e.g. [53]) whereas others have consistently failed to obtain any functional expression (e.g. [51,54]).

Changing the $\alpha1$ subunit to an allelic variant of the rat $\alpha6$ subunit (where a histidine residue at position 121 is replaced by glutamine), to form $\alpha6\beta1\gamma2S$ $GABA_A$ receptors, resulted in an increased single-channel conductance of 33 pS with an apparent subconductance state at 21 pS [55].

Kinetic analyses of the recombinant GABA$_A$ receptors containing $\alpha1\beta1$ or $\alpha1\beta1\gamma2S$ subunits indicated clear differences in their gating characteristics. Receptors composed of $\alpha1\beta1$ subunits exhibited only two open states for the main conductance state (15 pS), whereas both the subconductance (21 pS) and main conductance (29 pS) states for $\alpha1\beta1\gamma2S$ exhibited three open states [51]. The time constants describing the various open states for $\alpha1\beta1$ and $\alpha1\beta1\gamma2S$ receptors were quite similar. The main difference in their kinetics was due to the relative areas described by the time constants in the distribution of all open durations. An equal weighting of the two time constants was evident for $\alpha1\beta1$ receptors, which correlated with the shortest and intermediate open time constants for $\alpha1\beta1\gamma2S$-subunit-containing receptors. However, for $\alpha1\beta1\gamma2S$ main conductance states, greater weighting was attributed to the longest open-time constants [51]. This could account for the apparent absence of long-duration openings for $\alpha1\beta1$ compared with $\alpha1\beta1\gamma2S$ GABA$_A$ receptors. Analysis of the closed times proved more complicated, with up to five ($\alpha1\beta1$; 15 pS state) or six ($\alpha1\beta1\gamma2S$; 21 and 29 pS states) time constants necessary to account for the data. The single-channel data have provided a molecular distinction between the operation of the $\alpha1\beta1$ and $\alpha1\beta1\gamma2S$ GABA$_A$ receptor ion channels. Co-expression of $\alpha1$, $\beta1$ and $\gamma2S$ subunits appears to result in receptors composed of $\alpha1\beta1\gamma2S$ in preference to a mixture of receptors comprising $\alpha1\beta1$ and $\alpha1\beta1\gamma2S$ subunits. This is based partly on the kinetic analyses of single channels (i.e. no openings conforming to $\alpha1\beta1$ subunits are observed in the presence of $\gamma2S$ subunits [51]) and also on the much lower sensitivity of whole-cell GABA-activated currents to inhibition by $Zn^{2+}$, a feature conferred on the GABA$_A$ receptor by the presence of the $\gamma2$ subunit in the receptor complex [56–58].

## Pharmacology of GABA$_A$ receptor ion channels

The proposed transmembrane topography of the polypeptide subunits comprising a GABA$_A$ receptor is based largely on hydropathy profiles and the likely location (i.e. intracellular or extracellular) of key elements of the receptor structure, including N-linked glycosylation sites and various consensus sequences for phosphorylation by protein kinases. Each mature polypeptide subunit contains approx. 400–500 amino acids and, for a pentameric receptor, a total 2000–2500 residues. According to hydropathy plot analyses, up to two-thirds of the receptor, or more precisely 60% of the total number of residues, will reside in the extracellular compartment. The remaining 40% of amino acids will be equally divided between the transmembrane domains (20%) and the two putative intracellular loops (20%) (Fig. 7). Many of the amino acid residues will have charged side chains and/or the facility to form dipoles, giving considerable potential for the binding of external ligands that may affect receptor conformation and ultimately the ability of the ion channel to conduct $Cl^-$. It is therefore possible that GABA$_A$ receptors (and perhaps many other ligand-gated ion channels with a

**Fig. 7.**          **Schematic sectional representation of a GABA$_A$ receptor**

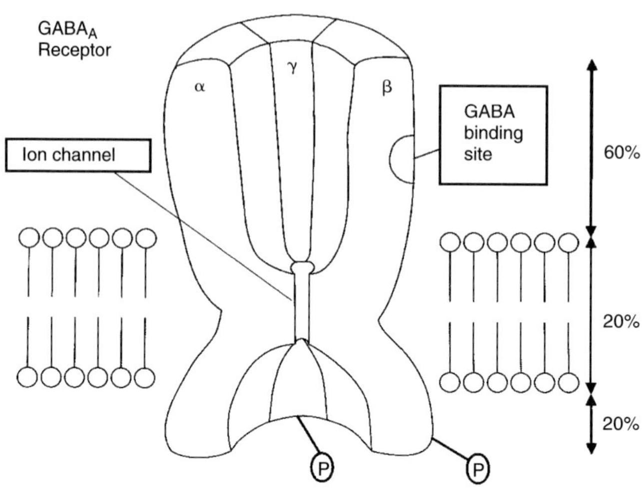



*Based on hydropathy plot analyses, the expected percentages of amino acid residues residing within or either side of the membrane are indicated. Only three out of a possible five subunits are shown. The identification of subunits within any GABA$_A$ receptor has so far proved elusive; however, most of the pharmacological properties of native GABA$_A$ receptors can be emulated by expressing α, β and γ subunits. The relative positions of the subunits and their immediate neighbours within the receptor structure are unresolved. Both β and γ subunits also possess consensus sequences for phosphorylation by protein kinases that can affect receptor function (see Chapter 10 of the present volume).*

similar distribution of amino acids) will be affected by many different classes of ligands. This could occur by opportunistic binding to charged amino acid residues on the receptor, in addition to targeting of compounds to naturally evolved binding sites of importance for endogenous ligands (e.g. the GABA recognition site).

There are numerous examples of GABA$_A$ receptor function being modulated by external ligands, causing enhanced or reduced regulation. In addition, there is also the facility for receptor function to be regulated by intracellular kinases that affect the phosphorylation state of the receptor (e.g. by protein kinase A and protein kinase C; see Chapter 10 of the present volume, by Raymond).

# The GABA$_A$ receptor agonist recognition site

Following the isolation of native bovine brain GABA$_A$ receptors, early biochemical studies [59] suggested that a 56–58 kDa polypeptide, defined as the 'β subunit', could be photoaffinity labelled with [$^3$H]muscimol [60] (and the same for native rat brain GABA$_A$ receptors [61]), and therefore might be responsible for the binding of GABA on to the receptor protein. Later studies, utilizing recombinant GABA$_A$ receptors expressed in *Xenopus* oocytes and site-directed mutagenesis, have attempted to establish the location of the GABA recognition site. Mutations of four amino acid residues (Tyr-157, Thr-160, Thr-202 and Tyr-205 [62]) near the N-terminus of a β subunit impaired the activation of a recombinant receptor (rat α1β2γ2) by GABA. This was evident from the increased EC$_{50}$ values apparent from dose–response curve analyses, and suggested the existence of two separate binding domains for GABA, or at least two domains involved with the activation of the ion channel (Fig. 8). However, additional amino acids in other subunits, particularly the α subunits, could also be important in defining the spatial geometry of the GABA recognition sites [63,64].

Apart from the natural transmitter, GABA, there are several compounds that have been utilized as agonists at GABA$_A$ receptors, including muscimol, 4,5,6,7-tetrahydroisoxazolo[5,4-*c*]pyridin-3-ol (THIP), isoguvacine, piperidine-4-sulphonic acid [65–67] and *trans*-4-aminocrotonic acid [68]. An additional agonist, 5-(4-piperidyl)isoxazol-3-ol (4-PIOL), behaves as a partial agonist on cultured hippocampal neurons [69]. Fluctuation analyses of membrane current noise induced by GABA agonists on rat neurons suggest that the elementary conductance of the ion channel does not vary, but agonists which are apparently more potent have longer mean open times and/or longer burst durations of channel activity [70,71]. However, in chick neurons, muscimol and isoguvacine preferentially activated subconductance states of the GABA-activated ion channel compared with GABA, although muscimol also induced a longer opening state of the channel that was not apparent with GABA [72]. The reasons for the apparent differences in agonist modulation of these GABA$_A$ receptor ion channels in the two species are unclear.

Studies using recombinant GABA$_A$ receptors composed of α1/α3/α5, β1–β3 and γ1–γ3 subunits have established that the EC$_{50}$ for GABA and other agonists is dependent on the subunit composition, with the lowest estimates obtained for GABA with α5β3γ2 and α5β3γ3 [73]. Moreover, the profile of some agonists was also dependent on subunit composition; for example, piperidine-4-sulphonic acid behaved as a partial agonist with recombinant receptors containing α1β2γ2 subunits whereas, with α3 or α5 in combination with β2γ2 subunits, this agent exhibited full agonist properties [73].

**Fig. 8.**          **GABA$_A$ receptor activation and the β subunit**

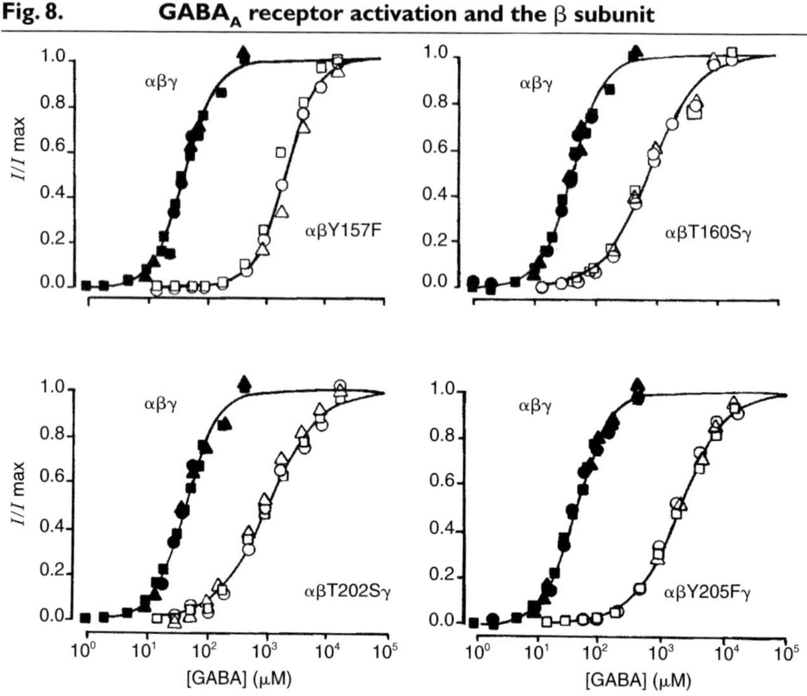



Site-directed mutational analysis of recombinant GABA$_A$ receptors expressed in Xenopus oocytes. Equilibrium concentration–response curves for GABA were constructed for the wild-type receptor (rat α1β2γ2) and for four different mutant receptors. The lateral shifts and increased EC$_{50}$ values for GABA illustrate the importance of four residues that form two separate domains in the β subunit (Tyr-157, Tyr-205, Thr-160 and Thr-202) for the activation of the receptor by GABA. Data obtained from [62]; reproduced from Nature [Amin, J. and Weiss, D.S. (1993) Nature (London) 366, 565–569], copyright 1993, Macmillan Magazines Limited.

# GABA$_A$ receptor antagonists

Bicuculline and picrotoxin are widely regarded as the archetypal GABA$_A$ receptor antagonists. Bicuculline, frequently used as a diagnostic indicator for the presence of GABA$_A$ receptors, acts as a competitive antagonist and binds to a discrete site on the receptor protein, unlike the non-competitive inhibitor picrotoxin [74]. The competitive inhibitory behaviour of bicuculline has been the subject of Schild analysis with native neuronal and recombinant GABA$_A$ receptors, with reported pA$_2$ values of 5.9 (cuneate nucleus slice) [75], 6.24, 6.1 (hippocampus) [76] and 6.4 (α1β1γ2L, where γ2L is the long form of γ2) [77]. For recombinant GABA$_A$ receptors, the pA$_2$ value is apparently consistent between α1β1 (6.0), α1β1γ2S

(5.96) and α1β1γ2L (5.99) receptors [78], demonstrating that inhibition is independent of these particular subunit compositions.

The sesquiterpene picrotoxin, or its active moiety picrotoxinin, is frequently regarded in the literature as an 'open channel' blocker, but the evidence for this has not been totally compelling (see [79–82]). Indeed, inhibition of GABA-activated responses by picrotoxin may result from this agent stabilizing the Cl⁻ ion channel in a closed conformation rather than an overt block of one or more channel open states [83–85]. Other compounds which may act mainly at a similar site(s) to picrotoxin or affect the binding of picrotoxin include cage convulsant compounds such as *t*-butylbicyclophosphorothionate [86,87], an analogue of diazepam, Ro 5-3663 [88], and alkyl-substituted γ-butyrolactones [89]. Interestingly, using recombinant GABA$_A$ receptors expressed in *Xenopus* oocytes, the IC$_{50}$ values for picrotoxin inhibition of GABA-activated conductances for α1β1-, α1β1γ2S- a nd α1β1γ2L-subunit-containing receptors were similar at 0.55, 0.64 and 0.46 μM respectively [78]. This suggests that picrotoxin-induced antagonism is also not sensitive to these receptor subunit combinations.

The site of action of picrotoxin on GABA receptors in insect neurons (*Drosophila melanogaster*) was addressed in a study of flies resistant to picrotoxin and to the cyclodiene insecticide dieldrin [90]. DNA sequencing revealed a mutation in the second transmembrane domain of the receptor protein, which is believed to line the ion channel, where a single amino acid, Ala-302, was mutated to serine. The position of this amino acid in the putative transmembrane topography of the receptor is compatible with the uncharged picrotoxin molecule binding within the ion channel lumen. This mutation also conferred resistance to inhibition by lindane (closely related to the cyclodienes) and by *t*-butylbicyclophosphoro-thionate [91].

Penicillin represents another type of GABA$_A$ receptor antagonist. It does not compete for the agonist recognition site but appears to preferentially block GABA-activated open ion channels [92–95]. The conductance of the ion channel was unaffected by penicillin, but the mean open duration was reduced in addition to an increased burst duration. These features are consistent with a simple physical occlusion by penicillin of current flow through the ion channel.

In addition to the major GABA antagonists already considered, other miscellaneous compounds capable of negatively modulating GABA$_A$ receptor function include d-tubocurarine [96], the neurosteroid dihydroepiandrosterone sulphate [97], pitrazepin [98] and m-sulphonate benzene diazonium chloride [98a].


## Modulators of GABA$_A$ receptors

### Benzodiazepines

These agents can modulate GABA$_A$ receptor function by binding to allosteric sites on the receptor protein. The modulation can result in either an enhancement of receptor function (increased Cl$^-$ ion flow) or a reduction. There are many examples of benzodiazepines (e.g. diazepam, flurazepam) that can enhance GABA-mediated responses on a variety of neuronal preparations [99]. These agents act mostly by increasing the opening frequency of single GABA-activated ion channels, with little effect on the ion channel open duration [100–102]. In contrast, the β-carboline inverse agonists, e.g. methyl-6,7-dimethoxyl-4-ethyl-β-carboline-3-carboxylic acid, reduce the opening frequency of GABA-operated ion channels, again without an effect on the open duration [102], while the benzodiazepine antagonist flumazenil has no effect on single GABA channel events [101]. For inhibitory synaptic currents, benzodiazepines often result in a prolongation of the current decay without increasing the IPSC amplitude. This modulation is reversed by flumazenil [13].

Studies using recombinant GABA$_A$ receptors and benzodiazepines have yielded important insights into the subunit requirements and possible amino acid residues necessary for receptor regulation. An absolute functional sensitivity to the benzodiazepines is conferred on the receptor by the γ subunit [103]. In addition, the identity of the α subunits within the receptor complex influences the type of benzodiazepine pharmacology [104]. Initially, benzodiazepine pharmacology was catergorized by classification into type I (or BZ1) GABA$_A$-benzodiazepine receptors (sensitive to the triazolopyridazine CL 218 872 and 2-oxoquazepam, based on ligand binding affinities) associated with α1-subunit-containing receptors in combination with β2 and γ2 subunits; and type II (or BZ2; reduced sensitivity to CL 218 872) receptors, which are associated with α2, α3 or α5 subunits. However, the receptors constituting the type II sites are likely to be heterogeneous and can be further subdivided based on ligand affinities. For example, the imidazolopyridine zolpidem, while being relatively selective for α1-subunit-containing receptors, also displays differential affinities for α2- and α5-subunit-containing receptors. GABA$_A$ receptors containing α5 are unaffected by zolpidem [105]. Interestingly, the identity of the α subunit can also affect not only the EC$_{50}$ values of the benzodiazepines for enhancing GABA-activated responses but also the maximum enhancement elicited by these agents [105–107]. In comparison, the α6 subunit has a discrete localization in maturing granule neurons in the cerebellum. Recombinant GABA$_A$ receptors containing this subunit exhibit responses to GABA that are insensitive to the traditional benzodiazepines such as diazepam, but can bind the partial inverse agonist ligand Ro 15-4513 [108].

The molecular determinants affecting type I or II properties in α1- and α2–α5-subunit-containing receptors, and the sensitivity to Ro 15-4513 of α6-subunit-containing receptors, have been addressed by forming chimaeric receptors by swopping domains between α1 and α3 subunits, and between α1 and α6 subunits. Using this strategy, it has been shown that the benzodiazepine pharmacology of GABA$_A$ receptors may reside in only a few amino acid residues. A single amino acid substitution of glutamate with glycine (position 225) in the α3 subunit, expressed in combination with β2 and γ2 subunits, results in high-affinity binding of the type I selective agent CL 218 872 [109]. In addition, by mutating an arginine residue at position 100 to histidine (present in most other $\alpha$ subunits), the mutated $\alpha$6 subunit in combination with $\beta$2 and $\gamma$2 subunits exhibited [$^3$H]flumazenil binding that could now be displaced by diazepam [108].

Variation of the $\beta$ subunit in recombinant GABA$_A$ receptors appears to have little effect on the EC$_{50}$ values and apparent efficacies of benzodiazepines [110] (cf. [106]). In contrast, variation of the $\gamma$ subunit can affect benzodiazepine modulation of GABA$_A$ receptor function. Substituting $\gamma$2 for $\gamma$1 in receptors containing $\alpha$2 and $\beta$1 subunits resulted in less potentiation of responses to GABA by benzodiazepine agonists. Furthermore, the inverse agonists Ro 15-4513 and methyl-6,7-dimethoxy-4-ethyl-$\beta$-carboline-3-carboxylate, unusually, did not inhibit but slightly enhanced GABA-activated responses [105,111]. The $\gamma$3 subunit has not been subjected to as much study as the other $\gamma$ subunits, but it would appear that $\gamma$3 can substitute for $\gamma$2 in GABA$_A$ receptors and maintain a conventional benzodiazepine pharmacology [112], although the apparent affinity for benzodiazepine agonists is reduced [113].

## Barbiturates

These compounds have multiple properties, allowing them to be designated as anticonvulsant, anaesthetic and sedative/hypnotic barbiturates. These agents have two clear effects on the GABA$_A$ receptor: an enhancement of GABA-activated responses at a site discrete from the benzodiazepine binding site, and a direct activation of the GABA$_A$ receptor in the absence of GABA [100,114,115]. Using fluctuation analysis of membrane current noise induced by GABA, the barbiturates enhanced the GABA response by increasing the mean open time of the ion channels without affecting single-channel conductance [100,115]. In hippocampal neurons, pentobarbitone also increased the decay time for monoexponential GABA-mediated IPSCs [116].

Single-channel analyses have confirmed the finding from earlier fluctuation analyses that pentobarbitone does not affect the GABA-activated single-channel conductance but results in an increased channel mean open time [117]. The time constants required to fit the frequency histograms of GABA$_A$ receptor open times were unaffected by pentobarbitone. The increased mean open time resulted from an increased relative frequency of occurrence for the longest open states [117]. Analysis of burst durations revealed that pentobarbitone also increased the frequency of the long-duration burst states, resulting in an increased mean burst duration [117].

With recombinant GABA$_A$ receptors, modulation by barbiturates does not appear to have a strict dependence on receptor subunit composition, unlike the effects of the benzodiazepines, although the degree of enhancement may depend subtly on the subunit composition (cf. [106]).

## Neurosteroids

The prospect of steroids being able to modulate $GABA_A$ receptor function was revealed following the discovery that the steroid anaesthetic, alphaxolone, enhanced responses to GABA [118]. Steroids can enhance $GABA_A$ receptor function in a manner similar to that of the barbiturates, although their sites of action are probably different [118–121]. A comparison of different steroids of the pregnan-20-one and androstan-17-one series, incorporating structural variations, has enabled a number of key elements to be identified that confer the ability to potentiate responses to GABA [122]; two of these are a 3α-hydroxy group and a keto group at position 20 (pregnanes) or 17 (androstanes). The site of action of steroids on the $GABA_A$ receptor is yet to be defined, but a general lipid-perturbing effect of the steroids seems unlikely following the report that intracellular alphaxolone is unable to enhance responses to GABA [123]. This suggests a site(s) on the extracellular face of the $GABA_A$ receptor for the binding of the steroids (cf. [124,125]). Moreover, some steroids can activate the $GABA_A$ receptor directly in the apparent absence of any GABA [126]. The natural occurrence of many steroids suggests that these compounds may be candidates for natural endogenous modulators of $GABA_A$ receptor function. In addition to the enhancement of $GABA_A$ receptor function, some steroids, including pregnanolone sulphate and dihydroepiandrosterone sulphate, conversely act in an inhibitory capacity, usually as non-competitive antagonists [125,127].

The mechanism of action of the steroids that enhance $GABA_A$ receptor function has been examined using fluctuation and single-channel analyses. The single GABA ion channel conductance is unaltered by the steroids, but the channel burst duration is increased [121,126]. This change probably results from an increased occurrence of long-duration channel openings, without an effect on the time constants that describe the open-state distributions, and as such has similarities with the mechanism of action of barbiturates that enhance $GABA_A$ receptor function [128].

Recombinant $GABA_A$ receptors have been used to investigate whether steroid modulation of receptor function is dependent on the polypeptide subunit composition. The overall view is so far unclear. Responses to GABA recorded from receptors composed of heteromeric α1β1γ2 or α1β1 subunits, or homomeric β1 subunits, were all enhanced by two pregan-20-one derivatives (3α-hydroxy- and 3α,21-dihydroxy-) [129], suggesting little dependence on subunit composition. However, in another study [130] the steroid potentiation of [³H]flunitrazepam binding was greater using α3β1γ2 constructs than α1β1γ2 or α2β1γ2. A dependence of steroid modulation on α and γ subunits was also evident from studies in *Xenopus* oocytes. GABA-activated membrane currents exhibited larger steroid-induced enhancements for α1β1 and α3β1 than α2β1 constructs [131]. Addition of the γ2 subunit to α1β1 and α2β1 constructs increased the steroid potency, but for α3β1γ2 the steroid potentiation was reduced [131]. Inhibition of responses to GABA by pregnanolone sulphate may also be dependent on the type of $\alpha$ subunit present in the receptor complex [132] (but see [133]).

## Ethanol

Ethanol shares many of the sedative and hypnotic properties usually associated with barbiturates and benzodiazepines and, not surprisingly, has a modulatory effect at the GABA$_A$ receptor. Ethanol can enhance GABA-activated responses at low concentrations (for ethanol) of 1–50 mM, an effect that is inhibited by the benzodiazepine partial inverse agonist Ro 15-4513 [134]. The level of enhancement can vary between different regions of the CNS [135,136] and might, therefore, be dependent on receptor subunit composition. Evidence supporting this view was provided from expression studies using *Xenopus* oocytes. Expression of GABA$_A$ receptors containing the long form of $\gamma$2 ($\gamma$2L) transduced responses to GABA that could be enhanced by ethanol; however, receptors containing the short form of $\gamma$2 ($\gamma$2S) were unaffected by ethanol [137]. $\gamma$2L differs from $\gamma$2S by the inclusion of eight extra amino acids containing a consensus sequence for phosphorylation by protein kinase C, and is thought to have an intracellular location [138,139]. Phosphorylation of the serine residue at position 343 (bovine sequence) appeared to be critical, since mutating Ser-343 to alanine prevented the ethanol-induced enhancement [140]. However, an additional study suggests that the importance of the state of phosphorylation of the GABA$_A$ receptor and its modulation by ethanol is controversial, as ethanol was unable to enhance GABA-activated responses at the following recombinant GABA$_A$ receptors: $\alpha$1$\beta$1$\gamma$2S, $\alpha$1$\beta$1$\gamma$2L and $\alpha$1$\beta$2$\gamma$2S [141].

## Ionic modulators: cations

Cations, particularly divalent cations, have been widely recognized as modulators of GABA receptor function [142–145]. The type of modulation can range from inhibition to potentiation of GABA-activated responses. For example, $Zn^{2+}$ can inhibit responses to GABA, particularly with cultured neurons [42,144,146–149]; however, in intact preparations such as brain slices, $Zn^{2+}$ either has no effect or can enhance GABA-activated responses [58,142,147]. The $Zn^{2+}$-induced inhibition of responses to GABA is usually non-competitive, and occurs at a discrete site not shared by barbiturates, benzodiazepines, steroids or picrotoxin [58] (Fig. 9).

In contrast to the inhibitory effects of $Zn^{2+}$ on cultured neurons, $La^{3+}$ and some members of the lanthanide series reversibly enhance responses to GABA by binding to another discrete site on the receptor protein not shared by benzodiazepines, barbiturates or picrotoxin [145,150] (Fig. 10). The main effect of $La^{3+}$ is a reduction in the GABA EC$_{50}$, resulting from a leftward shift of the dose–response curve, without a change in the maximum response to GABA. Interestingly, terbium ($Tb^{3+}$) induced membrane currents in the apparent absence of GABA which were enhanced by pentobarbitone and inhibited by bicuculline [150]. Whether this occurs from another site on the receptor is yet to be clarified; however, it is

**Fig. 9**        **Zn²⁺-induced inhibition of GABA-activated responses in cultured sympathetic neurons**

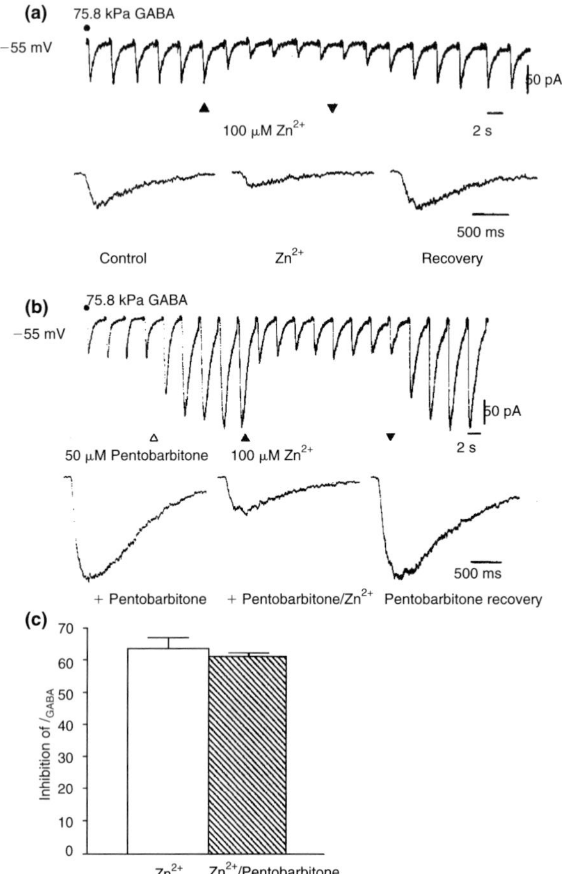



*(**a**) Whole-cell recordings at a holding potential of −55 mV. GABA was periodically applied from a juxtaposed pipette in the presence and absence of 100 μM Zn²⁺. (**b**) Responses to GABA were enhanced by pentobarbitone, but were still equally inhibited by co-application of Zn²⁺. (**c**) Inhibition induced by Zn²⁺ in the absence and presence of pentobarbitone. Reproduced from [42], with permission.*

intriguing to speculate that there could be many more discrete binding sites for cations on the $GABA_A$ receptor.

Using recombinant $GABA_A$ receptors, the inhibition of GABA-activated responses by $Zn^{2+}$ has been demonstrated to be dependent on receptor subunit composition. $GABA_A$ receptors containing a $\gamma$ subunit are much less sensitive to non-competitive inhibition ($IC_{50}$ values of approx. 500–600 μM) compared with $\alpha\beta$

**Fig. 10.** **Lanthanum enhances responses to GABA in dorsal root ganglion neurons**

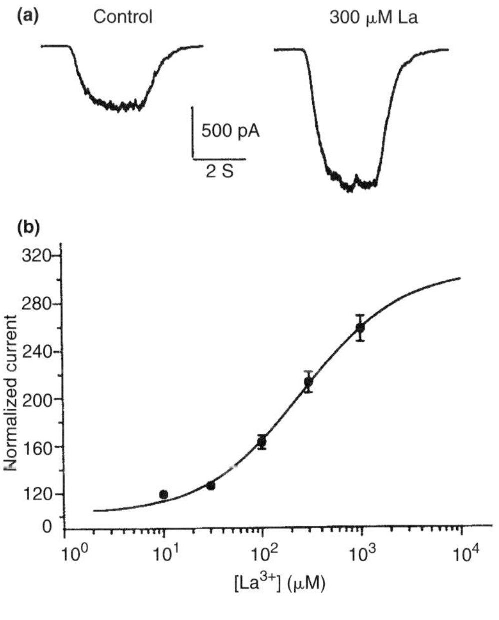


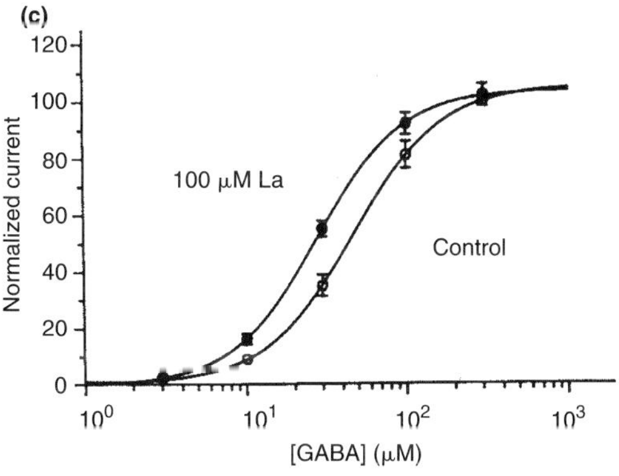


(*a*) Whole-cell recording of the GABA-activated current at a holding potential of −60 mV in the absence and presence of 300 μM La³⁺. (*b*) Concentration–response curve for La³⁺ enhancement of the response to 30 μM GABA. The estimated $K_J$ for La³⁺ was 231 μM. (*c*) La³⁺ (100 μM) induces a leftward shift in the GABA concentration–response curve without increasing the maximum response. Data reproduced from [145], with permission.

dimeric complexes ($IC_{50}$ approx. 1 $\mu$M) [56–58] (Fig. 11). In comparison, $GABA_A$ receptors containing the $\delta$ subunit in complexes of $\alpha1\beta1\delta$ were quite sensitive to inhibition by $Zn^{2+}$ (10 $\mu$M causing a 75% response inhibition), and inclusion of the $\gamma$2L subunit, forming $\alpha1\beta1\delta\gamma$2L, also resulted in greater sensitivity to $Zn^{2+}$ compared with receptors formed from $\alpha1\beta1\gamma$2L [151].

The enhancement induced by $La^{3+}$ was also apparently dependent on subunit composition. Using $\alpha1\beta$2-subunit-containing $GABA_A$ receptors, $La^{3+}$ ($EC_{50}$ approx. 200 $\mu$M) induced only a modest potentiation of the response to GABA (<70%), whereas with $\alpha1\beta2\gamma$2 receptors the potentiation was 3–4-fold greater ($EC_{50}$ 21 $\mu$M) [152]. Interestingly, addition of the $\delta$ subunit to $GABA_A$ receptors composed of $\alpha1\beta$1 subunits, giving the receptor complex $\alpha1\beta1\delta$, conferred an insensitivity to the potentiating effect of $La^{3+}$ up to 300 $\mu$M [151]. It is not known whether the potentiating effects of other lanthanide ions on GABA-activated responses also exhibit a dependence on subunit composition.


## Conclusions

This review has reported recent developments with regard to the physiological and pharmacological properties of $GABA_A$ receptors. The application of patch-clamp techniques to neurons in brain slices, allowing retention of some of their *in vivo* cytoarchitecture, has enabled an incisive resolution of inhibitory synaptic transmission. This has been useful in estimating how many $GABA_A$ receptor ion channels are opened by 'packets' of presynaptically released GABA, and whether or not the $GABA_A$ receptors are transiently saturated by the high concentrations present in the synaptic cleft. Moreover, analysing the fine structure of inhibitory transmission raises questions regarding the mechanisms of transmitter release (quantal versus non-quantal) and whether release can be synchronized by endogenous agents (e.g. $Zn^{2+}$).

The application of molecular biological techniques to the study of $GABA_A$ receptors has suggested the possibility of truly vast levels of receptor heterogeneity. One of the tasks for the future will be to assess the pharmacological and therapeutic impact of having so many different receptor subunits apparently able to form functional $GABA_A$ receptors. For example, how many are relevant? Do different subunit combinations really have distinct physiological and pharmacological roles in the CNS? Or are our techniques now of sufficient resolution for us to be looking at 'biological noise', demonstrating that the evolution of $GABA_A$ receptors, and perhaps of many ligand-gated ion channels, is not as precise as we once believed, and possibly that subunit composition is not so critical for function *in vivo*? However, if subunit composition should prove to be crucial, and this review has described some clear examples, then possible clinical benefit might accrue from having therapeutic compounds with increased selectivity that allows them to distinguish between subtypes of $GABA_A$ receptors (subtype-selective

## Fig. 11. Recombinant GABA$_A$ receptors are differentially sensitive to Zn$^{2+}$-induced inhibition

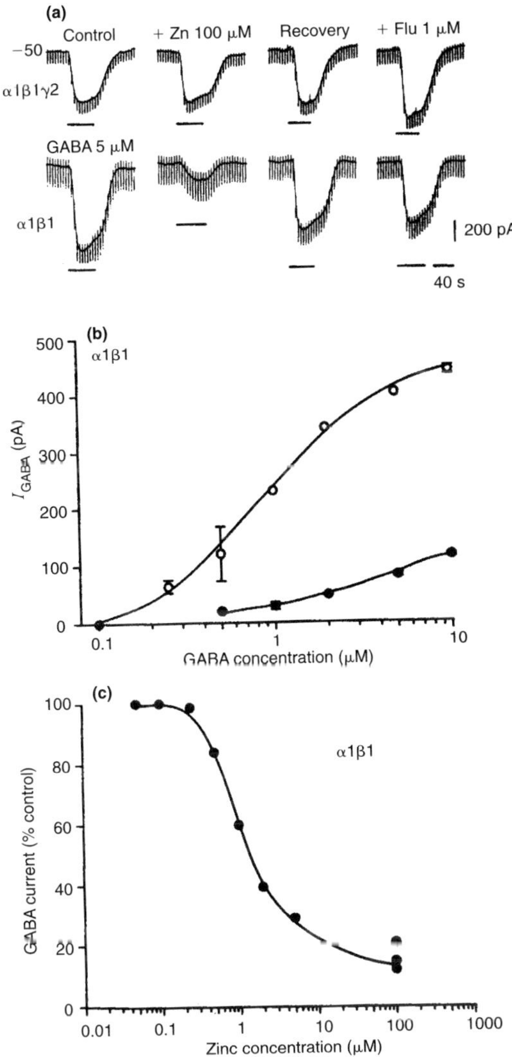



(a) Whole-cell recordings of GABA-activated currents from human embryonic kidney cells transfected with α1β1 (upper trace) or α1β1γ2S (lower trace) cDNAs. Zn$^{2+}$ was effective as an antagonist in cells expressing receptors devoid of the γ2 subunit. Conversely, the benzodiazepine agonist flurazepam (Flu) only enhanced responses to GABA when γ2 subunits were present. (b) Concentration–response curves for GABA in the absence (○) and presence (●) of 100 μM Zn$^{2+}$. (c) Concentration–inhibition curve for Zn$^{2+}$ measured against a response to 5 μM GABA. The IC$_{50}$ for Zn$^{2+}$ was 1.5 μM. Data reproduced from [57], with permission.

agents). For the future, such a notion may obviate many of the unwanted side effects commonly associated with many classes of drugs that act on CNS GABA$_A$ receptors.

## References


References included here are up until March 1995.
1.   Sivilotti, L. and Nistri, A. (1991) Prog. Neurobiol. **36**, 35–92
2.   Bowery, N.G. (1993) Annu. Rev. Pharmacol. Toxicol. **33**, 109–147
3.   Drew, C.A. and Johnston, G.A.R. (1992) J. Neurochem. **58**, 1087–1092
4.   Qian, H. and Dowling, J.E. (1993) Nature (London) **361**, 162–164
5.   Feigenspan, A., Wassle, H. and Bormann, J. (1993) Nature (London) **361**, 159–162
6.   Burt, D.R. and Kamatchi, G.L. (1991) FASEB J. **5**, 2916–2923
7.   Olsen, R.W. and Tobin, A.J. (1990) FASEB J. **4**, 1469–1480
8.   Nayeem, N., Green, T.P., Martin, I.L. and Barnard, E.A. (1994) J. Neurochem. **62**, 815–818
9.   Lacaille, J.-C. and Schwartzkroin, P.A. (1988) J. Neurosci. **8**, 1400–1410
10.  2.Lacaille, J.-C. and Schwartzkroin, P.A. (1988) J. Neurosci. **8**, 1411–1424
11.  Buhl, E.H., Halasy, K. and Somogyi, P. (1994) Nature (London) **368**, 823–828
12.  Edwards, F.A., Konnerth, A. and Sakmann, B. (1990) J. Physiol. (London) **430**, 213–249
13.  Otis, T.S. and Mody, I. (1992) Neuroscience **49**, 13–32
14.  Otis, T.S., Staley, K.J. and Mody, I. (1991) Brain Res. **545**, 142–150
15.  Kraszewski, K. and Grantyn, R. (1992) Neuroscience **47**, 555–570
16.  Vautrin, J., Schaffner, A.E. and Barker, J.L. (1992) Neurosci. Lett. **138**, 67–71
17.  Vautrin, J., Schaffner, A.E. and Barker, J.L. (1993) Hippocampus **3**, 93–102
18.  Staley, K.J. and Mody, I. (1992) J. Neurophysiol. **68**, 197–212
19.  Mody, I. and Otis, T.S. (1992) Neuroscience **49**, 13–32
20.  Mody, I., De Koninck, Y., Otis, T.S. and Soltesz, I. (1994) Trends Neurosci. **17**, 517–525
21.  Maconochie, D.J., Zempel, J.M. and Steinbach, J.H. (1994) Neuron **12**, 61–71
22.  Oh, D.J. and Dichter, M.A. (1992) Neuroscience **49**, 571–576
23.  Celentano, J.J. and Wong, R.K.S. (1994) Biophys. J. **66**, 1039–1050
23a. Thompson, S.M. and Gahwiler, B.H. (1992) J. Neurophysiol. **67**, 1698–1701
24.  Isaacson, J.S., Solis, J.M. and Nicoll, R.A. (1993) Neuron **10**, 165–175
25.  Perreault, P. and Avoli, M. (1989) J. Neurophysiol. **61**, 953–970
26.  Muller, W. and Misgeld, U. (1991) J. Neurophysiol. **65**, 141–147
27.  Aram, J.A., Michelson, H.B. and Wong, R.K.S. (1991) J. Neurophysiol. **65**, 1034–1041
28.  Xie, X. and Smart, T.G. (1991) Nature (London) **349**, 521–524
29.  Xie, X. and Smart, T.G. (1993) J. Physiol. (London) **460**, 503–523
30.  Lambert, N.A., Levitin, M. and Harrison, N.L. (1992) Neurosci. Lett. **135**, 215–218
31.  Hablitz, J.J. and Zhou, Z.M. (1992) J. Neurophysiol. **70**, 1264–1269
32.  Ben-Ari, Y., Cherubini, E., Corradetti, R. and Gaiarsa, J.L. (1989) J. Physiol. (London) **416**, 303–325
33.  Xie, X., Hider, R.C. and Smart, T.G. (1994) J. Physiol. (London) **478**, 75–86
34.  Michelson, H.B. and Wong, R.K.S. (1991) Science **253**, 1420–1423
35.  Cherubini, E., Gaiarsa, J.L. and Ben-Ari, Y. (1991) Trends Neurosci. **14**, 515–519
36.  Wang, J., Reichling, D.B., Kyrozis, A. and MacDermott, A.B. (1994) Eur. J. Neurosci. **6**, 1275–1280
36a. Jackson, M.B., Lecar, H., Mathers, D.A. and Barker, J.L. (1982) J. Neurosci. **2**, 889–894
37.  Hamill, O.P., Bormann, J. and Sakmann, B. (1983) Nature (London) **305**, 805–808
38.  Mathers, D.A., Grewal, A. and Wang, Y. (1989) Neurosci. Lett. **98**, 229–233
39.  Macdonald, R.L., Rogers, C. and Twyman, R.E. (1989) J. Physiol. (London) **410**, 479–499
40.  Smith, S.M., Zorec, R. and McBurney, R.N. (1989) J. Membr. Biol. **108**, 45–52
41.  Newland, C.F., Colquhoun, D.C. and Cull-Candy, S.G. (1991) J. Physiol. (London) **432**, 203–233
42.  Smart, T.G. (1992) J. Physiol. (London) **447**, 587–625
43.  Mathers, D.A. and Wang, Y. (1988) Synapse **2**, 627–632
44.  Weiss, D.S. (1988) J. Neurophysiol. **59**, 514–527

45. Weiss, D.S. and Magleby, K.L. (1989) J. Neurosci. **9**, 1314–1324
46. Twyman, R.E., Rogers, C.J. and Macdonald, R.L. (1990) J. Physiol. (London) **423**, 193–220
47. Bormann, J., Hamill, O.P. and Sakmann, B. (1987) J. Physiol. (London) **385**, 243–286
48. Araki, T., Ito, M. and Oscarsson, O. (1961) J. Physiol. (London) **159**, 410–435
49. Kelly, J.S., Krnjevic, K., Morris, M.E. and Yim, G.K.W. (1969) Exp. Brain Res. **7**, 11–31
49a. Kaila, K. (1994) Prog. Neurobiol. **42**, 489–537
50. Moss, S.M., Smart, T.G., Porter, N.M., Nayeem, N., Devine, J., Stephenson, F.A., Macdonald, R.L. and Barnard, E.A. (1990) Eur. J. Pharmacol. (Mol. Pharmacol.) **189**, 77–88
51. Angelotti, T.P. and Macdonald, R.L. (1993) J. Neurosci. **13**, 1429–1440
52. Porter, N.M., Angelotti, T.P., Twyman, R.E. and Macdonald, R.L. (1992) Mol. Pharmacol. **42**, 872–881
53. Verdoorn, T.A., Draguhn, A., Ymer, S., Seeburg, P.H. and Sakmann, B. (1990) Neuron **4**, 919–928
54. Krishek, B.J., Xie, X., Blackstone, C.D., Huganir, R.L. and Smart, T.G. (1994) Neuron **12**, 1081–1095
55. Angelotti, T.P., Tan, F., Chahine, K.G. and Macdonald, R.L. (1992) Mol. Brain Res. **16**, 173–178
56. Draguhn, A., Verdoorn, T.A., Ewert, M., Seeburg, P.H. and Sakmann, B. (1990) Neuron **5**, 781–788
57. Smart, T.G., Moss, S.M., Xie, X. and Huganir, R.L. (1991) Br. J. Pharmacol. **103**, 1837–1839
58. Smart, T.G., Xie, X. and Krishek, B.J. (1994) Prog. Neurobiol. **42**, 393–441
59. Sigel, E., Stephenson, F.A., Mamalaki, C. and Barnard, E.A. (1983) J. Biol. Chem. **258**, 6965–6971
60. Casalotti, S.O., Stephenson, F.A. and Barnard, E.A. (1986) J. Biol. Chem. **261**, 15013–15016
61. Bureau, M. and Olsen, R.W. (1990) Mol. Pharmacol. **37**, 497–502
62. Amin, J. and Weiss, D.S. (1993) Nature (London) **366**, 565–569
63. Sigel, E., Baur, R., Kellenberger, S. and Malherbe, P. (1992) EMBO J. **11**, 2017–2023
64. Smith, G.B. and Olsen, R.W. (1994) J. Biol. Chem. **269**, 20380–20387
65. Aprison, M.H. and Lipkowitz, K.B. (1989) J. Neurosci. Res. **23**, 129–135
66. Krogsgaard-Larsen, P. (1988) Med. Res. Rev. **8**, 27–56
67. Krogsgaard-Larsen, P., Frolund, B., Jorgensen, F.S. and Schousboe, A. (1994) J. Med. Chem. **37**, 2489–2505
68. Johnston, G.A.R., Curtis, D.R., Beart, P.M., Game, C.J.A., McCulloch, R.M. and Twitchin, B. (1975) J. Neurochem. **24**, 157–160
69. Kristiansen, U., Lambert, J.D.C., Falch, E. and Krogsgaard-Larsen, P. (1991) Br. J. Pharmacol. **104**, 85–90
70. Barker, J.L. and Mathers, D.A. (1981) Science **212**, 358–361
71. Segal, M. and Barker, J.L. (1984) J. Neurophysiol. **51**, 500–515
72. Mistry, D.K. and Hablitz, J.J. (1990) Pflugers Arch. **416**, 454–461
73. Ebert, B., Wafford, K.A., Whiting, P.J., Krogsgaard-Larsen, P. and Kemp, J.A. (1994). Mol. Pharmacol. **46**, 957–963
74. Simmonds, M.A. (1980) Neuropharmacology **19**, 39–45
75. Simmonds, M.A. (1982) Eur. J. Pharmacol. **80**, 347–358
76. Kemp, J.A., Marshall, G.R. and Woodruff, G.N. (1986) Br. J. Pharmacol. **87**, 677–684
77. Horne, A.L., Hadingham, K.L., Macaulay, A.J., Whiting, P.W. and Kemp. J.A. (1992) Br. J. Pharmacol. **107**, 732–737
78. Krishek, B.J. and Smart, T.G. (1995) Br. J. Pharmacol. **114**, 291P
79. Barker, J.L., McBurney, R.N. and Mathers, D.A. (1983) Br. J. Pharmacol. **80**, 619–629
80. Akaike, N., Oomura, H.Y. and Carpenter, D.O. (1985) Experientia **41**, 70–71
81. Yasui, S., Ishizuka, S. and Akaike, N. (1985) Brain Res. **344**, 176–180
82. Twyman, R.E., Rogers, C.J. and Macdonald, R.L. (1989) Neurosci. Lett. **96**, 89–95
83. Smart, T.G. and Constanti, A. (1986) Proc. R. Soc. London B **227**, 191–216
84. Newland, C.F. and Cull-Candy, S.G. (1992) J. Physiol. (London) **447**, 191–213
85. Yoon, K. W., Covey, D.F. and Rothman, S.M. (1993) J. Physiol. (London) **464**, 423–439
86. Squires, R.F., Casida, J.E., Richardson, M. and Saederup, E. (1983) Mol. Pharmacol. **23**, 326–336
87. Van Renterghem, C., Bilbe, G., Moss, S., Smart, T.G., Constanti, A., Brown, D.A. and Barnard, E.A. (1987) Mol. Brain Res. **2**, 21–31

88. Olsen, R.W. (1981) J. Neurochem. **37**, 1–13
89. Klunk, W.E., Kalman, B.L., Ferrendelli, J.A. and Covey, D.F. (1983) Mol. Pharmacol. **23**, 511–518
90. ffrench-Constant, R.H., Rocheleau, T.A., Steichen, J.C. and Chalmers, A.E. (1993) Nature (London) **363**, 449–451
91. Zhang, H.-G., ffrench-Constant, R.H. and Jackson, M.B. (1994) J. Physiol. (London) **479**, 65–75
92. Barker, J.L., McBurney, R.N., Mathers, D.A. and Vaughn, W. (1980) J. Physiol. (London) **308**, 18P
93. Chow, P. and Mathers, D.A. (1986) Br. J. Pharmacol. **88**, 541–547
94. Katayama, N., Tokutomi, N., Nabekura, J. and Akaike, N. (1992) Brain Res. **595**, 249–255
95. Twyman, R.E., Green, R.M. and Macdonald, R.L. (1992) J. Physiol. (London) **445**, 97–127
96. Lebeda, F.J., Hablitz, J.J. and Johnston, D. (1982) J. Neurophysiol. **48**, 622–632
97. Majewska, M.D., Demirgoren, S., Spivak, C.E. and London, E.D. (1990) Brain Res. **526**, 143–146
98. Gahwiler, B.H., Maurer, R. and Wuthrich, H.J. (1984) Neurosci. Lett. **45**, 311–316
98a. Krishek, B.J., Xie, X., Bouchet, M.J. and Smart, T.G. (1994) Neuropharmacology **33**, 1125–1130
99. Polz, P. (1988) Prog. Neurobiol. **31**, 349–424
100. Study, R.E. and Barker, J.L. (1981) Proc. Natl. Acad. Sci. U.S.A. **78**, 7180–7184
101. Vicini, S., Mienville, J.-M. and Costa, E. (1987) J. Pharmacol. Exp. Ther. **243**, 1195–1201
102. Rogers, C.J., Twyman, R.E. and Macdonald, R.L. (1994) J. Physiol. (London) **475**, 69–82
103. Pritchett, D.B., Sontheimer, H., Shivers, B.D., Ymer, S., Kettenmann, H., Schofield, P.R. and Seeburg, P.H. (1989) Nature (London) **338**, 582–585
104. Pritchett, D.B., Luddens, H. and Seeburg, P.H. (1989) Science **245**, 1389–1392
105. Puia, G., Vicini, S., Seeburg, P.H. and Costa, E. (1991) Mol. Pharmacol. **39**, 691–696
106. Sigel, E., Baur, R., Trube, G., Mohler, H. and Malherbe, P. (1990) Neuron **5**, 703–711
107. Wafford, K.A., Whiting, P.J. and Kemp, J.A. (1993) Mol. Pharmacol. **43**, 240–244
108. Wieland, H.A., Luddens, H. and Seeburg, P.H. (1992) J. Biol. Chem. **267**, 1426–1429
109. Pritchett, D.B. and Seeburg, P.H. (1991) Proc. Natl. Acad. Sci. U.S.A. **88**, 1421–1425
110. Hadingham, K.L., Wingrove, P.B., Wafford, K.A., Bain, C., Kemp, J.A., Palmer, K.J., Wilson, A.W., Wilcox, A.S., Sikela, J.M., Ragan, C.I. and Whiting, P.J. (1993) Mol. Pharmacol. **44**, 1211–1218
111. Wafford, K.A., Bain, C.J., Whiting, P.J. and Kemp, J.A. (1993) Mol. Pharmacol. **44**, 437–442
112. Knoflach, F., Rhyner, Th., Villa, M., Kellenberger, S., Drescher, U., Malherbe, P., Sigel, E. and Mohler, H. (1991) FEBS Lett. **293**, 191–194
113. Herb, A., Wisden, W., Luddens, H., Puia, G., Vicini, S. and Seeburg, P.H. (1992) Proc. Natl. Acad. Sci. U.S.A. **89**, 1433–1437
114. Mathers, D.A. and Barker, J.L. (1980) Science **209**, 507–509
115. Barker, J.L. and McBurney, R.N. (1979) Proc. R. Soc. London B **206**, 319–327
116. Segal, M. and Barker, J.L. (1984) J. Neurophysiol. **52**, 469–487
117. Macdonald, R.L., Rogers, C. and Twyman, R.E. (1989) J. Physiol. (London) **417**, 483–500
118. Harrison, N.L. and Simmonds, M.A. (1984) Brain Res. **323**, 287–292
119. Majewska, M.D., Harrison, N.L., Schwartz, R.D., Barker, J.L. and Paul, S.M. (1986) Science **232**, 1004–1007
120. Cottrell, G.A., Lambert, J.J. and Peters, J.A. (1987) Br. J. Pharmacol. **90**, 491–500
121. Barker, J.L., Harrison, N.L., Lange, G.D. and Owen, D.G. (1987) J. Physiol. (London) **386**, 485–501
122. Simmonds, M.A. (1991) Semin. Neurosci. **3**, 231–239
123. Lambert, J.J., Peters, J.A., Sturgess, N.C. and Hales, T.C. (1990) Ciba Found. Symp. **153**, 103–104
124. Gee, K.W. (1988) Mol. Neurobiol. **2**, 291–317
125. Majewska, M.D. (1992) Prog. Neurobiol. **38**, 379–395
126. Callachan, H., Cottrell, G.A., Hather, N.Y., Lambert, J.J., Nooney, J.M. and Peters, J.A. (1987) Proc. R. Soc. London B **231**, 359–369
127. Mienville, J.-M. and Vicini, S. (1989) Brain Res **489**, 190–194
128. Twyman, R.E. and Macdonald, R.L. (1992) J. Physiol. (London) **456**, 215–245

129. Puia, G., Santi, M., Vicini, S., Pritchett, D.B., Purdy, R.H., Paul, S.M., Seeburg, P.H. and Costa, E. (1990) Neuron **4**, 759–765
130. Lan, N.C., Gee, K.W., Bolger, M.B. and Chen, J.S. (1991) J. Neurochem. **57**, 1818–1821
131. Shingai, R., Sutherland, M.L. and Barnard, E.A. (1991) Eur. J. Pharmacol. **206**, 77–80
132. Zaman, S.H., Shingai, R., Harvey, R.J., Darlison, M.G. and Barnard, E.A. (1992) Eur. J. Pharmacol. **225**, 321–330
133. Puia, G., Ducic, I., Vicini, S. and Costa, E. (1993) Receptors Channels **1**, 135–142
134. Reynolds, J.N., Prasad, A. and MacDonald, J.F. (1992) Eur. J. Pharmacol. **224**, 173–181
135. Proctor, W.R., Soldo, B.L., Allan, A.M. and Dunwiddie, T.V. (1992) Brain Res. **595**, 220–227
136. Criswell, H.E., Simson, P.E., Duncan, G.E., McCown, T.J., Herbert, J.S., Morrow, A.L. and Breese, G.R. (1993) J. Pharmacol. Exp. Ther. **267**, 522–537
137. Wafford, K.A., Burnett, D.M., Leidenheimer, N.J., Burt, D.R., Wang, J.B., Kofuji, P., Dunwiddie, T.V., Harris, R.A. and Sikela, J.M. (1991) Neuron **7**, 27–33
138. Whiting, P.J., McKernan, R.M. and Iversen, L.L. (1990) Proc. Natl. Acad. Sci. U.S.A. **87**, 2. 9966–9970
139. Kofuji, P., Wang, J.B., Moss, S.J., Huganir, R.L. and Burt, D.B. (1991) J. Neurochem **56**, 713–715
140. Wafford, K.A. and Whiting, P.J. (1992) FEBS Lett. **313**, 113–117
141. Sigel, E. and Baur, R. (1993) FEBS Lett. **324**, 140–142
142. Smart, T.G. and Constanti, A. (1983) Neurosci. Lett. **40**, 205–211
143. Kaneko, A. and Tachibana, M. (1986) J. Physiol. (London) **373**, 463–479
144. Westbrook, G.L. and Mayer, M.L. (1987) Nature (London) **328**, 640–643
145. Ma, J.Y. and Narahashi, T. (1993) Brain Res. **607**, 222–232
146. Mayer, M.L. and Vyklicky, L. (1989) J. Physiol. (London) **415**, 351–365
147. Smart, T.G. and Constanti, A. (1990) Br. J. Pharmacol. **99**, 643–654
148. Aguayo, L.G. and Alarcon, J.M. (1993) J. Pharmacol. Exp. Ther. **267**, 1414–1422
149. Schwartz, R.D., Wagner, J.P., Yu, X. and Martin, D. (1994) J. Neurochem. **62**, 916–922
150. Ma, J.Y. and Narahashi, T. (1993) J. Neurosci. **13**, 4872–4879
151. Saxena, N.C. and Macdonald, R.L. (1994) J. Neurosci. **14**, 7077–7086
152. Im, M.S., Hamilton, B.J., Carter, D.B. and Im, W.B. (1992) Neurosci. Lett. **144**, 165–168

# Molecular structure of GABA$_A$ receptors

**F. Anne Stephenson**

School of Pharmacy, University of London,
29–39 Brunswick Square, London WC1N 1AX, U.K.

## Introduction

$\gamma$-Aminobutyric acid (GABA) is the major inhibitory neurotransmitter in the mammalian brain. It mediates its effects via specific interactions with integral membrane proteins, the GABA receptors. GABA receptors can be classified according to their respective transduction mechanisms following activation into GABA$_A$ and GABA$_B$ receptors. The focus of this chapter is the molecular properties of the GABA$_A$ receptors. GABA$_A$ receptors are fast-acting ligand-gated chloride ion channels. Thus receptor activation in the brain is followed within milliseconds by the gating or opening of an integral chloride ion channel which results, in general, in the hyperpolarization of the recipient neuronal cell (see Chapter 3 of this volume [1] for a more detailed discussion of the electrophysiological properties of GABA$_A$ receptors). The GABA$_A$ receptors are of importance both because of the pivotal role that they play in the regulation of brain excitability and the fact that their function is allosterically regulated by several distinct classes of therapeutic compounds. These include anxiolytic benzodiazepines such as valium, barbiturates, neurosteroids and some volatile anaesthetics.

The first GABA$_A$ receptor gene sequences were reported in 1987 by Schofield et al. [2]. Since that time, a plethora of related genes encoding GABA$_A$ receptor polypeptides have been identified. These polypeptides are thought to co assemble, predominantly in the brain, to form heteromeric receptor proteins with distinct biophysical and pharmacological properties. This chapter (adapted from [3]) focuses on the developments in this field since the GABA$_A$ receptor genes were first cloned, addressing in particular the structural and functional significance of mammalian GABA$_A$ receptor heterogeneity.

# GABA$_A$ receptor pharmacology

The rich pharmacology of the GABA$_A$ receptors has had a major impact on the elucidation of their biochemical properties. Before describing recent developments in the understanding of GABA$_A$ receptor structure, it is necessary to have an appreciation of their pharmacological properties. (Some of this background has been described already in Chapter 3 [1], but it is included again here as background to the molecular perspective.) Perhaps the most important of the pharmacological properties of the GABA$_A$ receptor with respect to the elucidation of its molecular structure was the realization that the anxiolytic benzodiazepine drugs (e.g. valium and librium) exerted their action by a facilitation of GABA neurotransmission (for detailed reviews, see [4,5]). Thus it was found that benzodiazepines bound with high affinity to an allosteric modulatory site of the GABA$_A$ receptors. In the presence of GABA, they potentiated the inhibitory response by an increase in the frequency of

**Fig. 1.**   **Schematic representation of the various binding sites associated with the GABA$_A$ receptors of mammalian brain**

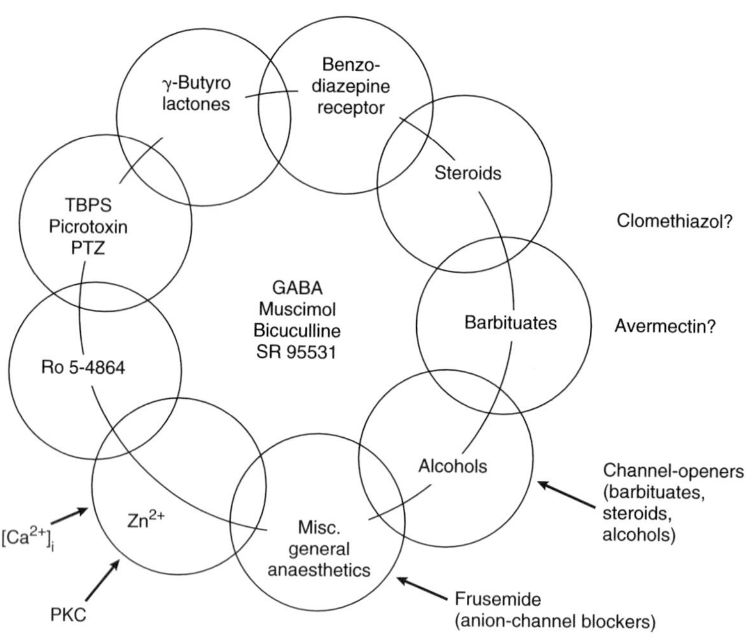


*Abbreviations: PTZ, pentyleneterazol; TBPS, t-butylbicyclophosphorothionate; [Ca$^{2+}$]$_i$, free cytosolic Ca$^{2+}$ concn; PKC, protein kinase C; misc., miscellaneous. Reprinted from [5], with the kind permission of F. Hoffmann-La Roche Ltd., Basel, Switzerland.*

chloride ion channel opening. The benzodiazepine mode of action permitted the purification of the GABA$_A$ receptors by benzodiazepine affinity chromatography (e.g. [6,7]). Benzodiazepines such as valium are anxiolytic. Although they are unable to activate GABA$_A$ receptors by themselves, by convention they are termed agonists. Other benzodiazepines, and related compounds thought to act at similar sites within the GABA$_A$ receptors, have been identified which are anxiogenic and, in the presence of GABA, decrease the frequency of chloride channel opening. These compounds are termed, by convention again, 'inverse agonists'. The benzodiazepine Ro 15-1788 has no or very low intrinsic efficacy, and is recognized as an antagonist. Further, although the classical benzodiazepines, valium and librium, bind to a single high-affinity site in the brain, other related compounds, which are again thought to act at the benzodiazepine site or at an adjacent, overlapping site, show shallow displacement curves in radioligand binding studies. It is now known that these compounds, e.g. the β-carbolines zolpidem and oxoquazepam, discriminate between GABA$_A$ receptor subtypes (see below).

Other allosteric modulators of GABA$_A$ receptors include the barbiturate drugs, non-competitive chloride channel blocking agents including picrotoxin and the cage convulsant compound $t$ butylbicyclophosphorothionate, certain neuro-steroids, the anthelminthic agents, the avermectins, $Zn^{2+}$, ethanol and the anti-convulsant drug loreclezole. Each of these compounds modulates GABA$_A$ receptor function by binding to a distinct site within the receptor complex. Fig. 1 is a diagrammatic representation of the complex pharmacology of the mammalian GABA$_A$ receptors.

## Molecular biology of GABA$_A$ receptors

At the current time, 16 mammalian GABA$_A$ receptor genes have been identified (for a review, see [8]). These are classified with respect to the conservation of the amino acid sequences of their gene products. Thus there are six GABA$_A$ receptor subunit types: α, β, γ, δ, ρ and π. Within four of these, isoforms exist with the now accepted nomenclature of α1–α6, β1–β3, γ1–γ3 and ρ1–ρ2. Isoforms of a single subunit type (e.g. comparing all the α subunits) have at least 70% amino acid sequence identity, whereas if the conservation in primary structure is compared across subunit classes (e.g. α versus β subunits), the percentage identity falls to within the range 30–40%. Some of the GABA$_A$ receptor genes undergo alternative splicing. The most prevalent of these is the γ2 subunit, which exists in two forms, $γ2_{Short}$ ($γ2_S$) and $γ2_{Long}$ ($γ2_L$) [9,10]. Alternative splicing of the human β3 [11] and the rat α6 [12] subunit genes has also been described. In the chick, the GABA$_A$ receptor subunit genes β2 and β4 have been found to undergo splicing to yield β2'β'4 [13,14]. (Note that, for the chick, β4 and γ4 GABA$_A$ receptor subunit genes have been reported, but it is probable that these are the avian homologues of the mammalian β1 and γ3 subunit genes respectively [15].) Two invertebrate GABA$_A$ receptor genes have

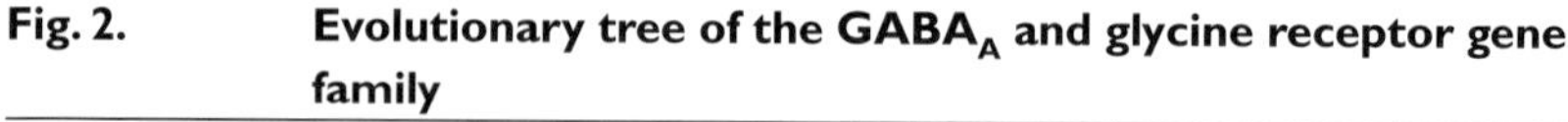

**Fig. 2.**          **Evolutionary tree of the GABA$_A$ and glycine receptor gene family**

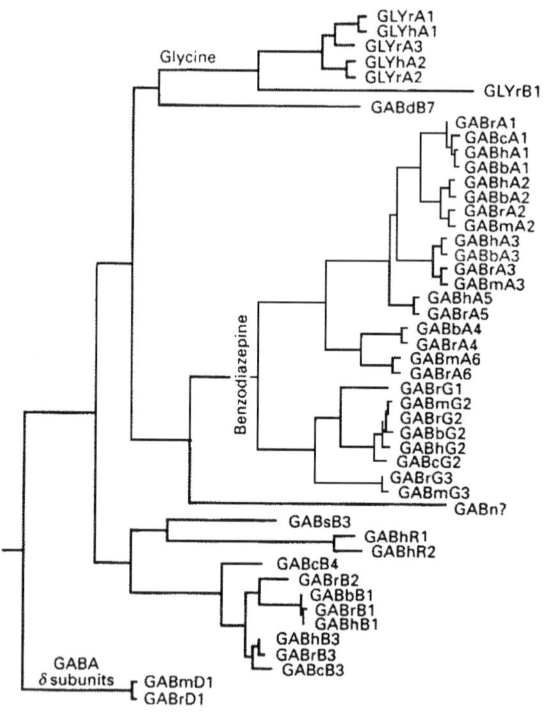


The figure shows an evolutionary tree of published GABA$_A$ and glycine receptor genes. It is constructed from the DNA sequences of 270 shared codons derived from an initial protein sequence alignment. The alignment included the majority of the extracellular domain and the transmembrane spanning regions, TM1–TM4. Gene sequences are labelled GAB for GABA$_A$ or GLY for glycine; and then organism, subunit type and number, abbreviated/designated as follows: b, bovine; c, chicken; d, Drosophila; h, human; m, mouse; n, nematode; s, snail; A, $\alpha$; B, $\beta$; D, $\delta$; G, $\gamma$; R, $\rho$. Thus, for example, GABcA1 is chicken GABA$_A$ receptor $\alpha$1 subunit, etc. Further details of the analysis can be found in [18]. Reproduced from [3] with permission.

been cloned. They are the *Drosophila Rdl* gene [16] and a GABA$_A$ receptor gene from the pond snail *Lymnaea stagnalis* [17]. Interestingly, in amino acid sequence comparisons with vertebrate GABA$_A$ receptor genes, these are both most closely related to the β subunits, with identities of the order of 28%. This low percentage conservation is in contrast with comparisons between the subunits of various vertebrate species, where values of greater than 90% are found. Fig. 2 shows the evolutionary tree of the GABA$_A$ receptor and the related glycine receptor subunit genes reported to date. The lineage is consistent with evolution from a single ancestral gene [18].

The mature GABA$_A$ receptor subunits have similar $M_r$ values, as deduced from their respective cDNA sequences. These range from 48000 ($\gamma$ subunits) to a maximum of 64000 ($\alpha$4 subunit). They all share a similar predicted domain structure, not only between themselves but also with other members of the ligand-gated ion channel superfamily, most notably the prototypic peripheral nicotinic acetylcholine receptor (cf. [19]). A schematic diagram of the pertinent features of a typical GABA$_A$ receptor subunit is shown in Fig. 3. Thus it is predicted that each subunit has an extended, extracellular, hydrophilic N-terminal domain of the order of 220 amino acids. Within this region are consensus sequences for N-glycosylation. There are four putative hydrophobic domains, TM1–TM4, which are predicted to

---

**Fig. 3.**  **Schematic representation of the pertinent features of GABA$_A$ receptors**

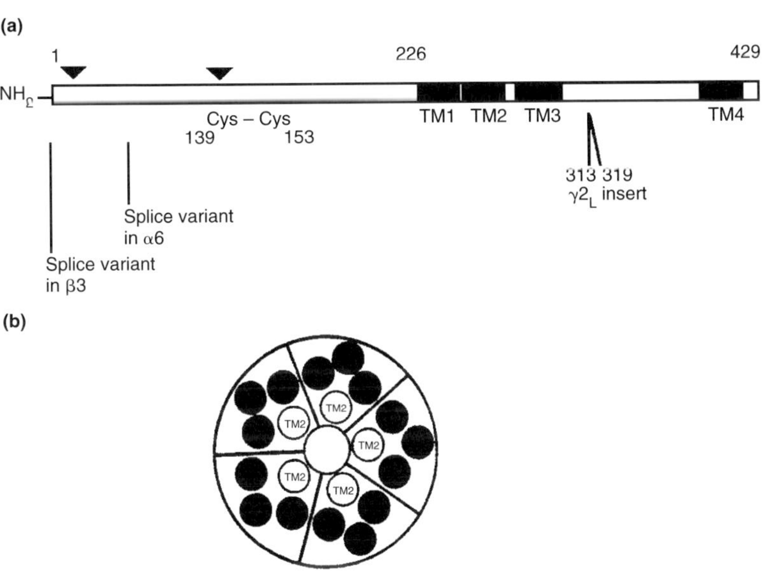


*(a) Numbered amino acid sequence of the bovine $\alpha$1 subunit, highlighting the domains conserved between all GABA$_A$ receptor polypeptides. ▼, Consensus sequences for N-glycosylation; TM 1–TM4, transmembrane domains; Cys–Cys, conserved extracellular motif common to all members of the ligand-gated ion channel superfamily. (b) Schematic view of the receptor oligomer, as viewed perpendicular to the plane of the membrane. Each of the five subunits of the receptor is represented as a segment within the annular structure, where the hole in the middle represents the chloride ion channel. The four membrane-spanning regions within each polypeptide are shown as filled circles, with the predicted $\alpha$ helix of TM2 lining the wall of the channel. The subunit complements and their ordering around the rosette are not definitively established; see the text for discussion. Reproduced from [3] with permission.*

span the membrane and form the chloride ion channel. TM2, again by analogy with nicotinic acetylcholine receptors and also as indicated by some mutagenesis studies (see below), is thought to form the inner lining of the channel. The TM1–TM3 regions are adjacent to the N-terminal domain, whereas TM4 is at the C-terminal end of the proteins. This transmembrane topology model predicts that both the N- and C-terminal regions are extracellular but, surprisingly, this remains unproven. Separating the TM3 and TM4 transmembrane spanning regions is a second hydrophilic region often referred to as the cytoplasmic or, alternatively, the intracellular loop domain. This region contains consensus sequences for phosphorylation by various protein kinases. It is also where the insertion of the eight amino acids in $\gamma2_L$ occurs. Splicing of the β3 and α6 subunit genes yields changes in the extracellular N-terminal domain. For the β3 subunit, variant signal peptides were found [11]; for the α6 subunit, a deletion of 10 amino acids (residues 57–66; rat numbering) was discovered [12].

Another feature of note of the GABA$_A$ receptor polypeptides, which is also highly conserved across all members of the ligand-gated ion channel superfamily, is a subdomain of the N-terminal region. This is a sequence of 15 amino acids that is predicted to form a Cys–Cys β loop. Modelling studies suggested that this domain may be involved in agonist binding ([20]; but see also section on localization of ligand binding domains within GABA$_A$ receptors, below).

When considering the different subunit types, the regions of amino acid conservation are found within the transmembrane spanning domains and also within defined regions of the N-terminal region. For the isoforms of one particular subunit type, on aligning the sequences the divergent regions are found at the extreme N- and C-terminal domains and, notably, within the respective cytoplasmic loops. Fig. 4 shows an alignment of mammalian α1–α6 subunit primary structures to illustrate the regions of divergence between subunit isoforms.


## Oligomeric structures of the GABA$_A$ receptors

It has been well established by classical protein chemical methods and electron microscopy studies that the peripheral nicotinic acetylcholine receptor has a pentameric quaternary structure, $\alpha_2\beta\gamma\delta$. More recently, other members of the ligand-gated ion channel superfamily, including the strychnine-sensitive glycine receptor and a cloned neuronal nicotinic acteylcholine receptor, were also shown to assemble as pentamers; the glycine receptor by chemical cross-linking [21] and the

---

**(Fig. 4. opposite)**

*The bovine α1, α2, α3 and α4 and rat α5 and α6 GABA$_A$ receptor subunit primary structures were aligned using the Multiple Sequence Alignment program of PC-GENE. Asterisks indicate amino acid sequence identity and dots indicate amino acid sequence similarity between all six subunits. The solid lines show the positions of the transmembrane regions TM 1–TM4. Reproduced from [3] with permission.*

**Fig. 4.**  **Comparison of the mammalian GABA_A receptor α subunit primary structures**

```
GAA1$BOVIN
GAA2$BOVIN
GAA3$BOVIN                                 QVESRRQEPGDFVKQDIGGLSP    22
GAA5$RAT                                   Q--------------------     1
GAA4$BOVIN                                 Q--------------------     1
GAA6$RAT                                   Q--------------------     1


GAA1$BOVIN  --QPSLQDFLKDNTTVFTRILDRLLDGYDNRLRPGLGERVTEVKTDIEVT    75
GAA2$BOVIN  ---NIQEDEAKNNITIFTRILDRLLDGYDNRLRPGLGDSITEVFTNIYVT    75
GAA3$BOVIN  KHAPDIPDDSTDNITIFTRILDRLLDGYDNRLRPGLGDAVTEVKTDIYVT   100
GAA5$RAT    MPTSSVQDETNDNITIFTRILDGLLDGYDNRLRPGLGERITQVRTDIYVT    82
GAA4$BOVIN  -----NQKEEKLCPENFTRILDSLLDGYDNRLRPGFGGPVTEVKTDIYVT    81
GAA6$RAT    -----LEDEGNFYSENVSRILDNLLEGYDNRLRPGFGGAVTEVKTDIYVT    65
              . .    .   **** **.********.*. .*.* *.*.**


GAA1$BOVIN  SFGPVSDHDMEYTIDVFFRQSWKDERLKFKGPMTVLRLNNLMASKIWTPD   125
GAA2$BOVIN  SFGPVSDTDMEYTIDVFFRQKWKDERLKFKGPMNILRLNNLMASKIWTPD   125
GAA3$BOVIN  SFGPVSDTIMEYTIDVFFRQTWHDERLKFDGPMKILPLNNLLASKIWTPD   150
GAA5$RAT    SFGPVSDTEMEYTIDVFFRQSWKDERLRFKGPMQRLPLNNLLASKIWTPD   132
GAA4$BOVIN  SFGPVSDVEMEYTMDVFFRQTWIDKRLKYDGPTEILRLNNMMVTKVWTPD   131
GAA6$RAT    SFGPVSDVEMEYTMDVFFRQTWTDERLKFKGPAEILSLNNLMVSKIWTPD   105
            *******  .****.******.*  *.**  .  **   *.***....*.****


GAA1$BOVIN  TFFHNGKKSVAHNMTMPNKLLRITEDGTLLYTMRLTVRAECPMHLEDFPM   175
GAA2$BOVIN  TFFHNGKKSVAHNMTMPNKLLRIQDDGTLLYTMRLTVQAECPMHLEDFPM   175
GAA3$BOVIN  TFFHNGKKSVAHNMTTPNKLLRLVDNGTLLYTMRLTIHAECPMHLEDFPM   200
GAA5$RAT    TFFHNGKKSIAHNMTTPNKLKRLEDDGTLIYTMRLTISAECPMQLEDFPM   182
GAA4$BOVIN  TFFRNGKKSVSHNMTAPNKLFRIMRNGTILYTMRLTISAECPMRLVDFPM   181
GAA6$RAT    TFFRNGKKSIAHNMTTPNKLFRLMHNGTILYTMRLTINADCPMRLVNFPM   155
            *** ***** ..**** ****.*. .*  ********. *.*** .***


GAA1$BOVIN  DAHACPLKFGSYAYTRAEVVYEWTREPARSVVVAEDGSRLNQYDLLGQTV   225
GAA2$BOVIN  DAHSCPLKFGSYAYTTSEVTYIWTYNASDSVQVAPDGSRLNQYDLPGQSI   225
GAA3$BOVIN  DVEACPLKFGSYAYTTAEVVYSWTLGKNKSVEVAQDGSRLNQYDLLGHVV   250
GAA5$RAT    DAHACPLKFGSYAYPNSEVVYVWTNGSTKSVVVAEDGSRLNQYHLMGQTV   232
GAA4$BOVIN  DGHACPLKFGSYAYPKSEMIYTWTKGPEKSVEVPKESSSLVQYDLIGQTV   231
GAA6$RAT    DGHACPLKFGSYAYPKSEIIYTWKKGPLYSVEVPEESSSLLQYDLIGQTV   205
            * *.********* .. .*..* **.   .   **   ** .*.* *.


GAA1$BOVIN  DSGIVQSSTGEYVVMTTHFHLKRKIGYFVIQTYLPCIMTVILSQVSFWLN   275
GAA2$BOVIN  GKETIKSSTGEYTVMTAHFHLKRKIGYFVIQTYLPCIMTVILSQVSFWLN   275
GAA3$BOVIN  GTEIIRSSTGEYVVMTTHFHLKRKIGYFVIQTYLPCIMTVILSQVSFWLN   300
GAA5$RAT    GTENISTSTGEYIMTAHFHLKRKIGYFVIQTYLPCIMTVILSQVSFWLN   282
GAA4$BOVIN  SSETIKSITGEYIVMTVYFHLRRKMGYFMIQTYIPCIMTVILSQVSFWIN   281
GAA6$RAT    SSETIKSNTGEYVIMTVYFHLQRKMGYFMIQIYTPCIMTVILSQVSFWIN   255
            . .. .***. .. ***.**. ** .  .* ******** **. ***


GAA1$BOVIN  RESVPARTVFGVTTVLTMTTLSISARN-SLPKVAYATAMDWFIAVCYAFV   324
GAA2$BOVIN  RESVPARTVFGVTTVLTMTTLSISARN-SLPKVAYATAMDWFIAVCYAFV   324
GAA3$BOVIN  RESVPARTVFGVTTVLTMTTLSISARN-SLPKVAYATAMDWFMAVCYAFV   349
GAA5$RAT    RESVPARTVFGVTTVLTMTTLSISARN-SLPKVAYATAMDWFIAVCYAFV   331
GAA4$BOVIN  KESVPARTVFGITTVLTMTTLSISARPISLPKVSYATAMDWFIAVCFAFV   331
GAA6$RAT    KESVPARTVFGITTVLTMTTLSISARH-SLPKVSYATAMDWFIAVCFAFV   304
            .********* **.*********.*   ****.******** *** .***


GAA1$BOVIN  FSALIEFATVNYFTKRGYAWDGKSVVPEKPKKVKDPLI---KKNNTYAPT   371
GAA2$BOVIN  FSALIEFATVNYFTKRGWAWDGKSVVNDK-KKEKASVM---IQNNAYAVA   370
GAA3$BOVIN  FSALIEFATVNYFTKRSWAWEGKKVPEALEMKKKTPAVPTKKTSTTFNIV   399
GAA5$RAT    FSALIEFATVNYFTKRGWAWDGKKALEAAKIKKKEREL IINKSTNAFTTG   381
GAA4$BOVIN  FSALIEFAAVNYFTNVQMEKAKRKTSKAPQEISAAPVLREKHPETPLQNT   381
GAA6$RAT    FSALIEFAAVNYFTNLQSQKAERQAQTAAKPPVAKSKTTESLEAEIVVHS   354
            ******** *****.          .            .  ..


GAA1$BOVIN  ATSYT-----------------------------PNLARG-----       382
GAA2$BOVIN  VANYA-----------------------------PNLSK------       380
GAA3$BOVIN  GTTYP---------------------------INLAKDTEFSA        415
GAA5$RAT    KLTHP-----------------------------PNIPKE-----       392
GAA4$BOVIN  NANLSMRKRANALVHSESDVGSRTDVGNISSKSSTVVQGSSEATPQSYLA   431
GAA6$RAT    DSKYHLKKRISSLTLP-----------------------------       370


GAA1$BOVIN  ---DPQLAYTARSATIEPREVRPETKPPEPKKTF----------------   413
GAA2$BOVIN  ---DPVLSTISKSATTPEPNKKPENKPAEAKKTF----------------   411
GAA3$BOVIN  ISKGAAPSTSSTPTIIASPKTTCVQDIPTETKTY----------------   449
GAA5$RAT    -----QLPGGTGNAVGTASTRASFFKTSFSKKTY----------------   421
GAA4$BOVIN  SSPNPFSRANAAETISAARAIPSALPSTPSRTGYVPRQVPVGSASTQHVF   481
GAA6$RAT    ------------------IVPSSEASKVLSRTPTLP--------------   388


GAA1$BOVIN  --------------------------NSVSKIDRLSRIAFPLL        430
GAA2$BOVIN  --------------------------NSVSKIDRMSRIVFPVL        428
GAA3$BOVIN  --------------------------NSVSKVDKISRIIFPVL        466
GAA5$RAT    ---------------------------NSISKIDKMSRIVFPTL       438
GAA4$BOVIN  GSRLQRIKTTVNSIGTSGKLSATTTPSAPPPSGSGTSKIDKYARILFPVT   531
GAA6$RAT    ---------------------------STPVTPPLLLPAIGGTSKIDQYSRILFPVA   418
                                        ...**.*. .** **.


GAA1$BOVIN  FGIFNLVYWATYLNREPQLKAPTPHQ        456
GAA2$BOVIN  FGTFNLVYWATYLNREPVLGVSP---        451
GAA3$BOVIN  FAIFNLVYWATYVNRESALKGMIRKQ        492
GAA5$RAT    FGTFNLVYWATYLNREPVINMATSPK        464
GAA4$BOVIN  FGAFNMVYWVVYLSKDT-MEKSESIM        556
GAA6$RAT    FAGFNLVYWIVYLSKDT-MEVSSTVE        443
            *. **.*** .*..... .
```

neuronal nicotinic acetylcholine receptor by mixing subtypes of wild-type and mutant subunits and studying their respective channel gating properties [22]. Again by analogy, it was presumed that the $GABA_A$ receptors were also pentameric. The $M_r$ of a heterogeneous population of $GABA_A$ receptors isolated from bovine cerebral cortex was determined by hydrodynamic analysis as 230000–240000 [23]. This, within experimental error, is consistent with the co-assembly of five glycosylated $GABA_A$ receptor subunits. Further, it has been shown by electron image analysis that a heterogeneous population of $GABA_A$ receptors, this time isolated from pig brain, has a rotational symmetry of five, again consistent with a pentameric structure [24]. More recently, Im et al. [25] used a novel approach to deduce the quaternary structures and subunit stoichiometries of two cloned $GABA_A$ receptor subtypes. Using a tandem construct of the α6–β2 $GABA_A$ receptor subunits, they showed that GABA-gated chloride ion channels were only formed following the co-expression of the α6–β2 construct with either an α6 or a γ2 clone. Thus it was deduced that these receptors were pentameric, with respective stoichiometries of $(\alpha6)_3(\beta2)_2$ and $(\alpha6)_2(\beta2)_2(\gamma2)_1$ ([25]; see also section on biochemical studies, below).

With 16 mammalian $GABA_A$ receptor genes, and not including variation due to the different splice forms of some subunits, Burt and Kamatchi [26] calculated that for pentameric structures there was the possibility of the existence of 151887 receptors, each with a different subunit permutation. It has been a major challenge to delineate the subunit complements of native $GABA_A$ receptors to investigate whether such extensive diversity is realistic. Several different approaches have been employed, with the most immediate progress being made utilizing the gene sequences. Thus, first, the electrophysiological and pharmacological properties of cloned receptors of defined subunit composition were characterized and compared with the properties of native $GABA_A$ receptors. Secondly, the spatial and temporal distributions of the respective $GABA_A$ receptor subunit mRNAs were determined by *in situ* hybridization. A complementary approach was the use of $GABA_A$-receptor-subunit-specific antibodies to determine subunit complements. Advances using each of these approaches are described in the following sections.

## Characterization of cloned $GABA_A$ receptors

For the study of the various cloned $GABA_A$ receptors, a variety of expression systems have been employed. These include the *Xenopus* oocyte (e.g. [27]), and transient (e.g. [28]) and stable (e.g. [29]) expression in a variety of mammalian cell types and in Sf9 insect cells using the baculovirus system (e.g. [30]). Generally, the *Xenopus* oocyte system has been used for electrophysiological characterization, whereas receptors expressed in the other systems are amenable to analysis by both electrophysiological and biochemical methods. The properties of the various cloned receptors seem to be independent of the expression system employed. However, it should be noted that, with the exception of single-channel recordings that have provided evidence for functional heterogeneity, these studies have characterized the properties of a *population* of cloned receptors. Detailed molecular comparisons of recombinant receptors expressed in the different systems has so far not been carried out. The physiological properties of the cloned GABA$_A$ receptor subtypes are described in more detail in Chapter 3 of this volume [1].

Each GABA$_A$ receptor polypeptide, when expressed alone in *Xenopus* oocytes, can form a GABA-gated homo-oligomeric chloride ion channel [31–33]. In general, however, the α1–α6, β1 and β2, γ1–γ3 and δ subunits expressed alone assemble at low efficiency, and it is probable that they do not form homo-oligomers *in vivo*. Two exceptions are the β3 and ρ subunits. The ρ subunit does assemble efficiently to form a functional GABA-gated ion channel [34]. Moreover, the ρ homomeric channels are reminiscent of invertebrate-type GABA receptors in that they are insensitive to the classical GABA$_A$ receptor antagonist, bicuculline [35]. They are also barbiturate- and benzodiazepine-insensitive [34]. The ρ subunit mRNAs are found predominantly in the retina [34]. It may be that they form a subset of GABA$_A$ receptors distinct from those found in the brain; they are sometimes termed GABA$_C$ receptors. Following expression in human embryonic kidney cells, the β3 subunit will also form homo-oligomers which bind [$^{35}$S] *t*-butylbicyclophosphorothionate with high affinity [36]. Further binding studies showed that these β3 homo-oligomers retained the barbiturate, neurosteroid and alphaxolone allosteric modulatory sites [36].

The co-expression of a single type of each of the α and β subunits results in the formation of robust GABA-gated ion channels which are potentiated by barbiturates and neurosteroids (e.g [2,37]). However, although there are some reports that these GABA currents are modulated by benzodiazepines [1,38], the consensus view currently is that, for a full spectrum of reproducible benzodiazepine responsiveness, the co-expression of an α, a β and either a γ2 or γ3 subunit is requisite [32,39,40]. Pairwise combinations such as αγ or βγ do not form channels with high efficiency. The δ subunit has not been identified so far with a particular pharmacological characteristic; thus it is more difficult to assess co-expression studies including this subunit (although see Chapter 4 of this volume [1] for a further discussion). Notwithstanding this caveat, reports are consistent with the idea that co-expressed δ-subunit-containing paired subunit combinations do not assemble with high efficiency [41]. The co-expression of α, β, γ and δ subunits has been reported to eliminate the benzodiazepine facilitation of GABA-gated chloride ion conductance [33], although a more recent report did observe diazepam potentiation of α1β1γ2$_L$δ receptors [41]. Further, single-channel properties and comparative pharmacological characterization of cells co-expressing these four subunits shows that they all do co-assemble rather than forming two distinct GABA$_A$ receptor subpopulations [41].

To summarize briefly, the results of the expression studies suggest that brain GABA$_A$ receptors comprise probably pentameric receptors of αβ, αβγ, possibly $\alpha\beta\delta$, $\alpha\beta\gamma\delta$ and $\rho$ subunit combinations. (No information is currently available regarding the $\pi$ subunit.) The ratios of the different subunits in each pentamer is not known, but will be discussed below. Similarly, account must be taken of the variants within a single subunit class, e.g. $\alpha1\beta\gamma$ compared with $\alpha2\beta\gamma$, etc. Again, expression studies have yielded insight into the characteristics endowed by the isoforms of a subunit type. These studies are described in the next section.

## Characteristic features of GABA$_A$ receptor subunit isoforms as determined by the study of recombinant receptors

As described above, it is now accepted from expression studies that the presence of both an $\alpha$ subunit and a $\gamma$ subunit is required for benzodiazepine potentiation of the various cloned GABA$_A$ receptors. However, the type of $\alpha$ or $\gamma$ subunit has been shown to influence the subtype of benzodiazepine pharmacology. The $\alpha1\beta x\gamma2$ receptors (where $\beta x$ is $\beta1$, $\beta2$ or $\beta3$) have a high affinity for compounds such as CL 218,872 and the $\beta$-carbolines, both of which were historically classified as benzodiazepine receptor type I ligands [28]. The $\alpha2\beta x\gamma2$, $\alpha3\beta x\gamma2$ and $\alpha5\beta x\gamma2$ receptors have a low affinity for these ligands, and therefore they correlate with benzodiazepine type II sites. A further subdivision has been made, since the $\alpha5\beta x\gamma2$ receptors have an even lower affinity for the drug zolpidem [42]. $\alpha4\beta x\gamma2$ and $\alpha6\beta x\gamma2$ receptors bind only the benzodiazepine partial inverse agonist Ro 15-4513 with high affinity, whereas classic benzodiazepines have very low affinity for these putative receptors [43,44]. This pharmacology correlates with the previously-described 'benzodiazepine-insensitive' or 'diazepam-insensitive' [$^3$H]Ro 15-4513 specific binding sites [45]. There have been reports of the co-expression of different $\alpha$ subunits together with $\beta\gamma$ which suggest that the four different types of subunit co-assemble: $\alpha1\alpha3$ by Verdoorn [46] and $\alpha1\alpha6$ by Mathews et al. [47]. However, the benzodiazepine pharmacological profile of these GABA$_A$ receptors has not been described.

Of the $\gamma$ subunits, the $\gamma2$ and $\gamma3$ polypeptides behave identically, in that their presence in $\alpha\beta\gamma$ constructs is required for a full benzodiazepine agonist and inverse agonist profile [32,39]. However, in these trimeric receptors containing $\gamma1$ subunits, the benzodiazepine inverse agonists behave as positive facilitators [48]. Also relevant is a report suggesting that the $\gamma3$ subunit could influence benzodiazepine pharmacology, since $\alpha1\beta x\gamma3$ receptors behaved like $\alpha5\beta x\gamma2$ recombinants, with a low affinity for zolpidem [49].

The presence of $\gamma$ subunits additionally decreases the sensitivity of the cloned receptors to antagonism by $Zn^{2+}$ [1,50,51]. Further, the presence of the $\gamma2_L$ splice variant renders the receptors susceptible to potentiation by ethanol [52]. It is not clear if this is a direct effect on the receptor itself or whether it occurs via a second messenger mechanism resulting in the activation of protein kinase C and

**Table 1.** **Summary of the characteristics of GABA$_A$ receptor polypeptides from studies of cloned receptor proteins**

Abbreviation: BZ, benzodiazepine. x denotes 1, 2 or 3 for β subunits or any number between 1 and 6 for α subunits.

| Subunit type | Comments |
| --- | --- |
| α1 | Type I BZ pharmacology in α1βxγ2 combinations |
| α2 | Type II BZ pharmacology in α2βxγ2 combinations |
| α3 | Type II BZ pharmacology in α3βxγ2 combinations |
| α4 | α4βxγ2 receptors have high affinity for Ro 15-4513 only |
| α5 | Type II BZ pharmacology and very low affinity for zolpidem |
| α6 | α6βxγ2 receptors have high affinity for Ro 15-4513 only; N-terminal splice variant in rat |
| β1 | αxβ1γ2 receptors are loreclezole-insensitive |
| β2 | αxβ2γ2 receptors are loreclezole-sensitive |
| β3 | αxβ3γ2 receptors are loreclezole-sensitive; N-terminal splice variant in human |
| γ1 | αxβxγ1 receptors bind BZ agonists only |
| γ2 | Confers a complete spectrum of BZ activities with α(1–3,5) βx combinations; reduces Zn$^{2+}$ sensitivity; splice variants γ2$_S$ and γ2$_L$; αxβxγ2$_L$ receptors are ethanol-sensitive |
| γ3 | Confers a complete spectrum of BZ activities with α(1–3,5) βx combinations; can confer low affinity for zolpidem in the absence of α5 |
| δ | No known characteristics |
| ρ | Forms robust homo-oligomers that are bicuculline insensitive |
| π | Expressed abundantly in reproductive tissues |

second messenger mechanism resulting in the activation of protein kinase C and subsequent phosphorylation of the $\gamma2_L$ subunit [52].

Until quite recently, it was thought that the presence of the variants of the β subunit did not influence the physiological or pharmacological properties of the cloned $GABA_A$ receptors. Interestingly, Wafford et al. [53] have shown that the anticonvulsant drug loreclezole binds to a novel site on the $GABA_A$ receptor. They have localized this site of action to the TM2 region of the β2 and β3 subunits.

A summary of the pertinent features of the $GABA_A$ receptor subunits is given in Table 1.

## Localization of $GABA_A$ receptor mRNAs by *in situ* hybridization

*In situ* hybridization with $GABA_A$-receptor-subunit-specific oligonucleotide probes has shown that each respective mRNA has a distinct pattern of expression. With the exceptions of the α6 subunit mRNA, which is localized almost exclusively in a single cell type (the mature cerebellar granule cell), the ρ subunits which are found predominantly in the retina and the π subunit which is expressed relatively abundantly in humans in the reproductive system but at comparatively low levels in brain [54], different $GABA_A$ receptor subunit mRNAs have a heterogeneous distribution across many different cell types in adult brain. Furthermore, each also undergoes developmental changes in its pattern of expression. There appears to be a more widespread heterogeneity in mRNAs expressed early in development compared with those expressed at the adult stage. *In situ* hybridization studies are not readily quantifiable, and there is the further caveat that, for heteromeric receptors, the levels of mRNA are not necessarily representative of functional receptors. Also, Primus et al. [55] reported that a significant proportion of rat β1 subunit mRNA is non-polyadenylated, thus introducing an extra variable. Nevertheless, the following generalizations and points can be made.

In adult brain, of the α subunit mRNAs, that for α1 is the most ubiquitously and abundantly expressed, followed by the α2 and α3 mRNAs. As stated above, the α6 mRNA is confined almost exclusively to the cerebellar granule cell; the α5 mRNA has a low overall distribution but is enriched particularly in the CA1 and CA3 regions of the hippocampus, while α4 transcripts are the rarest of the α subunits but are enriched in thalamic regions. Specific neuronal cell types have been identified which show the co-expression of at least two different α subunits, e.g. the mitral cells of the olfactory bulb (α1 and α3) and cerebellar granule cells (α1 and α6). The α2 subunit mRNA is notably found in both neuronal cells and the Bergmann glial cells of the cerebellum. The most abundant of the β subunit mRNAs are those for β2 and β3, which both show a widespread distribution; the β1 mRNA is present at lower levels. Again cell types have been identified in which the different β subunit mRNAs co-localize, an example being once more the mitral cells of the olfactory bulb. The γ2 subunit is the most abundant of the γ subunit mRNAs, followed by γ1; the γ3 transcript is rare. The δ subunit mRNA is enriched in cerebellar granule cells and the dorsolateral and ventrolateral geniculate nuclei of the thalamus. Thus, overall, the most ubiquitously expressed subunits are α1, α2, α3, β2, β3 and γ2. Other isoforms show more restricted patterns of distribution. Further, it is clear that some cell types express more than five different GABA$_A$ receptor subunit transcripts, thus implying the presence of more than one receptor subtype per cell. (For more detailed analyses and correlates of the *in situ* hybridization studies, see [56–58]; for a developmental profile, see [59].)

## Polypeptide complements and distributions of native GABA$_A$ receptors determined using subunit-specific antibodies

The distributions of the various subunit mRNAs are not necessarily representative of the localizations of functional, heteromeric receptor proteins. Thus a natural progression in the determination of the subunit complements of native GABA$_A$ receptors has been the production and use of subunit-specific antibodies. The antigens used for the production of such antibodies have been either synthetic peptides or bacterially-derived fusion proteins, both of which correspond to the divergent regions of the various GABA$_A$ receptor subunits. In general, the peptide sequences that have been employed successfully for GABA$_A$-receptor-subunit-specific antibody production are from the divergent N-terminal and C-terminal regions, whereas the bacterial fusion protein antigens are larger and encompass the respective cytoplasmic loops (cf. Fig. 4). The α1–α6, β1–β3, γ1–γ3 and δ subunits have all been identified in brain membrane preparations by immunoblotting by several different groups of workers. The $M_r$ values of the immunoreactive polypeptides agree fairly well with the values predicted from the cDNA sequences once account is taken of potential mass contributions by N-linked carbohydrates (summarized in [60]). Similarly, the overall regional distributions and abundances in brain of the different polypeptides determined qualitatively by immunoblotting and immunocytochemistry, and quantitatively by immunoprecipitation, agree well with the respective GABA$_A$ subunit mRNA localizations. However, the more interesting question is: which of the subunits co-assemble?

## Biochemical studies

Two biochemical approaches have been employed. In one approach, GABA$_A$ receptor subpopulations have been purified by immunoaffinity chromatography and the co-associating subunits identified by immunoblotting. Alternatively, receptor subpopulations have been immunoprecipitated from detergent extracts of

**Table 2.    Summary of the subunit complements of native GABA$_A$ receptors**

The 'primary antibody' column indicates the antibody specificity that was used for either immunoaffinity purification or immunoprecipitation studies. The co-associated subunits were then identified by immunoblotting with the specific antibodies as indicated. Positive results only are shown, i.e. since any one group of workers does not have the full repertoire of GABA$_A$ receptor subunit antibodies, not all polypeptides were screened in each case. ND indicates not detectable, and parentheses indicate a low concentration .

| Primary antibody | Experimental paradigm | Co-associated subunits | References |
|---|---|---|---|
| $\alpha$1 | Purification | $\alpha$1, $\alpha$2, $\alpha$3, $\beta$3, $\gamma$2 | 62,115 |
| | Purification | $\alpha$1, $\alpha$3 | 65 |
| | Precipitation | $\alpha$1, $\alpha$3 | 116 |
| | Precipitation | $\alpha$1, $\alpha$3, $\alpha$5, $\beta$2/3, $\gamma$2, $\delta$ | 67 |
| | Precipitation | $\alpha$1, $\beta$2/3, $\gamma$2$_L$, $\gamma$2$_S$, $\alpha$6 (implied) | 69 |
| $\alpha$2 | Purification | $\alpha$1, $\alpha$2, $\alpha$3, $\beta$3, $\gamma$2 | 62,115 |
| $\alpha$3 | Purification | $\alpha$1, $\alpha$2, $\alpha$3, $\beta$3, $\gamma$2 | 62,115 |
| | Purification | $\alpha$1, $\alpha$3, ($\alpha$5), $\beta$2/3, $\gamma$2, $\delta$ | 67 |
| | Precipitation | $\alpha$1, $\alpha$3 | 63 |
| $\alpha$4 | Purification | $\alpha$1, $\alpha$4, $\beta$2/3, $\gamma$2 | 66 |
| $\alpha$6 | Purification | $\alpha$1, $\alpha$6, $\beta$2/3, $\gamma$2<br>ND: $\alpha$2, $\alpha$3 | 68 |
| | Precipitation | $\alpha$6, $\gamma$2, $\delta$<br>ND: $\alpha$1, $\alpha$2, $\alpha$3 | 71 |
| $\beta$2 | Purification | $\alpha$1, ($\alpha$2), $\beta$2, $\gamma$2<br>ND: $\alpha$3, $\alpha$5, $\beta$1 | 74 |
| $\gamma$1 | Purification | $\alpha$2, $\gamma$1<br>ND: $\alpha$1, $\alpha$3, $\alpha$6, $\gamma$2, $\gamma$3, $\delta$ | 76 |
| | Purification | $\alpha$1, $\alpha$2, $\alpha$3, $\alpha$5, $\beta$2/3, $\gamma$1<br>ND: $\gamma$2, $\gamma$3 | 78 |
| $\gamma$2 | Purification | $\alpha$1, $\alpha$3, ($\alpha$5), $\beta$2/3, $\gamma$2, $\delta$ | 67 |
| $\gamma$3 | Purification | $\alpha$1, $\alpha$2, $\alpha$3, $\alpha$4, $\alpha$6, $\beta$2/3, $\gamma$2<br>ND: $\alpha$5, $\gamma$2 | 49 |
| | Purification | $\gamma$2, $\gamma$3<br>ND: $\gamma$1 | 76 |
| $\delta$ | Purification | $\alpha$1, $\alpha$3, $\beta$2/3, $\gamma$2, $\delta$ | 67 |

different brain regions and the immune pellets analysed for co-immunoprecipitated subunits. So far, most information has been forthcoming for the different $\alpha$ subunits of the GABA$_A$ receptors, because high-affinity specific antibodies were readily obtained. The salient features of these studies are discussed below. Table 2 summarizes the results from different laboratories in more detail.

With respect to the α subunits, there is universal agreement that the majority of native GABA$_A$ receptors contain a single type of α subunit (e.g. [61–63]). The most abundant of the α subunits in the adult rat brain is α1 [64]. The benzodiazepine pharmacology of the various α-subunit-enriched GABA$_A$ receptors was similar in profile to that found for the corresponding αβγ cloned receptors [63], and further correlation was thus with the type I and type II pharmacological subclasses (see earlier, and [28]). However, in some studies it was reported that two different α subunits could be detected in immunoaffinity-purified preparations or immune precipitates. For example, the coexistence of the α1 subunit with the α2, α3, α4, α5 or α6 subunit, and of the α2 with the α3 subunit, has been described ([62,65–68]; Table 2). GABA$_A$ receptors in detergent solution have a tendency to aggregate [23]. However, control experiments using antibody columns of different specificity in series demonstrated that the coexistence of two α subunit variants was not artefactual. For example, purification of GABA$_A$ receptors on an α2 and an α3 subunit antibody column in series yielded purified material containing α2 and α3 but not immunoreactivity for the most abundant α subunit (α1) [68]. Further, because only two different α subunit immunoreactivities have been reported as coexisting within a single receptor oligomer, this has formed the basis of the argument that there are probably two α subunits per receptor molecule [62,68]. It should be stressed, however, that there is no information as yet regarding α subunit stoichiometry in native GABA$_A$ receptors containing a single α subunit variant. The GABA$_A$ receptors that contain two types of α subunit have been identified as minority subpopulations, with the exception of the coexistence of the α1 and α6 subunits. Khan et al. [69] deduced that this population accounts for 50% of all the α6-subunit-containing GABA$_A$ receptors. This is in agreement with the direct findings of Pollard et al. [70], who demonstrated an α1/α6 subunit ratio of 1:1 in this receptor population. Notably, another group did not find evidence for the coexistence of the α1 and α6 subunits. This may be explained by the use of lower-avidity antibodies and reflects the difficulties in making comparisons between results from laboratories using different antibodies [71]. There have been reports whereby two different α subunits were coexpressed together with β γ subunits; the functional properties of these receptors were distinct from those of receptors containing a single α subunit, thus suggesting the co-assembly of all four subunit types ([46,47], and see above) The benzodiazepine pharmacologies of these mixed α subunit receptors is not known. For the α1α6βγ receptors, results are consistent with a benzodiazepine-insensitive pharmacology, i.e. dominance of the α6 subunit [69,70].

Such detailed studies have not been carried out for the β subunits, probably because it has been more difficult to raise useful isoform-specific antibodies. (The intracellular loops of the β subunits are more highly conserved than for the other subunit classes and, additionally, β subunits do not have a hydrophilic C-terminus.) Many studies have therefore utilized the monoclonal antibody bd-17, which recognizes both the $\beta 2$ and $\beta 3$ subunits but not $\beta 1$ [72,73]. Benke et al. [74], however, reported the production of specific antibodies against the $\beta 1$, $\beta 2$ and $\beta 3$ subunits. In this study, it was shown that the most prevalent $\beta$ subunit was $\beta 2$, which was most often associated with the $\alpha 1$ subunit [74]. Moreover, quantitative immunoprecipitation studies indicated the presence of a single type of $\beta$ subunit per receptor [74]. For all of the $GABA_A$ receptor subpopulations that have been identified, a $\beta$ subunit was always present (Table 2). This is consistent with the identification of the $\beta$ subunit as the high-affinity agonist binding site ([75], and see below).

Of the $\gamma$ subunits, the $\gamma 2$ subunit is the most prevalent. Quirk et al. [76] showed the co-association of the $\gamma 2$ and $\gamma 3$ subunits and, by analogy with the experiments carried out for the $\alpha$ subunits, proposed the existence of two $\gamma$ subunits per receptor molecule. This is in agreement with two other studies, but not that of Im et al. ([25], see below). Subunit proportions of $(\alpha 1)_2(\beta 1)_1(\gamma 2)_2$ were deduced in an elegant study by Backus et al. [77] by comparing the conductance properties of $\alpha 1\beta 2\gamma 2$ wild-type and mutant cloned receptors. Similarly, Khan et al. [69] deduced by immunoprecipitation studies that one subpopulation of receptors in the cerebellum comprised $\alpha 1\alpha 6\beta 2/3\gamma 2_L\gamma 2_S$. Conflictingly, however, Togel et al. [49] and Mossier et al. [78] purified $\gamma 3$- and $\gamma 1$-subunit-containing $GABA_A$ receptors respectively, but in these studies no evidence was found for the coexistence of the different $\gamma$ polypeptides. Note also that the elegant studies of Im et al. ([25], described above) imply the presence of only one $\gamma$ subunit per cloned receptor oligomer. In the $\gamma 3$-subunit-enriched receptors, no $\alpha 5$ subunit immunoreactivity was detected either, yet a low affinity for zolpidem was found ([49], and see above).

For the $\delta$ subunit, again there are inconsistencies between groups. Quirk et al. [71] found an exclusive co-localization of the $\delta$ subunit with the $\alpha 6$ polypeptide in the cerebellum. This $\alpha 6\delta$ subunit population did not bind benzo-diazepine agonists, antagonists or even inverse agonists with high affinity. No $\gamma 2$ immunoreactivity was detected in these cerebellar receptors. This is in contrast with the coexistence of $\alpha\beta\gamma\delta$ subunits as suggested by single-channel studies of cloned receptors (see above). Furthermore, Mertens et al. [67] found that the $\delta$ subunit was more widely expressed, and coexisted with $\alpha 1$, $\alpha 3$, $\beta 2/3$ and $\gamma 2$ subunits; the $\alpha 6$ subunit was not investigated.

It was mentioned earlier that Burt and Kamatchi [26] calculated the possible existence of 151887 different $GABA_A$ receptors. A subunit stoichiometry of $\alpha_2\beta_1\gamma_2$ is probably the most consistent with all published work on *native* receptors (see above; also, see below for further discussion). With the assumptions that within receptor oligomers different $\alpha$ subunits coexist, that different $\beta$ and $\gamma$ subunits do not, that the $\delta$ subunit substitutes for a $\gamma$ subunit and that $\rho$ subunits assemble as homo-oligomers, this value is reduced to a more manageable 137. [It is realized that the assumptions made are inconsistent with some reports and may give an underestimation (cf. Table 2); nevertheless, the characterization of the native receptors has significantly decreased the estimates of potential receptor numbers.]

## Immunocytochemical studies

There have been many reports of the distributions of specific GABA$_A$ receptor subunits in the developing and adult brain, and indeed in the periphery. These are, however, incomplete studies because no group has available a complete repertoire of antibodies which are amenable for use in immunocytochemistry. A further problem in interpretation is that, generally, single-subunit distributions are carried out, and this may not be representative of functional receptors. For example, significant intracellular pools of the GABA$_A$ receptor α1 subunit have been described which presumably are nascent polypeptides [79]. The most comprehensive study so far is that of Fritschy and Mohler [80], who described the regional and cellular distributions of seven major subunits, i.e. α1, α2, α3, α5, β2/3, γ2 and δ. Overall, the cellular distributions of the GABA$_A$ receptor subunits correlate well with *in situ* hybridization studies and the same overall generalizations can be made. As described above, each subunit has its own temporal and spatial pattern of distribution. Also, in one case at least five different subunits were found in the same neuron [81] and, in agreement with the biochemical studies, two α variants were localized in the same cell type. The findings of Fristchy and Mohler [80] identified the triplet combination α1β2/3γ2 as the most prevalent, in agreement with biochemical studies. Sometimes α2, α3 or δ was co-associated with these to suggest four subunit types per receptor [8]. Other triplet combinations identified were α2β2/3γ2, α3β2/3γ2 and α5β2/3γ2 [80]. Another minor subset of receptors was identified which lacked the co-association of the β2/3 subunit and included α1α2γ2, α2γ2, α3γ2, α2α3γ2 and α2α5δ [80]. Fig. 5 illustrates some of these points, based on work in my laboratory.

Some of the mapping studies have been extended to the electron microscope level in order to resolve the subcellular distributions of the GABA$_A$ receptor subtypes (e.g. [80,82,83]). One particular study focused on the subcellular localization of the α1 and α6 subunits in cerebellar granule cells using post-embedding fixation and immunogold labelling [82]. Points of interest found here were that α1 and α6 immunoreactivities were both found at synaptic and extra-synaptic extracellular sites within the same cell, but were enriched at the former. Further, in ultra-thin serial sections, synapses were found that contained either α1 or α6, or both α1 and α6, subunit immunoreactivities [83]. Taken in conjunction with the biochemical studies, this strongly suggests that the α1 and α6 subunits are present in the same synaptic GABA$_A$ receptor. Also, since all combinations with respect to these two α subunits were found, the functional significance of α subunit heterogeneity compared with the random association of actively transcribed α subunit genes is raised (see below). Fig. 6 shows the subcellular distributions of

## Fig. 5.  Immunocytochemical localization of GABA$_A$ receptor $\alpha$ subunits in adult rat brain

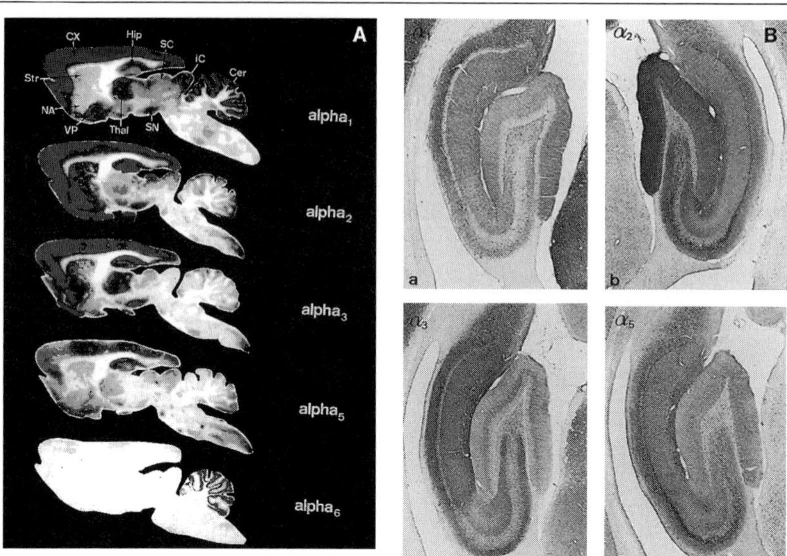



*(A) Global distribution of GABA$_A$ receptor $\alpha$ subunit immunoreactivity in the adult rat brain. Image analysis of saggital sections stained with each of the antibodies reveals the sites of most abundant immunoreaction. Note that the patterns are complementary in many brain regions. Abbreviations: Cer, cerebellum; CX, cortex; Hip, hippocampus; IC, inferior colliculus; NA, nucleus accumbens; SC, superior colliculus; SN, substantia nigra; Str, striatum; Thal, thalamus; VP, ventral pallidum. (B) Localization of the $\alpha$1, $\alpha$2, $\alpha$3 and $\alpha$5 subunits at higher magnification in the hippocampus. In (a), the $\alpha$1 subunit immunoreactivity decorates principally interneurons and, to a lesser extent, presumed pyramidal cell dendrites, mostly in the CA1 field. In (b), in contrast, the $\alpha$2 subunit immunoreactivity is most prominent in the molecular layer of the dentate gyrus, and on the somata of granule and pyramidal, especially CA3, cells. In (c) the $\alpha$3 subunit immunoreactivity appears most concentrated in the stratum oriens of all pyramidal fields, and in the stratum radiatum of CA1 (note also the graded staining in the dentate molecular layer). In (d), some of the most dense $\alpha$5 subunit immunoreactivity is associated with pyramidal somata, particularly in CA3. The scale bar represents 200 $\mu$m. Reprinted with permission from [117].*

**(Fig. 6. opposite)**

*(A) and (C) show labelling with an anti-$\alpha$6 antibody, (B) with an anti-GABA antibody and (D) with an anti-$\alpha$1 antibody. (A) and (B) show immunoparticles for the receptors located at synaptic junctions (arrows in A) between GABA-immunopositive Golgi cell terminals (Gt) and granule cell dendrites (d). Three immunoparticles are located between a Golgi cell terminal and a granule cell dendrite (d$_1$) at a site where the synaptic junction is not evident. However, on the next serial section (B), the synaptic specialization is clearly visible. (C) and (D) show the co-localization of the $\alpha$6 (C) and $\alpha$1 (D) GABA$_A$ receptor subunits in the same synaptic junction (arrows) made between a Golgi cell terminal (Gt) and a granule cell dendrite (d). Scale bar = 0.2 $\mu$m. Reprinted from [82], with permission.*

**Fig. 6.**      **Electron micrographs showing the co-localization of immunoreactivity for the GABA$_A$ receptor $\alpha$1 and $\alpha$6 subunits at the same synapse in serial ultra-thin sections of adult rat cerebellum using post-embedding immunogold labelling**

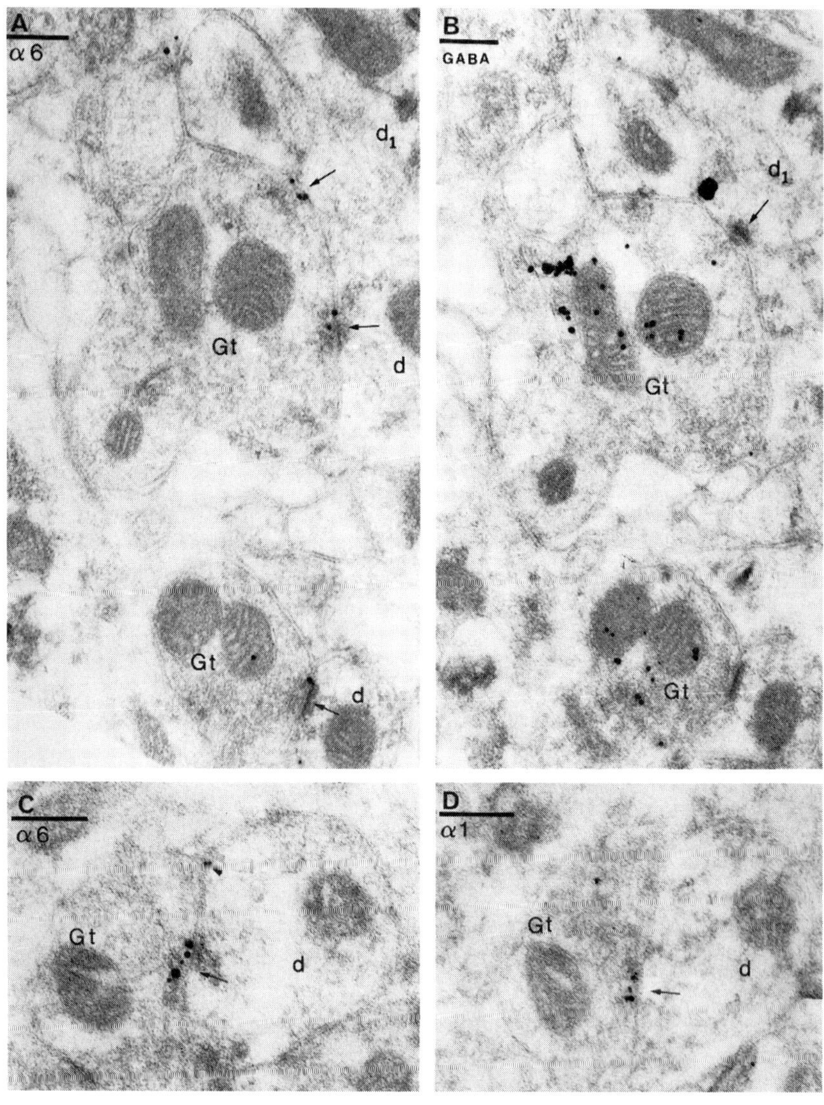

GABA$_A$ receptor α1 and α6 subunits in cerebellar granule cells. A more recent study has found, again by high resolution immunogold labelling, that, in hippocampal pyramidal cells, the α1 and α2 subunits are differentially targeted within the same cell [83]. Thus the α1 subunit was found in most GABAergic synapses on all post-synaptic domains of pyramidal cells. In contrast, the α2 subunit was located primarily in a subset of synapses on axon initial segments that receive GABAergic input from specialized axo-axonic cells [83]. In some cases, α1 and α2 subunits were co-localized in the same synapse [83].

## Functional significance of GABA$_A$ receptor heterogeneity

The question that immediately arises following the discovery of the multiple receptor subtypes is: why are there so many and what distinct functions, if any, do they subserve? It can be argued that all the different GABA$_A$ receptors serve the same function, in that they are all fast-acting, ligand-gated chloride ion channels and that, in fact, GABA$_A$ receptor heterogeneity has evolved with the conservation of the important part of the protein, the chloride ion channel, with all other changes being redundant. The diametrically opposed view, however, is that the GABA$_A$ receptor genes have indeed evolved to serve distinct functions. Support for this argument comes from the facts that each GABA$_A$ receptor gene shows a characteristic temporal and spatial pattern of expression (see above) and, additionally, that recombinant receptors do have subtly different properties. A complete understanding of the role played by GABA$_A$ receptor subtypes will only come from a fully integrated approach involving molecular properties through to behavioural neuroscience. Some points are made below.

Cloned receptors of defined subunit composition have differential sensitivities to activation by GABA and different desensitization rates (e.g. [1,26,84]). These properties may be important during development, where GABA acts both as a trophic agent and as a neurotransmitter (see [59] for a discussion). Recombinant GABA$_A$ receptors also have differential sensitivities to allosteric modulation by endogenous molecules found in the brain, which include the neurosteroids and Zn$^{2+}$ [1,37,50,51]. GABA$_A$ receptors which differ in their benzodiazepine pharmacological profiles cannot be discussed in the context of their physiological functions, since the current consensus view is that there are no endogenous benzodiazepine-like compounds. These binding sites, therefore, remain serendipitous both for the molecular dissection and classification of GABA$_A$ receptors and, importantly, for exploitation in the treatment of anxiety etc.

Many of the GABA$_A$ receptor subunits contain consensus amino acid sequences for phosphorylation by protein kinase A, protein kinase C and tyrosine kinase. Some of the subunits have been shown to be phosphorylated *in vitro*, with concomitant modulation of channel activity. Further, native GABA$_A$ receptors show differential responses following activation of different protein kinases {see, e.g., [85] for a more detailed discussion, and [1,86] (Chapters 3 and 9, respectively, of the current volume)}, thus suggesting that regulation of receptor subtype activity by intracellular mechanisms may be important.

In whole-animal studies, the GABA$_A$ receptor α6 subunit has been implicated in cerebellar motor control [87]. Interestingly, the β3 subunit gene is deleted in the neurological disorder Angelman's syndrome [88]. The mouse pink-eyed cleft-palate (p$^{cp}$) mutation has a deletion of a region that contains the α5, β3 and γ3 subunit genes [89]. This deletion is similar to the one found for Angelman's syndrome in humans, and these mice show neurological dysfunction. Interestingly, however, it was reported that one strain had deletions that resulted in no expression of the α5 and γ3 subunit genes, yet these mice were neurologically normal [90].

So far, the study of GABA$_A$ receptor subunit knock-out mice has yielded no useful information regarding the functional significance of particular subunits. Mice lacking the γ2 subunit predictably had a loss of benzodiazepine binding sites (>94% decrease compared with wild-type mice) [91]. However, the mice showed robust GABA responses, with the single-channel properties in the mutant mice consistent with the presence of αβ-subunit-containing GABA$_A$ receptors. Thus it was concluded that the γ2 subunit is not essential for the formation of functional GABA$_A$ receptors [91]. However, since the knock-out animals were not viable beyond a few days, it was proposed that an endogenous ligand active at the benzodiazepine site may be required for normal development [91]. α6-Subunit-deficient mice were found to survive, but they showed no distinguishing phenotype [92].

## Localization of ligand binding domains within GABA$_A$ receptors

The initial insight into the localization of drug binding sites within the GABA$_A$ receptors was gained from photoaffinity labelling studies. Thus it was shown that the benzodiazepine, [$^3$H]flunitrazepam, and the high-affinity GABA$_A$ receptor agonist, [$^3$H]muscimol, photoaffinity-labelled two different GABA$_A$ receptor subunits. These were distinguished on the basis of $M_r$ [93,94], and they were at the time widely recognized as the α ($M_r$ 51000; benzodiazepine binding) and β ($M_r$ 58000; GABA binding) subunits respectively, suggesting that these allosterically coupled binding sites were located on different subunits of GABA$_A$ receptors. Heterogeneity of polypeptides that were photoaffinity-labelled by [$^3$H]flunitrazepam was reported as early as 1980 [95]. This heterogeneity was particularly evident in the neonatal rat brain and, furthermore, showed a pharmacological specificity of photolabelling reminiscent of type I and type II receptors (summarized in [96]). It is now known that this was due to the presence of the minor α subunit isoforms. The discovery of the requirement for the γ2 subunit for benzodiazepine modulation of GABA responses led to a re-evaluation of the

[³H]flunitrazepam photoaffinity labelling studies. Using subunit-specific antibodies, it was confirmed that the α1, and not the γ2, subunit was the site of the photoaffinity labelling reaction [97]. Indeed, the site of labelling has been defined to be within the region of α1 comprising residues 59–148, by cyanogen bromide cleavage and antibody mapping experiments [98]. More recently, in agreement with these studies, Duncalfe et al. [99] further refined the site of [³H]flunitrazepam photoaffinity labelling in the bovine α1 subunit to His-102 (equivalent to rat residue 101). Thus, in order to explain the role of both the α and γ subunits in agonist benzodiazepine binding, it is proposed that this site is at the interface between these two subunits. Note that with the working model of $\alpha_2\beta_1\gamma_2$ for the most prevalent GABA$_A$ receptor of adult mammalian brain (see above), this would simplistically indicate the presence of two agonist benzodiazepine binding sites per receptor molecule. Specific activity determinations for GABA$_A$ receptors purified from mammalian brain by benzodiazepine affinity chromatography range from 0.8 to 3.0 benzodiazepine binding sites/mg of protein (summarized in [100]). These values are low for a pure protein with one binding site per 250000-$M_r$ receptor molecule, and are attributed to receptor inactivation during isolation. Two benzodiazepine binding sites per receptor would require an even higher theoretical specific activity for purification to homogeneity. Thus binding site stoichiometry remains an open question.

Specific amino acids within the α subunit primary structures that determine the type of benzodiazepine pharmacology have been identified by site-directed mutagenesis. Firstly, His-101 (rat amino acid numbering), which notably falls within the sequence identifed from the protein chemistry studies, is requisite for high-affinity agonist benzodiazepine binding [101]. This histidine is not conserved in the α4 and α6 subunit sequences, which have high affinity for the inverse agonist Ro 15-4513 only. The mutation Arg-100→His increased the affinity of α6(R100H)βxγ2 receptors for flumazenil and diazepam, but the $K_i$ values were still 6–30 fold lower than those found for the α1βxγ2 wild-type receptor [101]. Secondly, Pritchett and Seeburg [102] converted a cloned type II receptor into a receptor with type I benzodiazepine pharmacology by a single point mutation. Thus Gly-201 (α1 numbering) was identified as a determinant of high-affinity binding for type I-selective compounds [102]. More recently and interestingly, Mihic et al. [103] identified a residue in the γ2 subunit, Thr-142, which is important for the determination of benzodiazepine efficacy. Some previous publications identified this amino acid as Ser-142, but these now appear to be incorrect [103]. Comparison of the mutant α1β1γ2$_L$(T142S) with the wild-type α1β1γ2$_L$(T142) yielded receptors where the antagonist Ro 15-1788, and the partial inverse agonist Ro 15-4513, both acted as partial agonists; also, potentiation by conventional benzodiazepine agonists was increased 2-fold, but no effect was found on GABA affinity, efficacy or barbiturate or alphaxolone modulation of GABA responses [103].

As described above, [$^3$H]muscimol photoaffinity-labelled predominantly a 58000-$M_r$ subunit which was assumed to be the β subunit, although this has not actually been proven. [$^3$H]Muscimol reportedly also specifically photoaffinity-labelled the 53000-$M_r$ α1 subunit, albeit at a much decreased affinity compared with the 58000-$M_r$ β subunit. Thus again the experimental evidence available is most consistent with a high-affinity binding site at the interfaces between subunits, this time α and β. Amin and Weiss [75] showed that two similar domains of the β subunit in α1β2γ2 cloned receptors are required for activation of the receptors by GABA. These were identified as binding domain I [β2-(157–160); sequence Tyr-Gly-Tyr-Thr] and binding domain II [β2-(202–205); sequence Thr-Gly-Ser-Tyr]. The tyrosine and threonine residues were identified by mutagenesis as points of interaction with GABA. Binding domain I is completely conserved across the β subunits of rat and cow. The motif Thr-Gly-Xaa-Tyr of binding domain II is also conserved. Weiss and Amin [104] have further identified analogous amino acids in ρ1 homomeric receptors, i.e. Tyr-198, Thr-244 and Tyr-247, as being important for GABA activation. For the binding domain I-related sequence in the glycine receptor α1 subunit, Gly-160 has been identified as an important determinant of glycine and antagonist binding. Further and interestingly, mutations adjacent to Gly-160, i.e. Arg-159→Tyr and Tyr-161→Arg, result in a glycine α1 homomeric receptor that shows gating by GABA [105]. Sigel et al. [106], however, showed that the mutation Phe-64→Leu in the α1 subunit of α1β2γ2 receptors resulted in a 200-fold decrease in agonist affinity. This same residue, Phe-64, was shown to have incorporated radioactivity following [$^3$H]muscimol affinity labelling in the sequence Thr-Ile-Asp-Val-Phe, which is unique to the α1, α2, α3 and α5 subunits [107]. However, it is unlikely that this is the major amino acid that is photoaffinity-labelled by [$^3$H]muscimol, since predominant labelling is associated with a 58000-$M_r$ band and, although the α3 subunit has an $M_r$ similar to this, it is not a major component of benzodiazepine affinity-purified GABA$_A$ receptors.

Picrotoxinin is a classical non-competitive antagonist of GABA$_A$ receptors, and it has been proposed that it acts at the level of the chloride ion channel. Mutagenesis studies have identified the residues within the predicted TM2 domain that are important for picrotoxin-sensitivity. Thus a GABA$_A$ receptor subunit of a *Drosophila* mutant *Rdl* which was resistant to the insecticide dieldrin and to picrotoxin had a single point mutation, Ala-302→Ser, in the putative TM2 region [108]. Further, the co-expression of the glycine receptor β subunit with the α1 subunit renders the glycine-gated channel picrotixinin-insensitive [109]. Residues within the putative TM2 domain of the β subunit were subsequently identified as being important for the endowment of this property [109]. More recently, Gurley et al. [110] showed directly that the mutant receptor α1β2(T246F)γ2 is picrotoxin-insensitive. These reports thus suggest that the TM2 domain is important for picrotoxinin binding, and they can be extrapolated as providing evidence for the TM2 region forming the inner lining of the chloride ion channel.

**Fig. 7.    Schematic model of the major type of GABA$_A$ receptor expressed in adult mammalian brain, highlighting the amino acids implicated in ligand binding**

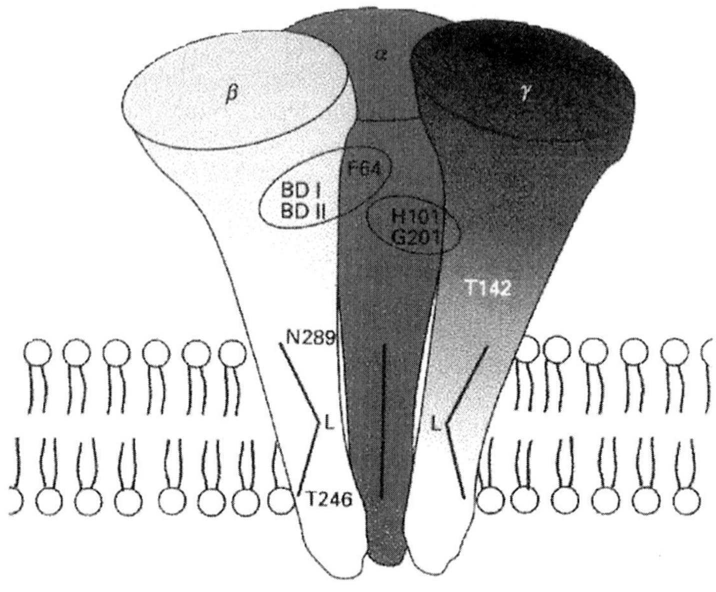


*The subunit stoichiometry most consistent with all published experimental results is $(\alpha 1)_2(\beta 2/3)_2(\gamma 2)_1$. The figure shows an $\alpha 1$ (white), $\beta 2$ (light grey) and $\gamma 2$ (dark grey) subunit, with the binding pocket for GABA at the interface of the $\alpha$ and $\beta$ subunits and residues important for benzodiazepine subpharmacology at the interface between the $\alpha$ and $\gamma$ subunits. Note that, in accordance with the studies of Unwin [19] on the peripheral nicotinic acetylcholine receptor, the GABA binding site is depicted as being approximately half-way down the extracellular region. The leucine residues (L) in the transmembrane region, TM2, are those conserved with the nicotinic receptor, which for the latter protein have been identified as the gate within the channel lumen [19]. This GABA$_A$ receptor will have type I benzodiazepine pharmacology, and it is also loreclezole-sensitive. Abbreviation: BD, binding domain. Reproduced from [3] with permission.*

The TM2 domain is also implicated in the binding of loreclezole to the GABA$_A$ receptor. As described above, loreclezole-sensitivity was found with β2- and β3-subunit-containing receptors [53]. Wingrove et al. [111] have further refined these studies, and identified Asn-289 (β2) and Asn-290 (β3), located at the C-terminal end of TM2, as the important determinant for high-affinity loreclezole binding.

The results of the mutagenesis studies are summarized in diagrammatic form in Fig. 7.

## Conclusions

Several years have elapsed since the first GABA$_A$ and glycine receptor gene sequences were reported. This finding led to the realization that there exists a superfamily of ligand-gated ion channels [2]. Since then, a wealth of information has been forthcoming on the molecular and pharmacological propertics of GABA$_A$ receptor gcnes and their products. The major challenge now, as indeed for many of the neurotransmitter receptor and indeed transporter proteins, is to integrate all this detail into a cohesive understanding of the functional significance of GABA$_A$ receptor heterogeneity in inhibitory amino acid neurotransmission. An additional future direction will include an understanding of the regulation of GABA$_A$ receptor gene expression. A developmental switch in the expression of GABA$_A$ receptor subtypes in cerebellar granule cells has been described [47,112]. Also, agents that elevate intracellular cAMP have been found to have long-term, bidirectional effects on the expression of the α1 and α6 subunits in primary cerebellar granule cells in primary culture, which may be important in mechanisms of receptor subtype plasticity [113]. Further, the protein brain-specific factor 1 has been identified as a neuronal-cell-type-enriched factor involved in the differential transcription of the GABA$_A$ receptor δ subunit gene [114]. Further investigation of these early results, together with the continued use of both standard GABA$_A$ receptor subunit transgenic mice and developmentally controlled transgenic mice, should provide insights into the function and regulation of GABA$_A$ receptor subunit gene expression.

A new subunit, the GABA$_A$ receptor ε was recently reported in Nature [118] This subunit reportedly confers anaesthetic insensitivity to (α β ε) receptors.

*I would like to thank the members of my group, Simon Pollard, Dr. Chris L. Thompson, Dr. Paul Chazot and Dr. Miroslav Cik, who have contributed to this review both practically (S.P. and C.L.T.) and with many useful discussions. I thank Professor G.G. Lunt and Dr. M. Ortells for kindly providing Fig. 3, Dr. Cik for help with Figs. 4 and 7, and Dr. Peter Somogyi and Dr Zolly Nusser for the use of Fig. 6 from our collaborative study. Referenced work from my laboratory was funded by the Medical Research Council (U.K.).*

## References

1. Smart, T.G. (1997) in Amino Acid Neurotransmission (Stephenson, F.A. and Turner, A.J., eds.), pp. 37–63, Portland Press, London
2. Schofield, P.R., Darlison, M.G., Fujita, N., Burt, D.R., Stephenson, F.A., Rodriguez, H, Ramachandran, J., Reale, V., Glencorse, T.A., Seeburg, P.H. and Barnard, E.A. (1987) Nature (London) 328, 221–227
3. Stephenson, F.A. (1995) Biochem. J. 310, 1–9
4. Olsen, R.W. and Venter, J.C. (eds.) (1986) Benzodiazepine/GABA-Receptors and Chloride Channels: Structural and Functional Properties, Alan R. Liss, New York
5. Haefely, W.E. (1994) in The Challenge of Neuropharmacology (Mohler, H. and Da Prada, M., eds.), pp. 15–39, Editiones Roche, Basel, Switzerland

6.  Sigel, E., Stephenson, F.A., Mamalaki, C. and Barnard, E.A. (1983) J. Biol. Chem. **258**, 7219–7223

7.  Stauber, G.B., Ransom, R.W., Dilber, A. and Olsen, R.W. (1987) Eur. J. Biochem. **167**, 125–133

8.  Olsen, R.W. and Tobin, A.J. (1990) FASEB J. **4**, 1469–1480

9.  Whiting, P.J., McKernan, R.M. and Iversen, L.L. (1990) Proc. Natl. Acad. Sci. U.S.A. **87**, 9966–9970

10. Kofuji, P., Wang, J.B., Moss, S.J., Huganir, R.L. and Burt, D.B. (1991) J. Neurochem. **56**, 713–715

11. Kirkness, E.F. and Fraser, C.M. (1993) J. Biol. Chem. **268**, 4420–4428

12. Korpi, E.R., Kuner, T., Kristo, P., Kohler, M., Herb, A., Luddens, H. and Seeburg, P.H. (1994) J. Neurochem. **63**, 1167–1170

13. Harvey, A.J., Chinchetru, M.A. and Darlison, M.G. (1994) J. Neurochem. **62**, 10–16

14. Bateson, A.N., Lasham, A. and Darlison, M.G. (1991) J. Neurochem. **56**, 1437–1440

15. Darlison, M.G. and Albrecht, B.E. (1995) Semin. Neurosci. **7**, 115–126

16. ffrench-Constant, R.H., Mortlock, D.P., Shaffer, C.D., MacIntyre, R.J. and Roush, R.T. (1991) Proc. Natl. Acad. Sci. U.S.A. **88**, 7209–7213

17. Harvey, R.J., Vreugdenhil, E., Zaman, S.H., Bhandal, N.S., Usherwood, P.N.R., Barnard, E.A. and Darlison, M.G. (1991) EMBO J. **10**, 3239–3245

18. Ortells, M.O. and Lunt, G.G. (1995) Trends Neurosci. **18**, 121–127

19. Unwin, N. (1993) J. Mol. Biol. **229**, 1101–1124

20. Cockcroft, V.B., Osguthorpe, D.J., Barnard, E.A. and Lunt, G.G. (1990) Proteins **8**, 386–397

21. Langosch, D., Thomas, L. and Betz, H. (1988) Proc. Natl. Acad. Sci. U.S.A. **85**, 7394–7398

22. Cooper, E., Couturier, S. and Ballivet, M. (1991) Nature (London) **350**, 235–238

23. Mamalaki, C., Barnard, E.A. and Stephenson, F.A. (1989) J. Neurochem. **52**, 124–134

24. Nayeem, N., Green, T.P., Martin, I.L. and Barnard, E.A. (1994) J. Neurochem. **62**, 815–818

25. Im, W.B., Pregenzer, J.F., Binder, J.A., Dillon, G.H. and Alberts, G.L. (1995) J. Biol. Chem. **270**, 26063–26066

26. Burt, D.R. and Kamatchi, G.L. (1991) FASEB J. **5**, 2916–2923

27. Levitan, E.S., Schofield, P.R., Burt, D.R., Rhee, L.M., Wisden, W., Kohler, M., Fujita, N., Rodriguez, H., Stephenson, F.A., Darlison, M.G., Barnard, E.A. and Seeburg, P.H. (1988) Nature (London) **335**, 76–79

28. Pritchett, D.B., Luddens, H. and Seeburg, P.H. (1989) Science **245**, 1389–1392

29. Hadingham, K.L., Harkness, P.C., McKernan, R.M., Quirk, K., Le Bourdelles, B., Horne, A.L., Kemp, J.A., Barnard, E.A., Ragan, C.I.A. and Whiting, P.J. (1992) Proc. Natl. Acad. Sci. U.S.A. **89**, 6378–6382

30. Pregenzer, J.F., Im, W.B., Carter, D.B. and Thomsen, D.R. (1993) Mol. Pharmacol. **43**, 801–806

31. Blair, L.A.C., Levitan, E.S., Marshall, J., Dionne, V.E. and Barnard, E.A. (1988) Science **242**, 577–579

32. Pritchett, D.B., Sontheimer, H., Shivers, B.D., Ymer, S., Kettenmann, H., Schofield, P.R. and Seeburg, P.H. (1989) Nature (London) **338**, 582–585

33. Shivers, B.D., Killisch, I., Sprengel, R., Sontheimer, H., Kohler, M., Schofield, P.R. and Seeburg, P.H. (1989) Neuron **3**, 327–337

34. Cutting, G.R., Lu, L., O'Hara, F., Kasch, L.M., Montrose-Rafizadeh, C., Donovan, D.M., Shimada, S., Antonarakis, S.E., Guggino, W.B., Uhl, G.R. and Kazazian, H.H. (1991) Proc. Natl. Acad. Sci. U.S.A. **88**, 2673–2677

35. Shimada, S., Cutting, G. and Uhl, G.R. (1992) Mol. Pharmacol. **41**, 683–687

36. Slany, A., Zezula, J., Tretter, V. and Sieghart, W. (1995) Mol. Pharmacol. **48**, 385–391

37. Puia, G., Vicini, S., Seeburg, P.H. and Costa, E. (1991) Mol. Pharmacol. **39**, 691–696

38. Sigel, E., Baur, R., Trube, G., Mohler, H. and Malherbe, P. (1990) Neuron **5**, 703–711

39. Herb, A., Wisden, W., Luddens, H., Puia, G., Vicini, S. and Seeburg, P.H. (1992) Proc. Natl. Acad. Sci. U.S.A. **89**, 1433–1437

40. Knoflach, F., Rhyner, Th., Villa, M., Kellenberger, S., Drescher, U., Malherbe, P., Sigel, E. and Mohler, H. (1991) FEBS Lett. **293**, 191–194

41. Saxena, N.C. and MacDonald, R.L. (1994) J. Neurosci. **14**, 7077–7086

42. Pritchett, D.B. and Seeburg, P.H. (1990) J. Neurochem. **54**, 1802–1804

43. Luddens, H., Pritchett, D.B., Kohler, M., Killisch, I., Keinanen, K., Monyer, H., Sprengel. R. and Seeburg, P.H. (1990) Nature (London) **346**, 648–651

44. Wisden, W., Herb, A., Weiland, H., Keinanen, K., Luddens, H. and Seeburg, P.H. (1991) FEBS Lett. **289**, 227–230
45. Sieghart, W., Eichinger, A., Richards, J.G. and Mohler, H. (1987) J. Neurochem. **48**, 46–52
46. Verdoorn, T. (1994) Mol. Pharmacol. **45**, 475–480
47. Mathews, G.C., Bolos-Sy, A.M., Holland, K.D., Isenberg, K.E., Covey, D.F., Ferrendelli, J.A. and Rothman, S.M. (1994) Neuron **13**, 149–158
48. Ymer, S., Draguhn, A., Wisden, W., Werner, P., Keinenen, K., Schofield, P.R., Sprengel, R., Pritchett, D.B. and Seeburg, P.H. (1990) EMBO J. **9**, 3261–3267
49. Togel, M., Mossier, B., Fuchs, K. and Sieghart, W. (1994) J. Biol. Chem. **269**, 12993–12998
50. Draguhn, A., Verdorn, T.A., Ewert, M., Seeburg, P.H. and Sakmann, B. (1990) Neuron **5**, 781–788
51. Smart, T.G., Moss, S.J., Xie, X. and Huganir, R.L. (1991) Br. J. Pharmacol. **103**, 1837–1839
52. Wafford, K.A., Burnett, D.M., Leidenheimer, N.J., Burt, D.R., Wang, J.B., Kofuji, P., Dunwiddie, T.V., Harris, R.A. and Sikela, J.M. (1991) Neuron **7**, 27–33
53. Wafford, K.A., Bain, C.J., Quirk, K., McKernan, R.M., Wingrove, P.B., Whiting, P.J. and Kemp, J.A. (1994) Neuron **12**, 775–782
54. Kirkness, E.F., Hedblom, E. and Fraser, C.M. (1995) Neurosci. Abstr. **21**, 849
55. Primus, R.J., Jacobs, A.A. and Gallager, D.W. (1992) Brain Res. **14**, 179–185
56. Wisden, W., Laurie, D.J., Monyer, H. and Seeburg, P.H. (1992) J. Neurosci. **12**, 1040–1062
57. Laurie, D.J., Seeburg, P.H. and Wisden, W. (1992) J. Neurosci. **12**, 1063–1076
58. Persohn, E., Malherbe, P. and Richards, J.G. (1992) J. Comp. Neurol. **326**, 193–216
59. Laurie, D.J., Wisden, W. and Seeburg, P.H. (1992) J. Neurosci. **12**, 4151–4172
60. McDonald, R.L. and Olsen, R.W. (1994) Annu. Rev. Neurosci. **17**, 569–602
61. Duggan, M.J. and Stephenson, F.A. (1990) J. Biol. Chem. **265**, 3831–3835
62. Duggan, M.J., Pollard, S. and Stephenson, F.A. (1991) J. Biol. Chem. **266**, 24778–24784
63. McKernan, R.M., Quirk, K., Prince, R., Cox, P.A., Gillard, N.P., Ragan, I.C. and Whiting, P.J. (1991) Neuron **7**, 667–676
64. Benke, D., Mertens, S., Trceziak, A., Gillessen, D. and Mohler, H. (1991) J. Biol. Chem. **266**, 4478–4483
65. Zezula, J. and Sieghart, W. (1991) FEBS Lett. **284**, 15–18
66. Kern, W. and Sieghart, W. (1994) J. Neurochem. **62**, 764–769
67. Mertens, S., Benke, D. and Mohler, H. (1993) J. Biol. Chem. **268**, 5965–5973
68. Pollard, S., Duggan, M.J. and Stephenson, F.A. (1993) J. Biol. Chem. **268**, 3753–3757
69. Khan, Z.U., Gutierrez, A. and De Blas, A.L. (1994) J. Neurochem. **63**, 371–374
70. Pollard, S., Thompson, C.L. and Stephenson, F.A. (1995) J. Biol. Chem. **270**, 21285–21290
71. Quirk, K., Gillard, N.P., Ragan, C.I., Whiting, P.J. and McKernan, R.M. (1994) J. Biol. Chem. **269**, 16020–16028
72. Schoch, P., Richards, J.G., Haring, P., Takacs, B., Stahli, C., Staehelin, T., Haefely, W. and Mohler, H. (1985) Nature (London) **314**, 168–171
73. Ewert, M., Shivers, B.D., Luddens, H., Mohler, H. and Seeburg, P.H. (1990) J. Cell Biol. **110**, 2043–2048
74. Benke, D., Fritschy, J.-M., Trzeciak, A., Bannwarth, W. and Mohler, H. (1994) J. Biol. Chem. **269**, 27100–27107
75. Amin, J. and Weiss, D.S. (1993) Nature (London) **366**, 565–569
76. Quirk, K., Gillard, N.P., Ragan, C.I., Whiting, P.J. and McKernan, R.M. (1994) Mol. Pharmacol. **45**, 1061–1070
77. Backus, K.H., Arigoni, M., Drescher, U., Scheurer, L., Malherbe, P. Mohler, H. and Benson, J.A. (1993) NeuroReport **5**, 285–288
78. Mossier, B., Togel, M., Fuchs, K. and Sieghart, W. (1994) J. Biol. Chem. **269**, 25777–25782
79. Nusser, Z., Roberts, D.B., Baude, A., Richards, J.G., Sieghart, W. and Somogyi, P. (1995) Eur. J. Neurosci. **7**, 630–646
80. Fritschy, J.-M. and Mohler, H. (1995) J. Comp. Neurol. **359**, 154–194
81. Fritschy, J.-M., Benke, D., Mertens, S., Oertel, W.H., Bachi, T. and Mohler, H. (1992) Proc. Natl. Acad. Sci. U.S.A. **89**, 6726–6730
82. Nusser, Z., Sieghart, W., Stephenson, F.A. and Somogyi, P. (1996) J. Neurosci. **16**, 103–114
83. Nusser, Z., Sieghart, W., Benke, D., Fritschy, J.-M. and Somogyi, P. (1996) Proc. Natl. Acad. Sci. U.S.A., in the press

84. Levitan, E.S., Blair, L.A.C., Dionne, V.E. and Barnard, E.A. (1988) Neuron **1**, 773–781
85. Krishek, B.J., Xie, X., Blackstone, C., Huganir, R.L., Moss, S.J. and Smart, T.G. (1994) Neuron **12**, 1081–1095
86. Raymond, L. (1997) in Amino Acid Neurotransmission (Stepenson, F.A. and Turner, A.J., eds.), pp. 177–194, Portland Press, London
87. Korpi, E.R., Kleingoor, C., Kettenmann, H. and Seeburg, P.H. (1993) Nature (London) **361**, 356–359
88. Sinnett, D., Wagstaff, J., Glatt, K., Woolf, F., Kirkness, E.F. and Lalande, M. (1993) Am. J. Hum. Genet. **52**, 1216–1229
89. Nakatsu, Y., Tyndale, R.F., DeLorey, T.M., Durham-Pierre, D., Gardner, J.M., McDanel, H.J., Nguyen. Q., Wagstaff, J., Lalande, M., Sikela, J., Olsen, R.W., Tobin, A.J. and Brilliant, M.H. (1993) Nature (London) **364**, 448–450
90. Culiat, C.T., Stubbs, I.J., Montgomery, C.S., Russell, I.B. and Rinchik, F.M. (1994) Proc. Natl. Acad. Sci. U.S.A. **91**, 2815–2818
91. Gunther, U., Benson, J., Benke, D., Fritschy, J.-M., Reyes, G., Knoflach, F., Crestani, F., Aguzzi, A., Arigoni, M., Lang, Y., Bluethmann, H., Mohler, H. and Luscher, B. (1995) Proc. Natl. Acad. Sci. U.S.A. **92**, 7749–7753
92. Jones, A., Korpi, E.R., McKernan, R.M., et al. (1997) J. Neurosci., in the press
93. Casalotti, S.O., Stephenson, F.A. and Barnard, E.A. (1986) J. Biol. Chem. **261**, 15013–15016
94. Deng, L.R., Ransom, R.W. and Olsen, R.W. (1986) Biochem. Biophys. Res. Commun. **138**, 1308–1314
95. Sieghart, W. and Karobath, M. (1980) Nature (London) **286**, 285–287
96. Sieghart, W. (1989) Trends Phramacol. Sci. **10**, 407–411
97. Stephenson, F.A., Duggan, M.J. and Pollard, S. (1990) J. Biol. Chem. **265**, 21160–21165
98. Stephenson, F.A. and Duggan, M.J. (1989) Biochem. J. **264**, 199–206
99. Duncalfe, L.L., Carpenter, M.R., Smillie, L.B., Martin, I.L. and Dunn, S.M.J. (1996) J. Biol. Chem. **271**, 9209–9214
100. Stephenson, F.A. (1988) Biochem. J. **249**, 21–32
101. Wieland, H.A., Luddens, H. and Seeburg, P.H. (1992) J. Biol. Chem. **267**, 1426–1429
102. Pritchett, D.B. and Seeburg, P.H. (1991) Proc. Natl. Acad. Sci. U.S.A. **88**, 1421–1425
103. Mihic, S.J., Whiting, P.J., Klein, R.L., Wafford, K.A. and Harris, R.A. (1994) J. Biol. Chem. **269**, 32768–32773
104. Weiss, D.S. and Amin, J. (1994) Soc. Neurosci. Abstr. **20**, 13.4
105. Schmieden, V., Kuhse, J. and Betz, H. (1993) Science **262**, 256–258
106. Sigel, E., Baur, R., Kellenberger, S. and Malherbe, P. (1992) EMBO J. **11**, 2017–2023
107. Smith, G.B. and Olsen, R.W. (1994) J. Biol. Chem. **269**, 20380–20387
108. ffrench-Constant, R., Rocheleau, T.A., Steichen, J.C. and Chalmers, A.E. (1993) Nature (London) **363**, 449–451
109. Pribilla, I., Takagi, T., Langosch, D., Bormann, J. and Betz, H. (1992) EMBO J. **11**, 4305–4311
110. Gurley, D.A., Amin, J., Ross, P., Weiss, D. and White, G. (1994) Soc. Neurosci. Abstr. **20**, 509
111. Wingrove, P.B., Wafford, K.A., Bain, C. and Whiting, P.J. (1994) Proc. Natl. Acad. Sci. U.S.A. **91**, 4569–4573
112. Thompson, C.L. and Stephenson, F.A. (1994) J. Neurochem. **62**, 2037–2044
113. Thompson, C.L., Pollard, S. and Stephenson, F.A. (1996) J. Neurochem. **67**, 434–437
114. Motejlek, K., Hauselmann, R., Leitgeb, S. and Luscher, B. (1994) J. Biol. Chem. **269**, 15265–15273
115. Pollard, S., Duggan, M.J. and Stephenson, F.A. (1991) FEBS Lett. **295**, 81–83
116. Luddens, H., Killisch, I. and Seeburg, P.H. (1991) J. Recept. Res. **11**, 535–551
117. Turner, J.D., Bodewitz, G., Thompson, C.L. and Stephenson, F.A. (1993) Psychopharmacol. Ser. **11**, 29–49
118. Davies, P.A., Hanna, M.C., Hales, T.G. and Kirkness, E.F. (1997) Nature (London) **385**, 820–823

# 5

# The inhibitory glycine receptor

Cord-Michael Becker*‡ and Dieter Langosch†

*Institut für Biochemie, Universität Erlangen-Nürnberg, Fahrstrasse 17,
D-91054 Erlangen, and †Institut für Neurobiologie, Universität Heidelberg,
Im Neuenheimer Feld 364, D-69120 Heidelberg, Germany

## Introduction

Glycine is the principal inhibitory transmitter in the spinal cord and brainstem. After vesicular release of glycine from presynaptic terminals, a depressant action on neuronal activity is exerted by its binding to post-synaptic receptors that exist in various isoforms. Glycine-mediated inhibition underlies the segmental regulation of spinal motoneurons by small interneurons including the Renshaw cells, but is also present in other areas of the central nervous system (CNS) (Krnjevic, 1981; Daly, 1990; Becker, 1992; Betz, 1992). The inhibitory action of glycine is distinct from its function as a co-agonist to glutamate at the NMDA (*N*-methyl-D-aspartate) receptor (Seeburg, 1994).

## Inhibitory glycine receptors are strychnine-sensitive

Glycine binds to post-synaptic receptors situated on the neuronal surface, where it induces the opening of an intrinsic anion channel. This results in chloride currents and post-synaptic hyperpolarization (Krnjevic, 1981; Takahashi, 1984; Takahashi et al., 1992). Moreover, a series of structurally related amino acids possesses glycine-like agonistic potencies decreasing in the order: glycine > β-alanine > taurine > α-alanine > serine. Receptor binding of glycine is efficiently antagonized by the convulsant alkaloid strychnine (Becker, 1992). Sublethal strychnine poisoning causes motor disturbances, e.g. increases in muscle tone and hyper-reflexia, including excessive startle responses. Higher doses of strychnine lead to convulsions and death (Becker, 1992). Molecular modelling has indeed identified structural similarities between glycine and a fragment of the strychnine molecule (Aprison and Lipkowitz, 1992; Galvez-Ruano et al., 1995). The use of [³H]strychnine as a specific probe of the inhibitory glycine receptor was introduced by Young and Snyder (1973). For the binding of [³H]strychnine to membrane preparations from spinal cord, a dissociation constant ($K_D$) of approx. 10 nM has

‡*To whom correspondence should be addressed.*

been determined (Young and Snyder, 1973; Becker, 1992). High-affinity binding sites for [³H]strychnine are present in the spinal cord, the sensory and acoustic ganglia of the brainstem and the retina (Zarbin et al., 1981; Probst et al., 1986). Increasing evidence indicates that glycine receptors are, in addition, expressed in the rat cerebral cortex (Naas et al., 1991; Becker et al., 1993). They may also contribute to the regulation of neuroendocrine functions (Becker et al., 1994) and the sperm acrosome reaction (Melendrez and Meisel, 1995).

## Glycine receptors: protein isoforms and subunit variants

Inhibitory glycine receptors are multisubunit transmembrane proteins that are assembled from ligand binding $\alpha$ and structural $\beta$ polypeptides (Betz, 1992). The glycine receptor complex, in addition, associates with gephyrin, a tubulin binding protein involved in the formation of post-synaptic receptor clusters (Kirsch et al., 1993). In the rodent CNS, two major isoforms of glycine receptor proteins are distinguishable that display characteristic developmental and regional expression patterns (Table 1). The neonatal isoform, $GlyR_N$, prevails in the spinal cord of the newborn rodent; it is replaced by the adult-type receptor ($GlyR_A$) 2 weeks after birth (Becker et al., 1988). $GlyR_N$ purified from spinal neurons appears to be a homo-oligomer composed of $\alpha2$ subunits of molecular mass 49 kDa (Becker et al., 1988, 1992; Hoch et al., 1989). During a short postnatal period, a glycine receptor variant is abundant in neurons of rat cerebral cortex which, by immunological criteria, resembles $GlyR_N$ (Table 1) (Naas et al., 1991; Becker et al., 1993). $GlyR_A$, the receptor isoform prevalent in adult rat spinal cord, is a complex glycoprotein of molecular mass >250 kDa (Pfeiffer et al., 1982; Becker et al., 1986; Langosch et al., 1990) composed of $\alpha1$ (48 kDa) and $\beta$ (58 kDa) polypeptides (Table 1, Fig. 1). The $\alpha1$ and $\beta$ subunits are integral membrane proteins which have been proposed to assemble into the heteropentameric chloride channel in a rosette-like arrangement (Langosch et al., 1990; Betz, 1992). The glycine and strychnine binding sites of $GlyR_A$ have been localized to the $\alpha1$ subunit (Pfeiffer et al., 1982; Ruiz-Gomez et al., 1990).

| Table 1. | Major protein isoforms of the glycine receptor | |
|---|---|---|
| | $GlyR_A$ | $GlyR_N$ |
| Subunits | $\alpha1$ (48 kDa), $\beta$ (58 kDa) | $\alpha2$ (49 kDa) |
| Complex structure proposed | Heteropentamer: $(\alpha1)_3(\beta)_2$ | Homopentamer (in cultured neurons) |
| Binding affinity for strychnine | High ($K_D$ ~10 nM) | Low |
| Expression pattern | Spinal cord and brainstem of adult animals | Spinal cord of neonates, cerebral cortex |

**Fig. 1.**        **Tentative structure of the glycine receptor isoform GlyRA**

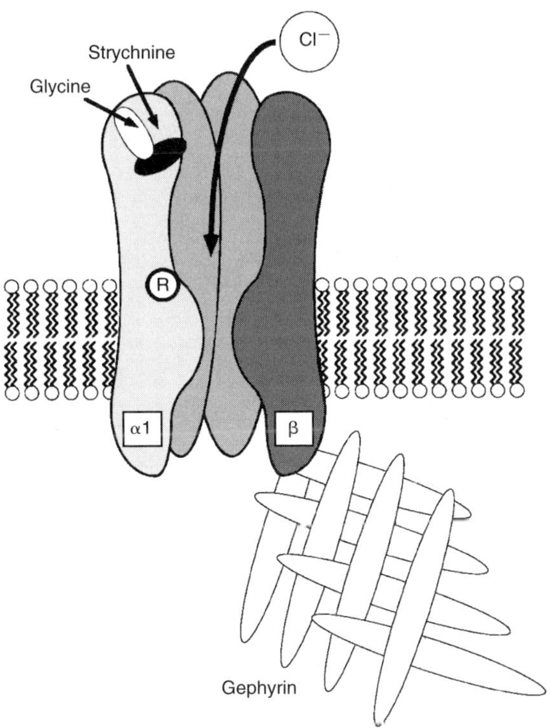


*A stoichiometry of $(\alpha 1)_3(\beta)_2$ has been proposed for this pentameric protein complex (Langosch et al., 1988; Kuhse et al., 1993). The $\alpha$ subunits carry non-identical, but overlapping, binding sites for glycine and strychnine. Gephyrin mediates the post-synaptic anchoring and clustering of the receptor channels.*

The number of glycine receptor subunits that have been cloned exceeds the heterogeneity of protein isoforms resolved by biochemical methods. In human, rat and mouse, four variants of the ligand binding subunit (α1–α4) have been identified (Grenningloh et al., 1987, 1990a; Kuhse et al., 1990a,b; Matzenbach et al., 1994), which are encoded by distinct genes (Table 2). Alternative splicing of α1 and α2 pre-mRNAs results in further complexity of α subunits (Kuhse et al., 1991; Malosio et al., 1991b). These α subunit variants (α1–α4) share a high degree of structural similarity, with sequence identities ranging from 80 to >90% (Matzenbach et al., 1994). In contrast, the β subunit diverges structurally from the α subunit variants, displaying an amino acid identity of approx. 47% when compared with the α1 subunit (Grenningloh et al., 1990b; Kingsmore et al., 1994a, Mülhardt et al., 1994; Handford et al., 1996).

**Table 2.        Glycine receptor subunit genes**

For murine loci, distances from the telomere are given in cM. n.d., not determined. See also Mouse Genome Database (1997); Online Mendelian Inheritance in Man, OMIM (1997).

| Subunit | Chromosomal locus | | References |
| | Mouse | Human | |
|---|---|---|---|
| $\alpha 1$ | *Glra1*: 11 (29 cM) | *GLRA1*: 5q31.2 | Shiang et al. (1993); Baker et al. (1994); Ryan et al. (1994) |
| $\alpha 2$ | *Glra2*: X (70 cM) | *GLRA2*: Xp21.2–p21.3 | Grenningloh et al. (1990a); Derry et al. (1991) |
| $\alpha 3$ | *Glra3*: 8 (25 cM) | n.d. | Kingsmore et al. (1994b) |
| $\alpha 4$ | *Glra4*: X (56 cM) | n.d. (Xq21–22?) | Matzenbach et al. (1994) |
| $\beta$ | *Glyrb*: 3 (36 cM) | *GLRB*: 4q32 | Kingsmore et al. (1994a); Mülhardt et al. (1994); Handford et al. (1996) |

Differential transcription of the glycine receptor subunit genes generates developmental (Grenningloh et al., 1987; Kuhse et al., 1990a,b; Akagi et al., 1991) and regional (Malosio et al., 1991b; Watanabe and Akagi, 1995) heterogeneity. Functional glycine receptors are not restricted to neurons, but are also present in glial cells at low levels. Correspondingly, glycine receptor $\alpha 1$ and $\beta$ subunits are expressed in glial cells of the rat spinal cord (Kirchhoff et al., 1996). Transcriptional activity is correlated with changes in glycine receptor isoform levels: postnatal $GlyR_A$ accumulation in the spinal cord coincides with $\alpha 1$ subunit mRNA levels. In contrast, $GlyR_N$ protein as well as the $\alpha 2$ transcripts are present around birth but display a drastic postnatal decrease (Becker et al., 1988, 1992). The developmental switch from $GlyR_N$ to $GlyR_A$ protein isoforms is matched by changes in the channel properties of glycinergic synaptic inhibition during maturation of the rat spinal cord (Takahashi et al., 1992, Bormann et al., 1993) and neuronal cultures (Lewis et al., 1991; St. John and Stephens, 1993). In the rat cerebral cortex, transient expression of $GlyR_N$ during a short postnatal period coincides with enhanced transcription of the $\alpha 2$ subunit gene (Malosio et al., 1991b; Naas et al., 1991, Becker et al., 1993). Transcription of the $\alpha 3$ and $\alpha 4$ subunit genes is low in the spinal cord (Kuhse et al., 1990b; Matzenbach et al., 1994). The widespread expression of the $\beta$ subunit throughout the rodent CNS remains to be explained (Malosio et al., 1991b).

## Functional domains of glycine receptor subunits

For the glycine receptor $\alpha$ and $\beta$ subunits, transmembrane topologies are predicted which are common to the superfamily of ligand-gated ion channels, including the

**Fig. 2.**     **Transmembrane topology of the glycine receptor α1 polypeptide**

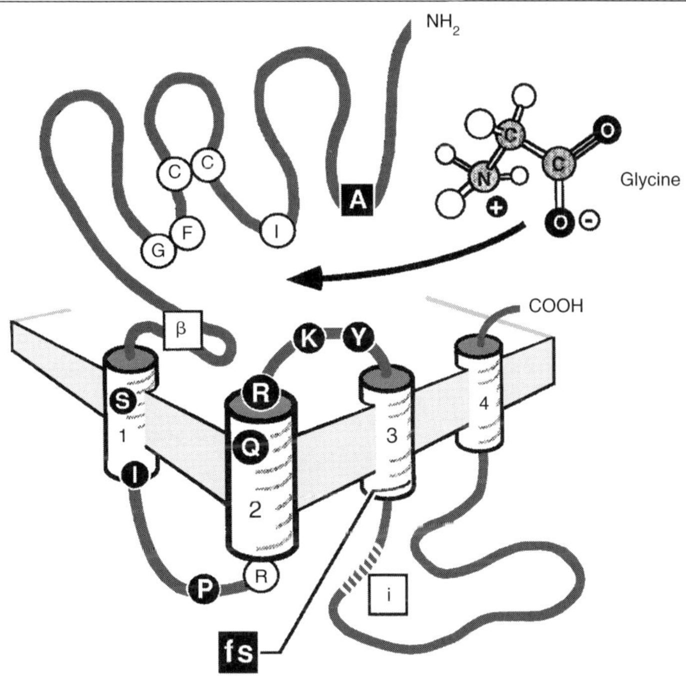



*Cylinders indicate transmembrane segments TM1–TM4. Given in the one-letter code, filled circles indicate sites of amino acid exchanges causing the human neurological disorder hyperekplexia (see Table 3). Filled boxes characterize murine mutations: mis-sense mutation (Ala-52→Ser) in* spasmodic; *fs, site of the frame-shift induced by a microdeletion in* oscillator. *Open circles depict the functionally important amino acid residues Ile-111, Cys-138, Cys-152, Phe-159, Gly-160 and Arg-252. Further structural details are indicated by: β, proposed β-sheet–β-turn structure (Vandenberg et al., 1992a, 1992b); i, insert generated by alternative splicing (Malosio et al., 1991a). Domains are not drawn to scale. Modified from Becker (1995), with permission, © The Neuroscientist.*

nicotinic acetylcholine and GABA$_A$ receptors (where GABA is γ-aminobutyric acid). In the mature α1 polypeptide, 220 amino acid residues form a large N-terminal domain that protrudes into the extracellular space (Grenningloh et al., 1987, 1990a; Betz, 1992). This domain is followed by four membrane-spanning segments (TM1–TM4) that are highly conserved between ligand-gated ion channels (Fig. 2). Consistent with the extracellular location of the N terminus, the mature protein is preceded by a hydrophobic cleavable signal peptide (Grenningloh et al., 1987, 1990a; Kuhse et al., 1990a,b, 1991, Akagi et al., 1991). Analysis of structure–function relationships has led to the identification of: (i) determinants of ligand binding and agonist–antagonist discrimination, (ii) domains involved in anion translocation, and (iii) sequences governing receptor biogenesis and assembly.

Recombinant glycine receptors faithfully reproduce many of the pharmacological and physiological characteristics of their native counterparts from the mammalian CNS or cultured neurons. These studies have taken advantage of the ability of glycine receptor α subunits to self-sufficiently create ligand-gated chloride channels upon expression in mammalian cells or *Xenopus laevis* oocytes (Schmieden et al., 1989; Sontheimer et al., 1989). Characteristic differences, however, exist between α1 and α2 homomeric receptors. While glycine is highly efficient in gating both types of channels, α1 subunit receptors are more responsive to β-alanine and taurine than α2 homomers (Schmieden et al., 1989, 1992; Sontheimer et al., 1989; Kuhse et al., 1990a; Akagi et al., 1991; Malosio et al., 1991a). In contrast, strychnine is an effective antagonist at both α1 and α2 homomeric receptors (Schmieden et al., 1989; Sontheimer et al., 1989; Grenningloh et al., 1990a; Kuhse et al., 1990a,b). From these observations, it is clearly evident that the α polypeptides contain agonist as well as antagonist binding sites (Pfeiffer et al., 1982; Kuhse et al., 1990a,b; Vandenberg et al., 1992a,b).

## Determinants of glycine receptor ligand binding

The initial assignment of the glycine receptor ligand binding domain to the N-terminal extracellular segment of its α1 subunit was based on incorporation studies using [$^3$H]strychnine and subsequent peptide mapping (Pfeiffer et al., 1982; Grenningloh et al., 1987; Ruiz-Gomez et al., 1990). A greater resolution of functional determinants has been achieved by comparative expression studies of α subunit variants either generated by site-directed mutagenesis or derived from human and murine motor disorders. From these studies, it is inferred that several discontinuous polypeptide segments contribute to ligand binding and discrimination. Accordingly, agonists and antagonists bind to partially overlapping, non-identical, determinants (Fig. 2).

(i) In the α1 subunit, Ala-52 is a determinant of agonist, but not strychnine, affinity. A mutation resulting in the substitution of this residue exists in the mutant mouse *spasmodic* (Ryan et al., 1994; Saul et al., 1994).

(ii) The isoleucine residue at position 111 of the α1 subunit is a mediator of taurine binding. In contrast, α2 receptors carry a smaller valine residue in the analogous position and are barely responsive to taurine (Schmieden et al., 1992).

(iii) Determinants of both agonist and high-affinity strychnine binding have been located at amino acid positions 159–161 (α1) (Kuhse et al., 1990a; Schmieden et al., 1992, 1993; Vandenberg et al., 1992b). Phe-159 is important for agonist recognition, as it restricts glycine receptor responses to small-sized amino acid agonists. Upon substitution of this phenylalanine residue by tyrosine, as found in GABA$_A$ receptor polypeptides, α1 channels are generated which also respond to

GABA (Schmieden et al., 1993). A glycine residue corresponding to position ($\alpha$1) 160 ($\alpha$1) is conserved in other $\alpha$ subunit variants, except the $\alpha$2* variant of the neonatal rat spinal cord. In this polypeptide, substitution of the glycine residue decreases affinities for both glycine and strychnine (Kuhse et al., 1990a; Vandenberg et al., 1992a). Functional groups complementary to these recognition sites have been identified in agonistic and antagonistic receptor ligands (Galvez-Ruano et al., 1995; Schmieden and Betz, 1995).

(iv) A unique extracellular loop structure, formed by putative disulphide bridging of the two cysteine residues at positions 198 and 209, has been proposed to confer agonist–antagonist discrimination on the $\alpha$1 subunit (Schmieden et al., 1992; Vandenberg et al., 1992a; Rajendra et al., 1995a). Glycine affinity is almost completely abolished by substitution of Tyr-204 ($\alpha$1), whereas strychnine binding is unaffected (Vandenberg et al., 1992a,b). Additional determinants have been identified among the flanking amino acid residues that are crucial for both agonist and antagonist binding to the glycine receptor. Interestingly, alteration of these determinants did not profoundly lower the maximal ion translocation obtained at glycine concentrations saturating the receptor. This observation supports the concept of separate, yet interdependent, domains within the individual subunit polypeptide structure (Laube et al., 1995a; Rajendra et al., 1995a). Additionally, two precisely conserved cysteine residues (Cys-138 and Cys-152) exist which are conserved in analogous positions in all glycine, $GABA_A$ and nicotinic acetylcholine receptor polypeptides known (Grenningloh et al., 1987). In contrast to earlier concepts, however, this motif does not contribute to the agonist site of the receptor (Vandenberg et al., 1993).

(v) The external vestibule of the integral ion channel is marked by an arginine residue at position ($\alpha$1) 271, corresponding to the extracellular border of transmembrane segment 2. Substitutions of this residue underlie the hereditary motor disorder hyperekplexia (Shiang et al., 1993). While these substitutions reduce receptor affinities for glycinergic ligands, they also convert the partial agonists $\beta$-alanine and taurine into competitive antagonists that are unable to initiate chloride currents (Langosch et al., 1994; Rajendra et al., 1994, 1995b; Laube et al., 1995a). Two explanations have been offered for this which are not necessarily mutually exclusive: mutations of Arg-271 may result in the uncoupling of the agonist binding process from the channel gating mechanism (Rajendra et al., 1995b), or may unmask an inhibitory subsite for $\beta$-amino acids (Laube et al., 1995a). These mutations also convert the competitive antagonist picrotoxin into an allosteric potentiator of the glycine receptor, suggesting that Arg-271 acts as a functional integration point for influx from various extracellular ligand domains (Lynch et al., 1995).

From these data, it is inferred that at least three to four stretches of amino acids contribute to form the binding sites for the different agonists and antagonists on the glycine receptor. This situation is reminiscent of results obtained for the nicotinic acetylcholine receptor (Galzi et al., 1991) and the $GABA_A$ receptor (Amin and Weiss, 1993). Moreover, it underscores the evolutionary relationship among the members of the ligand-gated ion channel superfamily (Ortells and Lunt, 1995).

An additional ligand site that modulates the efficacy of glycinergic receptor activity has recently been described. At micromolar concentrations, the bivalent cation $Zn^{2+}$ potentiates apparent agonist affinities, whereas higher concentrations exert a negative modulation. The determinants of $Zn^{2+}$ potentiation have been assigned to an aspartate-rich motif (residues 74–86 of α1) (Laube et al., 1995b). Positive modulation of glycine receptor function has also been observed for compounds initially described as 5-hydroxytryptamine receptor-3 antagonists, but this still awaits characterization of its structural basis (Chesnoy-Marchais, 1996).

## Biophysics of chloride channel function

Pharmacological and single-channel analyses have been applied to characterize the channel function of recombinant glycine receptors. By comparing the characteristics of the different homo- and hetero-oligomeric receptor isoforms, and by partially interconverting the subunit variants by site-directed mutagenesis, various structural determinants of glycine receptor channel function have been defined.

Recombinant homo-oligomeric α1 receptors are sensitive to the noncompetitive channel blocker picrotoxinin, the active ingredient of the alkaloid picrotoxin (Pribilla et al., 1994). Hetero-oligomeric α1/β glycine receptors, in contrast, are insensitive to this drug. Receptors consisting of α1 and a β subunit mutant, whose transmembrane segment TM2 was made identical to the α1 sequence, proved to be picrotoxinin-sensitive. This points to the TM2 segment as being the site of blocker binding and, thus, the channel-forming polypeptide motif (Pribilla et al., 1994). As the TM2 segment had previously been shown to form the channel of the related cation-selective nicotinic acetylcholine receptors (Hucho, 1986), this finding revealed an architectural feature common to ligand-gated anion and cation channels.

The TM2 segment is most likely to traverse the membrane in an α-helical secondary structure, and contains multiple serine/threonine residues in the different receptors. Hydroxylated side chains are well suited to facilitate ion passage by partially substituting for the ion's hydration shell. Synthetic peptides corresponding to the TM2 segment of the glycine receptor α1 subunit were found to induce the formation of randomly gated, i.e. agonist-independent, channels upon incorporation into planar lipid bilayers (Langosch et al., 1991; Reddy et al., 1993). The majority of elementary conductances formed in this way did not correspond to the conductance of the α1 glycine receptor channels. Nevertheless, these observations attest to the principal capability of the TM2 polypeptide to form ion-conducting aggregates in membranes.

Triphenylmethylphosphonium, a large organic cation, efficiently blocks the channel of the nicotinic acetylcholine receptor by binding to its TM2 segment

(Hucho, 1986), but does not block the glycine receptor (Rundström et al., 1994). Conversely, an anionic molecule of similar size and general structure, cyanophenylborate, was found to block the channel of homo-oligomeric α1 glycine receptors but not of recombinant nicotinic receptors (Rundström et al., 1994). It appears, therefore, that the two channels are rather similar in their overall dimensions, irrespective of their ion selectivity. The basis of cation-selectivity has been resolved for homo-oligomeric receptors composed of polypeptides derived from the neuronal nicotinic α7 subunit. In these polypeptide constructs, several residues of the glycine receptor α1 subunit were simultaneously inserted at multiple positions within the TM2 segment and the cytoplasmic flanking region (Galzi et al., 1992). The converse experiment, i.e. rendering the glycine receptor cation-selective, has so far not been achieved, however (D. Langosch, unpublished work; B. Saul, T. Kuner and C.-M. Becker, unpublished work). At the glycine receptor, cyanotriphenylborate blocks α1 but not α2 homo-oligomeric channels (Rundström et al., 1994). This difference could be traced to a glycine–alanine exchange at position 254 (α1), which is close to the cytoplasmic terminus of TM2. This result corroborates the pivotal role of this segment in glycine receptor channel function, as originally assessed by picrotoxinin blockage.

The different glycine receptor isoforms are distinguishable not only by pharmacological criteria but also by single-channel properties. Analysis showed that α1 homo-oligomers display a main elementary chloride conductance of 86 pS at a chloride concentration of 145 mM, which is accompanied by four less abundant subconductance states down to 18 pS. In contrast, α2 and α3 receptors show an additional conductance of 105–111 pS, which is the main state. In this case the 86 pS state appears only as a subconductance (Bormann et al., 1993). Mutation of Gly-254 (α1) into an alanine, the corresponding residue of the α2 subunit, generated a conductance pattern in the α1 mutant similar to that of α2 or α3 channels. Similar results were obtained with α1/β hetero-oligomers. These receptor channels adopt only the three lowest conductances observed in α1 homo-oligomers, with 44 pS being the highest and most abundant one. Again, this difference could be attributed to residues of the β subunit that reside either within the TM2 segment itself or in the extracellular region linking segments TM2 and TM3 (Bormann et al., 1993). While the different conductances are interpreted to correspond to various conformations of the open channel, these findings show that the probability of open channel conformations critically depends on the amino acid sequence within or near segment TM2. In addition to the pharmacological analyses with channel blockers, this provides independent proof for the channel-forming function of this region of the receptor polypeptide. Recordings from recombinant receptors with defined subunit compositions are not only instrumental in understanding the biophysics of channel function, but also provide insights into the composition of glycine receptors expressed in native nerve cells. GlyR$_A$ has been reported to be show low sensitivity to picrotoxin (Akaike and Kaneda, 1989). This is consistent with a hetero-oligomeric α1/β subunit model, as derived independently from biochemical and immunological approaches. Glycine-gated single channels recorded from cultivated spinal neurons or slices from adult spinal cord are characterized by main conductances of 42–48 pS (Hamill et al., 1983; Bormann et al., 1987; Takahashi and Momiyama, 1991; Takahashi et al., 1992). This is reminiscent of the main-state conductance of recombinant $\alpha1/\beta$ glycine receptors, and again points to a hetero-oligomeric structure of the underlying protein complexes. On the other hand, larger conductances ranging from 64 to 94 pS have been found at low frequency at all stages of spinal cord development examined (Takahashi and Momiyama, 1991). This indicates that homo-oligomeric glycine receptor channels are present that are capable of adopting these larger conductances.

## Glycine receptor biosynthesis and post-translational modification

The biogenesis of functional glycine receptor complexes is still poorly understood, but appears to comprise several distinct stages. As evident from pulse–chase labelling studies with cultured neurons and recombinant expression experiments, oligomerization of glycine receptor subunits follows translation within a few minutes (W. Hoch, F. Holzinger, H. Betz and C.-M. Becker, unpublished work). This process is guided by sequence motifs present within the N-terminal domains of the $\alpha1$ and $\beta$ subunits (Kuhse et al., 1993). Subcellular sorting and compartmentalization of newly synthesized receptors are crucially dependent on an arginine residue at position 252 ($\alpha1$). This amino acid residue flanks the cytoplasmic border of segment TM2, and its substitution results in the retention of receptor subunits within intracellular compartments (Langosch et al., 1993). It has been hypothesized that the positively charged arginine residue retains segment TM2 within its hydrophobic environment as the nascent polypeptide gyrates through the membrane (Langosch et al., 1993). Oligomerization is followed by a slower process of transition into a fully functional state. As shown for native $GlyR_N$ of cultured neurons (Hoch et al., 1989) and recombinant $\alpha1$ homo-oligomers expressed in mammalian cells (W. Hoch, F. Holzinger, H. Betz and C.-M. Becker, unpublished work), newly synthesized glycine receptors still lack antagonist binding properties. The ligand binding capacity is acquired after maturation periods of $\geqslant20$ min, which might correspond to as yet unidentified post-translational modifications of the receptor polypeptide. Post-translational modifications also appear to modulate synaptic glycine receptor channels. Both protein kinases A and C phosphorylate the $\alpha1$ subunit at sites within the large cytoplasmic domain between transmembrane segments TM3 and TM4 (Vaello et al., 1994). Activation of protein kinase C appears to enhance, and that of protein kinase A to decrease, glycine-induced currents (Song and Huang, 1990; Vaello et al., 1994; Nishizaki and Ikeuchi, 1995; Schonrock and Bormann, 1995).

# Gephyrin

Glycine receptors are associated with the peripheral membrane protein gephyrin (Pfeiffer et al., 1982), which exists in several splice variants that are widely expressed throughout the CNS (Prior et al., 1992; Kirsch et al., 1993). Gephyrin resides at the cytoplasmic face of the post-synaptic membrane (Triller et al., 1985). It is thought to mediate the formation of post-synaptic glycine receptor clusters by linking the transmembrane complex to the neuronal cytoskeleton. Indeed, inhibition of gephyrin synthesis by use of antisense oligonucleotides prevents receptor clustering in cultured neurons (Kirsch et al., 1993). The case for gephyrin as a constituent of a subsynaptic meshwork that immobilizes glycine receptors is further strengthened by the identification of a site of gephyrin attachment to the glycine receptor $\beta$ subunit. This attachment site has been confined to a 19-amino-acid motif within the cytoplasmic loop connecting transmembrane segments TM3 and TM4 (Kirsch et al., 1995; Meyer et al., 1995). The widespread co-distribution throughout the CNS of gephyrin and the glycine receptor $\beta$ polypeptide has been attributed to them forming a functional pair of proteins that represents the synaptic anchoring mechanism (Kirsch et al., 1995). During cellular maturation, glycine receptors do not undergo any major redistribution from diffuse to clustered, but form clusters upon insertion into the neuronal plasma membrane. This contrasts with the maturation pattern of nicotinic acetylcholine receptors at the neuromuscular junction (St. John et al., 1986; Nicola et al., 1992). Rather, the gephyrin binding $\beta$ polypeptide appears to funnel newly synthesized glycine receptor complexes into an intracellular, gephyrin-dependent, transport route. This transport would then directly target the hetero-oligomeric receptor channel to its post-synaptic destination (Kirsch et al., 1995). As gephyrin binds to polymerized tubulin, it is believed to immobilize glycine receptors at synaptic sites via linkage to the neuronal cytoskeleton (Kirsch et al., 1993, 1995). Indeed, treatment of cultured neurons with agents that depolymerize cytoskeletal microtubules and microfilaments affects both the morphology of glycine receptor clusters and the desensitization behaviour of the currents recorded (Delon and Legendre, 1995; Kirsch and Betz, 1995).

Recent observations, however, indicate that the distribution of gephyrin is neither confined to, nor essential for, functional glycine receptors. Rather, gephyrin also appears to accumulate at non-glycinergic synapses. In rat retina and spinal cord, gephyrin and the $GABA_A$ receptor subunits $\alpha2$ and $\beta3$ were found to occur at the same synapses (Sassoe-Pognetto et al., 1995; Koulen et al., 1996; Todd et al., 1996). This matches the observation that, upon recombinant expression in a human cell line, the $GABA_A$ receptor $\beta3$ subunit was partially targeted to intracellular gephyrin aggregates. Thus gephyrin may indeed underlie post-synaptic clustering of those $GABA_A$ receptor isoforms containing the $\beta3$ polypeptide (Kirsch et al., 1995). On the other hand, functional glycine receptors also exist in cells from which gephyrin is virtually absent. Retinal bipolar cells express a glycine receptor isoform resembling $GlyR_A$, but lack detectable amounts of gephyrin mRNA. This indicates that the anchor protein is not required for glycine receptor function in rod bipolar cells (Enz and Bormann, 1995). In short, these observations suggest a broader role for gephyrin than previously anticipated: while gephyrin induces the post-synaptic aggregation of glycine receptors, it also appears to interact with $GABA_A$ receptors. The physiological role of this process remains to be elucidated.

## Organization of glycine receptor subunit genes

Consistent with phylogenetic gene duplications (Matzenbach et al., 1994), human and murine glycine receptor subunits constitute a family of genes that share a similar intron–exon organization (Shiang et al., 1993; Kingsmore et al., 1994a; Matzenbach et al., 1994; Mülhardt et al., 1994). The coding regions of the human *GLRA1* and the murine *Glra1*, *Glra2* and *Glyrb* genes (Table 2) are distributed on nine exons spread over large stretches of genomic DNA. For individual genes, deviations from this scheme exist: the *Glra1* gene possesses an additional acceptor site within exon 9 that results in the alternative splice variants α1 and α1$_{ins}$ (Malosio et al., 1991a; Matzenbach et al., 1994). In the *Glra2* gene, the alternative exons 3a and 3b underlie the mature transcripts α2A and α2B respectively (Kuhse et al., 1991; Matzenbach et al., 1994). The human and murine glycine receptor α1, α3 and β subunit genes are autosomally located, while the human and murine α2, as well as the murine α4, subunit genes reside on the X chromosome (Table 2).

## Glycine receptors in neurological disease

Based on its clinical resemblance to subconvulsive strychnine poisoning, glycine receptor dysfunction has long been considered as a candidate mechanism for hypertonic motor disorders in human and animal models (for a review, see Becker, 1990). Indeed, recessive mutations of the glycine receptor α1 and β subunit genes underlie the pathological murine phenotypes *spastic* and *spasmodic*, as well as the lethal condition *oscillator* (Buckwalter et al., 1994; Kingsmore et al., 1994a; Mülhardt et al., 1994; Ryan et al., 1994; Saul et al., 1994; Kling et al., 1997). Tremor, startle reactions and myoclonic jerks are the most prominent signs of these phenotypes. These mutations show that genetic defects that affect distinct subunits of the same neurotransmitter receptor, i.e. $GlyR_A$, result in similar disorders. In cattle and horses, a hereditary deficit of glycine receptors results in myoclonus and cerebral hyperexcitability (Gundlach et al., 1988, 1993). In these cases, however, the genes responsible for the hypertonic condition remain to be identified. In the human, mutations of the glycine receptor α1 subunit gene cause the hereditary disorder hyperekplexia (startle stiff baby syndrome). In addition, evidence exists for secondary glycine receptor alterations in a variety of metabolic and neurodegenerative diseases in humans and animals (reviewed in Betz and Becker, 1988).

**Fig. 3.**   **Glycine receptor mutants in the mouse**

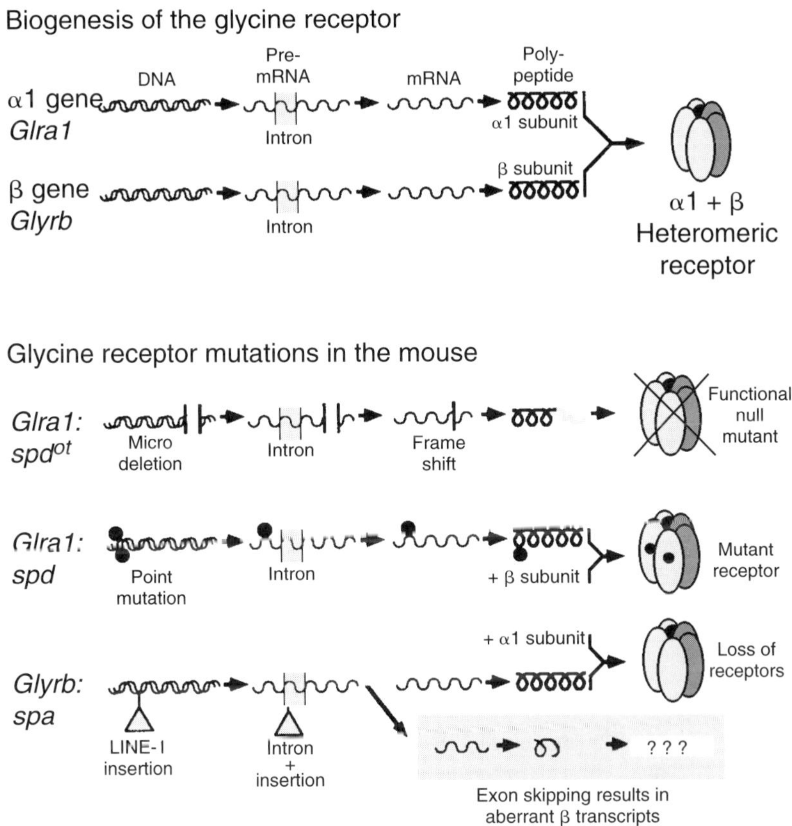



*Mutations of the glycine receptor α1 and β subunit genes result in hypertonic motor disorders. (i) The mutant mouse oscillator carries the null allele* $Glra1^{spd\text{-}ot}$, *in which a microdeletion causes a translational frame shift. This results in the complete loss of* $GlyR_A$ *protein from the CNS of homozygous mutants. (ii) In the spasmodic mouse (* $Glra1^{spd}$*), a mis-sense mutation (Ala-52→Ser) diminishes agonist affinity (increased* $K_D$*). (III) In the spastic mouse, the mutant allele* $Glrb^{spa}$ *is characterized by the intronic insertion of a LINE-1 element causing a splice defect. This results in a decrease in receptor numbers (decreased* $B_{max}$*), without affecting the receptor protein structure. Modified from Becker (1995), with permission,© The Neuroscientist.*

## The *spastic* gene of the mouse: *Glyrb*^spa

The *spastic* (*spa*) mouse carries a recessive mutation on chromosome 3 which becomes phenotypic approx. 2 weeks after birth. Muscle rigidity, tremor, exaggerated startle reactions and myoclonic jerks are easily detected in homozygous mutants (Becker, 1990). The *spastic* gene is a mutant allele of the *Glyrb* gene in which a LINE-1 (or L-1) element has been inserted into intron 5. LINE-1 elements are transposable elements that are mobile in the mammalian genome and are structurally related to retroviruses (Kingsmore et al., 1994a; Mülhardt et al., 1994). In the *spastic* allele, the DNA sequence inserted into intron 5 interferes with the normal splicing of β subunit pre-mRNA (Fig. 3), causing exon skipping (Mülhardt et al., 1994). As a consequence, truncated mRNAs are formed that code for non-functional polypeptides. Skipping of exons, however, is an incomplete process in the mutant. Supported by some residual formation of wild-type-like, full-length β transcripts, synthesis of $GlyR_A$ still proceeds at low levels (Mülhardt et al., 1994). Thus the $GlyR_A$ content in the spinal cord and retina of homozygous mutants is decreased to 20–30% of the wild-type level (White and Heller, 1982; Becker et al., 1986, 1992; Pinto et al., 1994). Despite this numerical reduction, the polypeptide composition of the glycine receptors remaining in the CNS of *spastic* homozygotes is not affected. Comprising α1 and β subunits in addition to the post-synaptic anchoring protein gephyrin, these receptors are indistinguishable from the wild-type receptor complex $GlyR_A$ (Becker et al., 1986). Consistent with the numerical receptor defect, the *spastic* phenotype is rescued by a transgene expressing β subunit mRNA (Hartenstein et al., 1996).

Differences in the subunit compositions of glycine receptor isoforms offer an explanation for the development of *spastic* symptoms: $GlyR_N$ is thought to be devoid of β subunits. Thus the postnatal appearance of the mutant phenotype coincides with the switch from the unaffected neonatal to the diminished adult receptor isoform, $GlyR_A$ (Becker et al., 1992). Recordings from the spinal cord of homozygous mutants show that this deficit in $GlyR_A$ is matched by the loss of glycine-mediated, chloride-dependent synaptic inhibition and the appearance of synchronous oscillations in motoneurons (Biscoe and Duchen, 1986; Simon, 1995). As pointed out previously, recombinant α1 subunits suffice to generate functional receptor homomers in heterologous expression experiments (Schmieden et al., 1989; Sontheimer et al., 1989). Thus the mechanism by which diminished β polypeptide expression results in the loss of glycine-gated chloride channels remains to be elucidated. One might speculate that α1 homomers, not properly targeted at the subsynaptic gephyrin network, are prone to accelerated protein degradation.

## Mutant alleles of the murine *Glra1* gene

The *spasmodic* (*spd*) mouse (Lane et al., 1987) carries a mutant allele of the *Glra1* gene (chromosome 11) in which a mis-sense mutation encodes the substitution of Ala-52 (Figs. 2 and 3). Mutant receptors display reduced agonist, but conserved strychnine, binding affinities, consistent with a loss of neuromuscular inhibition (Ryan et al., 1994; Saul et al., 1994). The *spasmodic* locus is linked by synteny homology to the human chromosomal region 5q31.3, which also contains the human *GLRA1* gene (Shiang et al., 1993; Baker et al., 1994).

The recessive *Glra1^spd-ot* allele underlies the *oscillator* phenotype of the mouse, which is lethal during the third postnatal week (Buckwalter et al., 1994). Interestingly, action tremor is more pronounced than exaggerated startle responses in homozygous animals. In the *Glra1^spd-ot* allele, a microdeletion within exon 8 predicts a translational frame shift. Depending on alternative exon use, the *Glra1^spd-ot* allele encodes two mutant transcripts (Buckwalter et al., 1994): (i) in the predominant α1 mRNA species, the site of the microdeletion corresponding to transmembrane segment TM3 is followed by 127 mis-sense amino acids, and (ii) in the alternative transcript, the polypeptide chain is terminated after 19 residues due to an early stop codon. [$^3$H]Strychnine binding is dramatically reduced (Buckwalter et al., 1994; Kling et al., 1997), and biochemical analysis confirms the complete absence of GlyR$_A$ protein from the CNS of homozygous *oscillator* mice (Fig. 3). Thus the *oscillator* gene is a functional null allele of the *Glra1* gene (Kling et al., 1997).

## Hyperekplexia mutations of the *GLRA1* gene: clues to glycine receptor function and regulation in the human

Hyperekplexia (startle disease; stiff baby syndrome) is a congenital motor disorder that displays autosomal-recessive as well as dominant modes of inheritance. A variety of case reports indicate that the symptoms of this disorder change with age. Affected neonates exhibit exaggerated startle responses and episodic muscle stiffness which may result in fatal apnoea. In early childhood, muscle tone returns to normal, but excessive startling persists throughout life. Startle responses may even trigger a loss of postural control and result in immediate, unprotected, falling (Andermann et al., 1980; Heller and Hallett, 1982; Shiang et al., 1993, Brune et al., 1996). Following extensive genomic mapping studies, Shiang et al. (1993) identified mis-sense mutations of the *GLRA1* gene that cause hereditary hyperekplexia. The mutations reported encode substitutions of (α1) Arg-271, which delineates the extracellular border of transmembrane segment TM2 (Fig. 2). In both the recessive and dominant forms of hyperekplexia, additional mis-sense mutations have been identified (Table 3), all of which give rise to substitutions of amino acid residues located from the TM1 segment through the extracellular loop following segment TM2. Heterologous expression studies show that these mutations impair agonist affinities and/or reduce conductance states of expressed mutant channels, resulting in a partial loss of function (Langosch et al., 1994; Rajendra et al., 1994, 1995b; Laube et al., 1995a, Lynch et al., 1995). Two conclusions can be drawn from these studies: (i) hyperekplexia mutations may affect intrinsic activities of glycine receptor ligands, as β-alanine and taurine are converted from partial agonists into competitive antagonists (Laube et al., 1995a; Rajendra et al., 1995b), and (ii) hyperekplexia mutations may result in the uncoupling of the agonist binding process from the channel activation mechanism of the receptor (Rajendra et al., 1995b). In a case of recessive hyperekplexia, Brune et al. (1996) identified a null allele of the *GLRA1* gene characterized by a deletion of exons 1–6. Born to consanguineous parents, the affected child was homozygous for this *GLRA1^null* allele, consistent with a complete loss of gene function. Despite this 'knockout' situation, the child displayed relatively mild symptoms at the age of 6 years, including a disinhibition of vestigial brainstem reflexes, i.e. exaggerated startle responses, head retraction jerks (Brune et al., 1996). Thus, in contrast to the lethal effect of the *null* allele *Glra1*<sup>spd-ot</sup> in the *oscillator* mouse, the complete loss of the glycine receptor α1 subunit is tolerated in the human.

**Table 3.    Mutant alleles of the human glycine receptor α1 subunit gene *GLRA1***

See also Online Mendelian Inheritance in Man, OMIM (1997).

| Allelic variant | Phenotype | Mode of inheritance | References |
| --- | --- | --- | --- |
| Ser-231→Arg | Hyperekplexia | Recessive | * |
| Ile-244→Asn | Hyperekplexia | Recessive | Rees et al. (1994) |
| Pro-250→Thr | Hyperekplexia | Dominant | † |
| Gln-266→His | Hyperekplexia | Dominant | Milani et al. (1996) |
| Arg-271→Leu | Hyperekplexia | Dominant | Shiang et al. (1993) |
| Arg-271→Gln | Hyperekplexia | Dominant | Shiang et al. (1993) |
| Lys-276→Glu | Hyperekplexia, spastic paraparesis | Dominant | Elmslie et al. (1996) |
| Tyr-279→Cys | Hyperekplexia | Dominant | Shiang et al. (1995) |
| *GLRA 1*<sup>null</sup> | Hyperekplexia | Recessive | Brune et al. (1996) |

*K. Reuter, S. Özbey, S. Jafari, U. Stephani and C.-M. Becker, unpublished work.

†B. Saul, T. Kuner, D. Sobetzko, W. Brune, F. Hanefield, H.-M. Meinck and C.-M. Becker, unpublished work.

In conclusion, these clinical studies may offer clues as to the function and regulation of inhibitory glycine receptors in the human. (i) Human hyperekplexia and the *spasmodic* syndrome of the mouse both result from mis-sense mutations of α1 subunit genes. In the human, muscle stiffness is most prominent around birth but is alleviated with age (Andermann et al., 1980; Shiang et al., 1993). The murine phenotype, however, appears postnatally, matching the developmental expression pattern of GlyR$_A$ (Becker et al., 1988, 1992; Saul et al., 1994). This suggests that the regulation of human *GLRA1* gene expression is disparate from the developmental profile seen in rodents. (ii) While generalized muscular hypertonia in 'stiff babies' is consistent with an alteration of spinal glycine receptors, the symptoms of juvenile and adult patients are compatible with a loss of inhibition that is confined to the brainstem (Brune et al., 1996). Indeed, glycine receptor levels in brainstem nuclei of adult individuals appear to exceed those of the spinal cord anterior horn (Probst et al., 1986; Naas et al., 1991). One might hypothesize that changes in clinical status are a consequence of the regional redistribution of glycine receptor α1 subunits during human development. (iii) While homozygosity for a null allele of the α1 subunit gene is tolerated in the human (Brune et al., 1996), it is lethal in the mouse (Kling et al., 1997). This observation also suggests either that glycine receptor subunit regulation substantially differs between these species or that the loss of this polypeptide is effectively compensated in the human.

## Outlook for a therapeutical orphan

Inhibitory glycine receptors represent a well characterized family of ligand-gated ion channels. Their involvement in hereditary motor disorders has been recognized in the human, as well as in animal models, and they still serve as candidate genes in a variety of disorders associated with CNS hyperexcitability. Glycine receptors appear, however, as 'therapeutical orphans' when compared with the sophisticated pharmacology of their first-degree cousins, the $GABA_A$ receptors. The recent discovery of positive modulation of glycine receptors by $Zn^{2+}$ and 5-hydroxytryptamine receptor-3 antagonists offers a favourable opportunity that may allow glycine receptors to leave their long-held territory of strychnine toxicology.

*Work in the authors' laboratories is supported by the Deutsche Forschungsgemeinschaft (SFB 317 and Heisenberg-Programm), the Bundesministerium für Bildung und Forschung, and the Fonds der Chemischen Industrie.*

## References

Akagi, H., Hirai, K. and Hishinuma, F. (1991) FEBS Lett. **281**, 160–166

Akaike, N. and Kaneda, M. (1989) J. Neurophysiol. **62**, 1400–1409

Amin, J. and Weiss, D.S. (1993) Nature (London) **366**, 565–569

Andermann, F., Keene, D.L., Andermann, E. and Quesney, L.F. (1980) Brain **103**, 985–997

Aprison, M. and Lipkowitz, K. (1992) J. Neurosci. Res. **31**, 166–174

Baker, E., Sutherland, G.R. and Schofield, P.R. (1994) Genomics **22**, 491–493

Becker, C.-M. (1990) FASEB J. **4**, 2767–2774

Becker, C.-M. (1992) Handb. Exp. Pharmacol. **102**, 539–575

Becker, C.-M. (1995) Neuroscientist **1**, 130–141

Becker, C.-M., Hermans-Borgmeyer, I., Schmitt, B. and Betz, H. (1986) J. Neurosci. **6**, 1358–1364

Becker, C.-M., Hoch, W. and Betz, H. (1988) EMBO J. **7**, 3717–3726

Becker, C.-M., Schmieden, V., Tarroni, P., Strasser, U. and Betz, H. (1992) Neuron **8**, 283–289

Becker, C.-M., Betz, H. and Schröder, H. (1993) Brain Res. **606**, 220–226

Becker, C.-M., Kling, C., Mülhardt, C., Saul, B. and Kuhse, J. (1994) Ann. N.Y. Acad. Sci. **733**, 155–162

Betz, H. (1992) Q. Rev. Biophys. **25**, 381–394

Betz, H. and Becker, C.-M. (1988) Neurochem. Int. **13**, 137–146

Biscoe, T.J. and Duchen, M.R. (1986) J. Physiol. (London) **379**, 275–292

Bormann, J., Hamill, O.P. and Sakmann (1987) J. Physiol. (London) **385**, 243–286

Bormann, J., Rundström, N., Betz, H. and Langosch, D. (1993) EMBO J. **12**, 3729–3737

Brune, W., Weber, R.G., Saul, B., von Knebel-Doeberitz, M., Grond-Grinsbach, C., Kellermann, K., Meinck, H.-M. and Becker, C.-M. (1996) Am. J. Hum. Genet. **58**, 989–997

Buckwalter, M.S., Cook, S.A. and Davisson, M.T. (1994) Hum. Mol. Genet. **3**, 2025–2030

Chesnoy-Marchais, D. (1996) Br. J. Pharmacol. **118**, 2115–2125

Daly, E.C. (1990) in Glycine Neurotransmission (Ottersen O. and Storm-Mathisen, V., eds.), pp. 25–66, Wiley, Chichester

Delon, J. and Legendre, P. (1995) NeuroReport **6**, 1932–1936

Derry, J.M. and Barnard, P.J. (1991) Genomics **10**, 593–597

Elmslie, F.V., Hutchings, S.M., Spencer, V., Curtis, A., Covanis, T., Gardiner, R.M. and Rees, M. (1996) J. Med. Genet. **33**, 435–436

Enz, R. and Bormann, J. (1995) Vis. Neurosci. **12**, 501–507

Galvez-Ruano, E., Lipkowitz, K.B. and Aprison, M.H. (1995) J. Neurosci. Res. **41**, 775–781

Galzi, J.L., Bertrand, D., Devillers-Thiery, A., Revah, F., Bertrand, S. and Changeux, J.P. (1991) FEBS Lett. **293**, 198–202

Galzi, J.L., Devillers, T.A., Hussy, N., Bertrand, S., Changeux, J.P. and Bertrand, D. (1992) Nature (London) **359**, 500–505

Grenningloh, G., Rienitz, A., Schmitt, B., Methfessel, C., Zensen, M., Beyreuther, K., Gundelfinger, E.D. and Betz, H. (1987) Nature (London) **328**, 215–220

Grenningloh, G., Schmieden, V., Schofield, P.R., Seeburg, P.H., Siddique, T., Mohandas, T.K., Becker, C.-M. and Betz, H. (1990a) EMBO J. **9**, 771–776

Grenningloh, G., Pribilla, I., Prior, P., Multhaup, G., Beyreuther, K., Taleb, O. and Betz, H. (1990b) Neuron **4**, 963–970

Gundlach, A.L., Dodd, P.R., Grabara, C.S.G., Watson, W.E.J., Johnston, G.A.R., Haper, P.A.W., Dennis, J.A. and Healy, P.J. (1988) Science **241**, 1807–1809

Gundlach, A.L., Kortz, G., Burazin, T.C., Madigan, J. and Higgins, R.J. (1993) Brain Res. **628**, 263–270

Hamill, O.P., Bormann, J. and Sakmann, B. (1983) Nature (London) **305**, 805–808

Handford, C.A., Lynch, J.W., Baker, E., Webb, G.C., Ford, J.H., Sutherland, G.R. and Schofield, P.R. (1996) Mol. Brain Res. **35**, 211–219

Hartenstein, B., Schenkel, J., Kuhse, J., Besenbeck, B., Kling, C., Becker, C.-M., Betz, H. and Weiher, H. (1996) EMBO J. **15**, 1275–1282

Heller, A.H. and Hallett, M. (1982) Brain. Res. **234**, 299–308

Hoch, W., Betz, H. and Becker, C.-M. (1989) Neuron **3**, 339–348

Hucho, F. (1986) Eur. J. Biochem. **158**, 221–226

Kingsmore, S.F., Giros, B., Suh, D., Bieniarz, M., Caron, M.G. and Seldin, M.F. (1994a) Nature Genet. **7**, 136–142

Kingsmore, S.F., Suh, D. and Seldin, M.F. (1994b) Mamm. Genome **5**, 831–832

Kirchhoff, F., Mülhardt, C., Pastor, A., Becker, C.-M. and Kettenmann, H. (1996) J. Neurochem. **66**, 1383–1390

Kirsch, J. and Betz, H. (1995) J. Neurosci. **15**, 4148–4156

Kirsch, J., Wolters, I., Triller, A. and Betz, H. (1993) Nature (London) **366**, 745–748

Kirsch, J., Kuhse, J. and Betz, H. (1995) Mol. Cell. Neurosci. **6**, 450–461

Kling, C., Koch, M., Saul, B. and Becker, C.-M. (1997) Neuroscience, in the press

Koulen, P., Sassoe-Pognetto, M., Grunert, U. and Wässle, H. (1996) J. Neurosci. **16**, 2127–2140

Krnjevic, K. (1981) in Handbook of Physiolgy (Geiger, S.R., ed.), pp. 107–154, Am. Soc. Physiol, Baltimore

Kuhse, J., Schmieden, V. and Betz H. (1990a) Neuron **5**, 867–873

Kuhse, J., Schmieden, V. and Betz, H. (1990b) J. Biol. Chem. **265**, 22317–22320

Kuhse, J., Kuryatov, A., Maulet, Y., Malosio, M.L., Schmieden, V. and Betz, H. (1991) FEBS Lett. **283**, 73–77

Kuhse, J., Laube, B., Magalei, D. and Betz, H. (1993) Neuron **11**, 1049–1056

Lane, P.W., Ganser, A.L., Kerner, A.L. and White, W.F. (1987) J. Hered. **78**, 353–356

Langosch, D., Thomas, L. and Betz, H. (1988) Proc. Natl. Acad. Sci. U.S.A. **85**, 7394–7398

Langosch, D., Betz, H. and Becker, C.-M. (1990) in Glycine Neurotransmission (Ottersen, O. and Storm-Mathisen, V., eds.), pp. 67–82, Wiley, Chichester

Langosch, D., Hartung, K., Grell, E., Bamberg, E. and Betz, H. (1991) Biochim. Biophys. Acta **1063**, 36–44

Langosch, D., Herbold, A., Schmieden, V., Borman, J. and Kirsch, J. (1993) FEBS Lett. **336**, 540–544

Langosch, D., Laube, B., Rundström, N., Schmieden, V., Bormann, J. and Betz H. (1994) EMBO J. **13**, 4223–4228

Laube, B., Langosch, D., Betz, H. and Schmieden, V. (1995a) NeuroReport **206**, 897–900

Laube, B., Kuhse, J., Rundström, N., Kirsch, J., Schmieden, V. and Betz, H. (1995b) J. Physiol. (London) **483**, 613–619

Lewis, C.A., Ahmed, Z. and Faber, D.S. (1991) J. Neurophysiol. **66**, 1291–1303

Lynch, J.W., Rajendra, S., Barry, P.H. and Schofield, P.R. (1995) J. Biol. Chem. **270**, 13799–13806

Malosio, M.L., Grenningloh, G., Kuhse, J., Schmitt, B., Prior, P. and Betz, H. (1991a) J. Biol. Chem. **266**, 2048–2053

Malosio, M.L., Marqueze, P.B., Kuhse, J. and Betz, H. (1991b) EMBO J. **10**, 2401–2409

Matzenbach, B., Maulet, Y., Sefton, L., Courtier, B., Avner, P., Guenet, J.L. and Betz, H. (1994) J. Biol. Chem. **269**, 2607–2612

Melendrez, C.S. and Meisel, S. (1995) Biol. Reprod. **53**, 676–683

Meyer, G., Kirsch, J., Betz, H. and Langosch, D. (1995) Neuron **15**, 563–572

Milani, N., Dalprá, L., Del Prete, A., Zanini, R. and Larizza, L. (1996) Am. J. Hum. Genet. **58**, 420–422

Mouse Genome Database (1997) Mouse genome informatics, The Jackson Laboratory, Bar Harbor, Maine, U.S.A.; World Wide Web http://www.informatics.jax. org/

Mülhardt, C., Fischer, M., Gass, P., Simon-Chazottes, D., Guénet, J.-L., Kuhse, J., Betz, H. and Becker, C.-M. (1994) Neuron **13**, 1003–1015

Naas, E., Zilles, K., Gnahn, H., Betz, H., Becker, C.-M. and Schröder, H. (1991) Brain Res. **561**, 139–146

Nicola, M.A, Becker, C.-M. and Triller, A. (1992) Neurosci. Lett. **138**, 173–178

Nishizaki, T. and Ikeuchi, Y. (1995) Brain Res. **687**, 214–216

Online Mendelian Inheritance in Man, OMIM (1997) National Center for Biotechnology Information, National Library of Medicine, Bethesda, MD, U.S.A.; World Wide Web http://www3.ncbi.nlm. nih. gov/omim/

Ortells, M.O. and Lunt, G.G. (1995) Trends Neurosci. **18**, 121–127

Pfeiffer, F., Graham, D. and Betz, H. (1982) J. Biol. Chem. **257**, 9389–9393

Pinto, L.H, Grünert, U., Studholm, K., Yazulla, S., Kirsch, J. and Becker, C.-M. (1994) Invest. Ophthalmol. Vis. Sci. **13**, 3633–3639

Pribilla, I., Takagi, T., Langosch, D., Bormann, J. and Betz, H. (1994) EMBO J. **11**, 4305–4311

Prior, P., Schmitt, B, Grenningloh, G., Pribilla, I., Multhaup, G., Beyreuther, K., Maulet, Y., Werner, P., Langosch, D., Kirsch, J. and Betz, H. (1992) Neuron **8**, 1161–1170

Probst, A., Cortes, R. and Palacios, J.M. (1986) Neuroscience **17**, 11–35

Rajendra, S., Lynch, J.W, Pierce, K.D., French, C.R., Barry, P.H. and Schofield, P.R. (1994) J. Biol. Chem. **269**, 18739–18742

Rajendra, S., Vandenberg, R.J., Pierce, K.D., Cunningham, A.M., French, P.W., Barry, P.H. and Schofield, P.R. (1995a) EMBO J. **14**, 2987–2998

Rajendra, S., Lynch, J.W., Pierce, K D., French, C.R., Barry, P.H. and Schofield, P.R. (1995b) Neuron **14**, 169–175

Reddy, G.L., Iwamoto, T., Tomich, J.M. and Montal, M. (1993) J. Biol. Chem. **268**, 14608–14615

Rees, M.I., Andrew, M., Jawad, S. and Owen, M.J. (1994) Hum. Mol. Genet. **3**, 2175–2179

Ruiz-Gomez, A., Morato, E., Garcia, C.M., Valdivieso, F. and Mayor, F.J. (1990) Biochemistry **29**, 7033–7040

Rundström, N., Schmieden, V., Betz, H., Bormann, J. and Langosch, D. (1994) Proc. Natl. Acad. Sci. U.S.A. **91**, 8950–8954

Ryan, S.G., Buckwalter, M.S., Lynch, J.W., Handford, C.A., Segura, L., Shiang, R., Wasmuth, J.J., Camper, S.A., Schofield, P. and O'Connell, P. (1994) Nature Genet. **7**, 131–135

Sassoe-Pognetto, M., Kirsch, J., Grunert, U., Greferath, U., Fritschy, J.M., Möhler, H., Betz, H. and Wässle, H. (1995) J. Comp. Neurol. **357**, 1–14

Saul, B., Schmieden, V., Mülhardt, C., Gass, P., Kuhse, J. and Becker, C.-M. (1994) FEBS Lett. **350**, 71–76

Schmieden, V. and Betz, H. (1995) Mol. Pharmacol. **48**, 919–927

Schmieden, V., Grenningloh, G., Schofield, P.R. and Betz, H. (1989) EMBO J. **8**, 695–700

Schmieden, V., Kuhse, J. and Betz, H. (1992) EMBO J. **11**, 2025–2032

Schmieden, V., Kuhse, J. and Betz, H. (1993) Science **262**, 256–258

Schonrock, B. and Bormann, J. (1995) NeuroReport **6**, 301 304

Seeburg, P.H. (1994) Trends Neurosci. **19**, 359–365

Shiang, R., Ryan, S.G., Zhu, Y.Z., Hahn, A.F., O'Connell, P. and Wasmuth, J.J. (1993) Nature Genet. **5**, 351–358

Shiang, R., Ryan, S.G., Zhu, Y.Z., Fielder, T J., Allen, R.J., Fryer, A., Yamashita, S., O'Connell, P. and Wasmuth, J.J. (1995) Ann. Neurol. **38**, 85–91

Simon, E.S. (1995) Neurology **45**, 1883–1892

Song, Y.M. and Huang, L.Y. (1990) Nature (London) **348**, 242–245

Sontheimer, H., Becker, C.-M., Pritchett, D.B., Schofield, P.R., Grenningloh, G., Kettenmann, H., Betz, H. and Seeburg, P.H. (1989) Neuron **2**, 1491–1497

St. John, P.A. and Stephens, S.L. (1993) J. Neurosci. **13**, 2749–2757

St. John, P.A., Wayne, M.K., Mazetta, J.S., Lange, G.D. and Barker, J.L. (1986) J. Neurosci. **6**, 1492–1512

Takahashi, T. (1984) Proc. R. Soc. London B **221**, 103–109

Takahashi, T. and Momiyama, A. (1991) Neuron **7**, 965–969

Takahashi, T., Momiyama, A., Hirai, K., Hishinuma, F. and Akagi, H. (1992) Neuron **9**, 1155–1161

Todd, A.J., Watt, C., Spike, R.C. and Sieghart, W. (1996) J. Neurosci. **16**, 974–982

Triller, A., Cluzeaud, F., Pfeiffer, F., Betz, H. and Korn, H. (1985) J. Cell Biol. **101**, 683–688

Vaello, M.L., Ruiz, G.A., Lerma, J. and Mayor, F.J. (1994) J. Biol. Chem. **269**, 2002–2008

Vandenberg, R.J., Handford, C.A. and Schofield, P.R. (1992a) Neuron **9**, 491–496

Vandenberg, R.J., French, C.R, Barry, P.H., Shine, J. and Schofield, P.R. (1992b) Proc. Natl. Acad. Sci. U.S.A. **89**, 1765–1769

Vandenberg, R.J., Rajendra, S., French, C.R., Barry, P.H. and Schofield, P.R. (1993) Mol. Pharmacol. **44**, 198–203

Watanabe, E. and Akagi, H. (1995) Neurosci. Res. **23**, 377–382

White, W.F. and Heller, A.H. (1982) Nature (London) **298**, 655–657

Young, A.B. and Snyder, S.H. (1973) Proc. Natl. Acad. Sci. U.S.A. **70**, 2832–2836

Zarbin, M.A., Wamsley, J.K. and Kuhar, M.J. (1981) J. Neurosci. **1**, 532–547

# 6

# Metabotropic glutamate receptors

Jeremy M. Henley* , Rachel Bruton† and Saul A. Richmond†

Department of Anatomy, School of Medical Sciences,
University of Bristol, University Walk, Bristol BS8 1TD, U.K., and
†Department of Pharmacology, Medical School,
University of Birmingham, Edgbaston, Birmingham B15 2TT, U.K.

## Introduction

The acidic amino acid L-glutamate activates the major class of excitatory neurotransmitter receptors (excitatory amino acid receptors or glutamate receptors) in the vertebrate central nervous system. Because glutamatergic synapses have such a widespread distribution, glutamate receptors have the potential of being involved in nearly every aspect of central nervous system function. Thus, even though glutamate was first suggested as a neurotransmitter only in the late 1950s, the receptors which mediate its actions have already been shown to be important in processes as varied as the control of motor, cardiovascular and visual functions; synapse formation, stabilization and elimination; learning and memory; and epilepsy, neuronal degenerative disorders and the mediation of neuronal cell death.

Glutamate receptors can be characterized into two broad groups: those that contain an integral ion channel, the ionotropic glutamate receptors, and those that are G-protein-coupled, the metabotropic glutamate receptors (mGluRs). The subtypes of ionotropic glutamate receptors are dealt with in other chapters in this volume; here we will deal exclusively with the mGluRs.

## Discovery of mGluRs

The first mGluR responses were discovered in the mid-1980s following the observation that glutamate and quisqualate activated phosphoinositide hydrolysis in a variety of neuronal preparations [1–4]. The first direct evidence that mGluRs are coupled to phospholipase C via G-proteins came from *Xenopus* oocyte expression studies [5,6]. Phosphoinositide hydrolysis results in the liberation of inositol 1,4,5-trisphosphate (Ins$P_3$), which in turn causes the mobilization of intracellular calcium. In *Xenopus* oocytes, intracellular calcium activates a calcium-

* To whom correspondence should be addressed.

sensitive oscillatory chloride current via a pertussis-toxin-sensitive G-protein-dependent mechanism. Sugiyama et al. [5,6] injected oocytes with rat brain mRNA and showed that L-glutamate, quisqualate and ibotenate, but not $N$-methyl-D-aspartate (NMDA), α-amino-3-hydroxy-5-methylisoxazolepropionate (AMPA) or kainate, were able to elicit pertussis-toxin-sensitive oscillatory chloride currents.

In addition to the direct observations that quisqualate and ibotenate activate PtdIns hydrolysis, pharmacological studies of ionotropic receptors also supported the existence of metabotropic receptors. Thus ibotenate is an effective agonist at NMDA receptors, but its effects on PtdIns hydrolysis are not blocked by NMDA receptor antagonists such as D-2-amino-5-phosphonopentanoic acid or MK-801 [7]. Similarly, the metabotropic effects of quisqualate are not blocked by the specific ionotropic receptor antagonists 6-cyano-7-nitroquinoxaline-2,3-dione or 6,7-dinitroquinoxaline-2,3-dione [8–10].

Since these initial observations, rapid progress has been achieved in the identification, characterization and isolation of the receptors that directly mediate phosphoinositide hydrolysis. As will be discussed in more detail below, it is now clear that a family of mGluRs exists and that these receptors can couple, either directly or indirectly, to a range of second messenger systems including phospholipase C, adenylate cyclase, phospholipase D, regulation of cGMP and arachidonic acid release [11].

## Molecular biology of mGluRs

The first cDNA encoding a mGluR (mGluR1a) was cloned independently by two groups [12,13]. Since then, cDNAs encoding alternatively spliced and truncated versions of the receptor (mGluR1b and mGluR1c) have also been described [14,15]. The mGluR1a gene has a large open reading frame of 3597 bp in a total length of 4282 bp encoding a predicted polypeptide of 1199 amino acids with a calculated $M_r$ of 133229. This protein is significantly larger than other known members of the G-protein-linked receptor superfamily (Fig. 1). Furthermore, there is little identity with other G-protein-linked receptors for which sequences are known, and almost none of the conserved regions shared between other G-protein-linked receptors are retained in the mGluR family.

The cDNA encoding mGluR1a was used to screen cDNA libraries for additional mGluR subtypes [15–18]. To date, there are at least seven reported distinct mGluR subtypes, some of which can exist in multiple splice variants (Table 1) [19].

### mGluR subgroups

The family of mGluRs has been divided into three groups based on the amino acid sequences, the agonist profiles and the effector mechanism of the recombinant receptors.

**Table 1.** **Properties of m GluRs**

The abbreviations are given in the text, except for: Glu, L-glutamate; Quis, quisqualate; Ibo, ibotenate.

| Group | No. of residues | Agonist rank order | Effector system |
|---|---|---|---|
| Group 1 | | | |
| mGluR1a | 1199 | | |
| mGluR1b | 907 | | |
| mGluR1c | 897 | Quis > Glu > Ibo > 1 S,3R-ACPD | Stimulates phospholipase C |
| mGluR5a | 1171 | | |
| mGluR5b | 1193 | | |
| | | | |
| Group 2 | | | |
| mGluR2 | 872 | L-CCG-I > Glu = 1 S,3R-ACPD > Ibo > Quis | Inhibits adenylate cyclase |
| mGluR3 | 879 | | |
| | | | |
| Group 3 | | | |
| mGluR4a | 912 | | |
| mGluR4b | 983 | L-AP4 > Glu » 1 S,3R-ACPD > Quis > Ibo | Inhibits adenylate cyclase |
| mGluR6 | 871 | | |
| *mGluR7* | *914* | | |

**Fig. 1.**          **Schematic diagram of the putative structure of a mGluR**

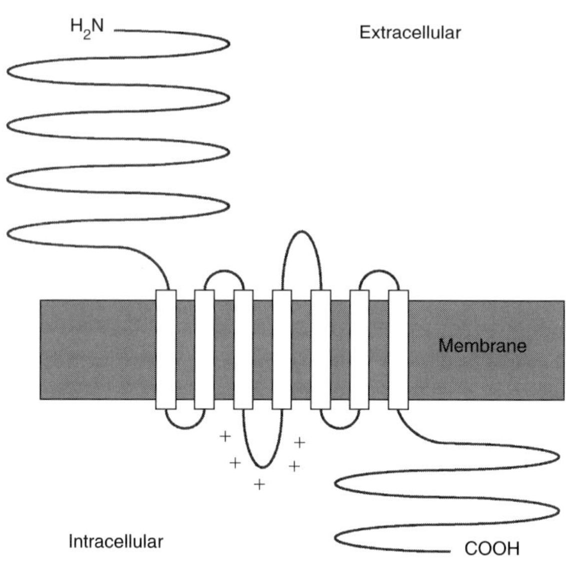


## Group 1 (mGluR1 and mGluR5)

Group 1 receptors share about 70% sequence similarity (Table 2), and stimulate the production of $InsP_3$ and diacylglycerol from phosphatidylinositol 4,5-bis-phosphate. Diacylglycerol then activates protein kinase C, while $InsP_3$ mobilizes calcium from intracellular stores [13,16]. The pharmacology of group 1 mGluRs is unique. When expressed in *Xenopus* oocytes or in Chinese hamster ovary (CHO) [20] or baby hamster kidney (BHK) [21] cells, mGluR1 is activated most potently by quisqualate and L-glutamate (low micromolar concentrations) and relatively weakly by 1*S*,3*R*-aminocyclopentane dicarboxylic acid (1*S*,3*R*-ACPD) (high micromolar to millimolar concentrations).

## Group 2 and group 3 mGluRs

These receptors do not evoke currents when expressed in *Xenopus* oocytes, indicating that they do not readily couple to phosphoinositide turnover. Rather, they are negatively coupled to adenylate cyclase when expressed in non-neuronal cells, since they mediate the agonist-induced inhibition of forskolin-stimulated cAMP formation. Group 2 mGluRs (mGluR2 and mGluR3) are approx. 70% similar (Table 2); L-glutamate and 1*S*,3*R*-ACPD are the best agonists, while quisqualate is comparatively ineffective [15,22]. Group 3 mGluRs (mGluR4, mGluR6) are insensitive to 1*S*,3*R*-ACPD but are potently activated by L-2-amino-4-phosphonobutyrate (L-AP4) [23].

**Table 2.**  **Amino acid sequence similarity between mGluR subtypes**

The values represent the percentage of amino acid residues that are identical or conserved among the mGluR subtypes. The data are taken from the review by Suzdak et al. [19].

| Similarity (%) | | | | | | |
| --- | --- | --- | --- | --- | --- | --- |
| mGluR1 | mGluR2 | mGluR3 | mGluR4 | mGluR5 | mGluR6 | mGluR7 |
| mGluR1 100 | 46 | 44 | 43 | 61 | 40 | 38 |
| mGluR2 | 100 | 70 | 43 | 43 | 46 | 45 |
| mGluR3 | | 100 | 47 | 42 | 45 | 43 |
| mGluR4 | | | 100 | 40 | 69 | 69 |
| mGluR5 | | | | 100 | 39 | 38 |
| mGluR6 | | | | | 100 | 67 |
| mGluR7 | | | | | | 100 |

## General structural features of mGluRs

Hydrophobicity plots for all of the deduced sequences reveal at least eight hydrophobic regions of 20 or more uncharged amino acid residues. For mGluR1a there are also two large hydrophilic regions, one of 570 residues at the N-terminal end of the polypeptide and the other of approx. 360 residues at the C-terminus [12,13]. In all cases the putative hydrophobic signal peptide is followed by a large N-terminal domain of more than 500 amino acid residues. However, the mGluR1c splice variant is truncated by 302 amino acid residues at the C-terminus. Interestingly, despite this difference, recombinant mGluR1c activates phospholipase C in *Xenopus* oocytes with a pharmacological profile identical with that of expressed mGluR1a [14].

All of the known sequences for G-protein-linked receptors that are activated by large glycoprotein hormones possess a large N-terminal extracellular domain and clusters of cysteine residues immediately preceding the putative seven transmembrane domains. On the other hand, G-protein-linked receptors for small neurotransmitter molecules appear to lack this large N-terminal extracellular domain and the cysteine clusters. Interestingly, mGluRs have both sequence and structural characteristics more in common with the former class of receptors than the latter, despite the small size of their agonist [24].

Sequence analysis algorithms to search for identity between the mGluR N-terminal domain and other proteins have revealed that the extracellular domains of the mGluR family are related to bacterial periplasmic proteins [25]. Models of protein folding based on this similarity yielded a predicted agonist binding site that includes amino acid residues Ser-165 and Thr-188. Site-directed mutagenisis to replace Ser-165, Thr-188 or both residues with Ala has been performed to test whether these residues are indeed important for agonist binding. The affinities for L-glutamate and quisqualate were markedly reduced when either residue was changed alone, and when both residues were changed the affinities were reduced at least 10000-fold. These results suggest that, similar to bacterial periplasmic proteins, agonist binding to mGluRs is stabilized by hydrogen bonding, and that Ser-165 and Thr-188 are involved in this process [25].

The structure and post-translational processing of mGluR1a have been analysed by *in vitro* cell-free translation, protease protection and deglycosylation [26]. In the rabbit-reticulocyte translation system, mGluR1a yielded a predominant 142000-$M_r$ polypeptide, which was processed to an apparent $M_r$ of 147000 in the presence of dog pancreatic microsomes due to N-linked glycosylation. Proteinase K treatment resulted in a protected fragment of approx. $M_r$ 92000, while a C-terminal deletion resulted in almost complete protection from protease action. Consistent with the hydrophobicity plots from the deduced amino acid sequences, these data show that the N-terminus of mGluR1a is translocated into the lumen of the endoplasmic reticulum and will consequently be located extracellularly when targeted to the plasma membrane [26].

## mGluR gene deletion studies

A major advance for the study of the roles of a wide range of proteins, including receptors, has been the development of the molecular biological techniques required to produce transgenic animals. Transgenic mice lacking the mGluR1 gene have been generated by two separate groups, one based in Europe [27] the other in the U.S.A. [28,29], and overall there is good agreement between the two groups over the effects of mGluR1 knockout.

The European group inserted a *lac*Z/neo[r] expression unit into the mGluR1 gene locus in-frame with the coding sequence at the level of the second intracellular loop. Since all seven transmembrane domains and the intracellular regions of the receptor are required for function, *lac*Z insertion destroyed mGluR1 activity. Furthermore, *lac*Z allowed β-galactosidase activity to be used as a marker for gene expression [27]. The mutant mice had motor co-ordination and spatial learning deficits, but showed no major anatomical changes. They were found to have normal basic excitatory synaptic transmission responses in both the hippocampus and the cerebellum. The mutants did, however, display impaired long-term depression in the cerebellum, and long term potentiation (LTP) in the hippocampal mossy fibres, but not CA1, was also reduced [27].

The U.S. group independently generated a strain of mutant mice with a deletion in the mGluR1 gene [28]. These workers also found that the gross anatomy of the hippocampus, synaptic transmission, long-term depression and short-term potentiation in the CA1 region were all unaffected in the mutants compared with wild-types. However, in contrast with the European group, they did find significant reductions in LTP in the CA1, and some specific aspects of associative learning were impaired. While the reasons for the differences in the impairment of LTP in CA1 need to be resolved, it is clear from the results of both groups that mGluR1 may modulate synaptic plasticity [28].

The mGluR1 gene deletion had most profound effects in the cerebellum although, here again, no gross anatomical changes were observed [29]. The mutant mice showed typical cerebellar motor impairment characteristics such as ataxic gait and intention tremor. In addition, although excitatory transmission from the climbing fibres or the parallel fibres to the Purkinje cells was present, long-term depression was severely deficient and a learned motor response (conditioned eyeblink) was also impaired [29].

The field of transgenic animals and gene targeting as applied to neuroscience is expanding rapidly, with many proteins representing interesting candidates for deletion. Transgenic animals will provide a particularly valuable model for the study of synaptic plasticity and it is only a matter of time before more mutants become available that lack other individual mGluRs, and possibly even lack multiple subtypes of mGluRs. Such animals will undoubtedly give considerable insight into the physiological roles of these receptors.

# Pharmacology of mGluRs

L-Glutamate is believed to be the endogenous agonist for all ionotropic glutamate receptors and mGluRs. The flexible structure of the molecule, which allows it to adopt a variety of low-energy conformations, probably accounts for its non-selective agonist activity. L-Aspartate is another endogenous acidic amino acid that has been proposed as a neurotransmitter in certain synapses. L-Aspartate is a potent agonist at NMDA receptors [8] and displays weak activity at mGluRs [30]. Similarly, quisqualate stimulates both ionotropic (AMPA-type) glutamate receptors and mGluRs. While quisqualate is relatively potent at the group 1 mGluRs compared with group 2 and 3 mGluRs, it is by no means specific. Furthermore, its potent action at AMPA receptors greatly limits its use as a pharmacological tool.

Ibotenate was one of the first agonists of mGluR-evoked PtdIns hydrolysis in brain slices to be recognized [1,31]. Unfortunately, ibotenate is also a potent agonist at NMDA receptors. Its ionotropic activity is blocked by NMDA receptor antagonists such as MK-801, but its metabotropic activity is not. Ibotenate shows no selectivity between the mGluR subtypes and stimulates a wide range of metabotropic responses [7,30,32–34]. However, despite being a well characterized full agonist at mGluRs, its use is compromised by its quite low potency, lack of subtype selectivity and potent action at NMDA receptors.

*trans*-ACPD, a conformationally restricted analogue of L-glutamate, was the first mGluR-selective agonist to be described. Four stereoisomers of ACPD exist, and historically the 1*S*,3*R* and 1*R*,3*S* isomers were called *trans*-ACPD and the 1*S*,3*S* and 1*R*,3*R* isomers were called *cis*-ACPD. Racemic mixtures of the 1*S*,3*R* and 1*R*,3*S* isomers were termed (±)-*trans*-ACPD, and those of the 1*S*,3*S* and 1*R*,3*R* isomers (±)-*cis*-ACPD. Due to the confusion caused by this somewhat complex nomenclature, it is now conventional to use only the *S* and *R* designations and not the *cis* and *trans* terminology [35]. 1*S*,3*R*-ACPD activates both group 1 and group 2/3 mGluRs with micromolar potencies, but has no effects on ionotropic glutamate receptors [7].

Members of a new family of antagonists based on phenylglycines have recently been synthesized and are currently under investigation. These compounds show considerable promise as the first potentially specific mGluR- and mGluR-subtype-specific antagonists. The chemistry and pharmacology of these substances has been reviewed [36]. The agonist and antagonist activities of a series of phenylglycine derivatives have been tested on the signal transduction of representative mGluR1, mGluR2 and mGluR4 subtypes expressed individually in CHO cells [37].

Ten phenylglycine derivatives were examined, including (*S*)- and (*R*)-forms of 3-hydroxyphenylglycine (3HPG), 4-carboxyphenylglycine (4CPG), 4-carboxy-3-hydroxyphenylglycine (4C3HPG), 3-carboxy-4-hydroxyphenylglycine (3C4HPG) and (+)- and (−)-α-methyl-4-carboxyphenylglycine (MCPG or αM4CPG). (*S*)-3HPG acted as an agonist for mGluR1, while (*S*)-4C3HPG, (*S*)-3C4HPG and (*S*)-4CPG served as effective agonists for mGluR2. The rank order of agonist potencies for mGluR2 was L-glutamate > (*S*)-4C3HPG > (*S*)-3C4HPG > (*S*)-4CPG, while the compounds showed no agonist activity on either mGluR1 or mGluR2. (*S*)-4C3HPG, (*S*)-3C4HPG, (*S*)-4CPG and (+)-MCPG effectively antagonized the action of L-glutamate on mGluR1. The rank order of antagonism was (*S*)-4C3HPG ≥ (*S*)-4CPG ≥ (+)-MCPG > (*S*)-3C4HPG. (+)-MCPG was also an effective competitive antagonist for mGluR2. None of the phenylglycine derivatives tested showed any agonist or antagonist activity for mGluR4. Thus (*S*)-4C3HPG, (*S*)-4CPG and (*S*)-3C4HPG act both as agonists at mGluR2 and as antagonists at mGluR1. Significantly, however, (+)-MCPG is a pure antagonist for mGluR1 and mGluR2.

In the neonatal rat spinal cord, two other phenylglycine derivatives have been shown to display marked selectivity for presynaptic responses [38]. The presynaptic depressant action of L-AP4 can be distinguished from the similar action of 1*S*,3*R*-ACPD and (2*S*,3*S*,4*S*)-α-(carboxycyclopropyl)glycine (L-CCG-I) and from the post-synaptic effects of 1*S*,3*R*-ACPD by the novel antagonists α-methyl-L-AP4 and α-methyl-L-CCG-I. This degree of selectivity represents a considerable improvement over previous-generation antagonists such as (+)-MCPG, and neither antagonist had much effect on ionotropic glutamate receptor responses. α-Methyl-L-AP4 is a potent and selective antagonist for presynaptic receptors preferentially activated by L-AP4, while α-methyl-L-CCG-I is a relatively selective antagonist for different presynaptic receptors which are preferentially activated by 1*S*,3*R*-ACPD.

The presynaptic effects of 1*S*,3*R*-ACPD and L-CCG-I in the neonatal rat cord are likely to be mediated by mGluR2, since the pharmacology matches well with that observed in CHO cells expressing only mGluR2 [37]. However, L-AP4 is not an agonist at mGluR2, so the L-AP4-evoked presynaptic depression observed in the neonatal rat cord cannot be mediated by that subtype [23]. Thus it seems that at least two presynaptic receptors are involved, one of which is mGluR2. Since α-methyl-L-CCG-I blocks 1*S*,3*R*-ACPD- and L-CCG-I-evoked, but not L-AP4-evoked, presynaptic depressions, this compound appears to be a selective antagonist for mGluR2 [38].

The receptor mediating the L-AP4-evoked presynaptic depression remains unclear, but it is unlikely to be either of the L-AP4-sensitive subtypes mGluR4 or mGluR6. In contrast with the response to L-AP4 in the neonatal spinal cord, mGluR4 is insensitive to (+)-MCPG [37], while mGluR6 is localized exclusively in the retina [39]. It is possible that the response to L-AP4 is mediated by the recently described L-AP4-sensitive mGluR7, but it has been suggested that inhibition of adenylate cyclase may not be the predominant transduction mechanism for this receptor in neurons [40]. In any event, since the receptor that mediates L-AP4-evoked presynaptic depression is potently and selectively blocked by α-methyl-L-AP4, the tools are now available for its detailed characterization.

**Table 3.**      **Distribution of mGluR1–mGluR5 mRNAs in rat brain**

The greater the number of ticks, the higher the level of gene expression. The data are taken from the review by Gallagher and co-workers [50].

| | mGluR1 | mGluR2 | mGluR3 | mGluR4 | mGluR5 |
|---|---|---|---|---|---|
| Cortex | ✔ | ✔✔ | ✔✔ | – | ✔✔✔ |
| Cerebellum | | | | | |
|   Purkinje cells | ✔✔✔✔ | – | – | – | – |
|   Granule cells | ✔ | – | – | ✔✔✔✔ | – |
|   Golgi cells | ✔✔ | ✔✔✔✔ | ✔✔ | – | ✔ |
| Hippocampus | | | | | |
|   CA1 pyramidal cells | ✔ | – | – | – | ✔✔✔ |
|   CA3 pyramidal cells | ✔✔✔ | – | – | – | ✔✔✔ |
| Dentate gyrus | | | | | |
|   Granule cells | ✔✔ | ✔✔ | ✔ | ✔ | ✔✔✔ |
|   CA4 pyramidal cells | ✔✔✔ | – | – | ✔ | ✔✔✔ |
| Basal ganglia | | | | | |
|   Striatum | ✔✔ | ✔ | ✔ | – | ✔✔✔ |
|   Nucleus accumbens | ✔ | ✔ | ✔ | – | ✔✔✔ |
| Amygdala | | | | | |
|   Basolateral | ✔ | ✔✔✔ | ✔✔ | – | – |
|   Medial | ✔ | ✔✔✔ | – | – | – |
|   Central | ✔ | – | – | – | – |
| Main olfactory bulb | | | | | |
|   Mitral cells | ✔✔✔ | – | – | – | – |
|   Tufted cells | ✔✔✔ | – | – | – | – |
|   Interior granule cells | ✔ | ✔ | – | ✔✔✔ | ✔✔✔ |
| Accessory olfactory bulb | | | | | |
|   Mitral cells | ✔✔✔ | ✔✔✔ | – | – | – |
|   Granule cells | ✔ | ✔✔ | – | – | – |

# Distribution of mGluRs

## Gene expression

The localization of mRNAs encoding the different mGluR subtypes has been investigated using *in situ* hybridization, and each subtype displays distinct gene expression patterns in rat brain (Table 3).

## Group 1 mGluRs

mRNAs encoding mGluR1s are prevalent in cerebellar Purkinje cells, in the granule and CA3 pyramidal cells of the hippocampus, in the mitral and tufted cells of the olfactory bulb and in the thalamus [41,42]. The splice variant mGluR1c mRNAs are generally less abundant than the mRNAs coding for mGluR1a and mGluR1b but, in contrast with mGluR1a and mGluR1b, the mRNAs for mGluR1c are more highly expressed in cerebellar granule cells than in Purkinje cells. This finding indicates that the splicing of mGluR1 is not identical in all neuronal cell types [14].

The distribution of mGluR5 gene expression is anatomically distinct from that of mGluR1. Thus in the cerebellum there are low levels of mRNA encoding mGluR5 but high levels encoding mGluR1; the converse is observed in the striatum. The mRNAs encoding mGluR5 and mGluR1 also display markedly different cellular distributions in the hippocampus [16].

## Group 2 and group 3 mGluRs

mGluR2 mRNA is highly expressed in the olfactory bulb and cerebral cortex, with lower levels of expression in the striatum and cerebellar cortex [15]. mGluR3 mRNA is prevalent in the cerebral cortex, the striatum and the granule cells of the hippocampus [22]. mGluR4 gene expression is distributed widely throughout the rat brain, but is most highly expressed in cerebellar granule cells [22].

## Radioligand binding

Because of the lack of specific radioligands of sufficient affinity, direct autoradiography or membrane binding studies are technically complex and somewhat difficult to interpret. However, there have been reports using quisqualate-sensitive L-[$^3$H]glutamate binding in the presence of AMPA and NMDA to block ionotropic receptor binding [43–45]. These studies have not, as yet, been able to address the distributions of individual mGluR subtypes.

## Immunohistochemistry

The subcellular distribution of mGluR1a has been investigated in the hippocampus and cerebellum at the electron microscopic level using a highly sensitive immunogold labelling technique [46]. It was shown that mGluR1a is located preferentially in areas immediately adjacent to the post-synaptic density. In the climbing fibres of the cerebellum, double-labelling techniques have indicated that, whereas AMPA receptors are located in the post-synaptic density directly beneath the presynaptic terminal, mGluR1a subunits are at the edges around the post-synaptic density [47]. These observations suggest that AMPA receptors may be stimulated by every release of L-glutamate from the presynaptic terminal, whereas mGluR1a may only be activated under conditions which produce a very strong neurotransmitter release. At present the physiological implications of such a system remain unclear.

The localization of mGluR5 has been investigated using an antibody raised against a fusion protein containing the C-terminal sequence [48]. High levels of immunoreactivity were detected in the olfactory bulb, cerebral cortex, hippocampus, septum and nucleus accumbens, consistent with the *in situ* hybridization data for mGluR5 mRNAs. Electron microscopic analysis of the striatum showed the mGluR5 immunoreactivity to be located at asymmetrical synapses, suggesting a predominantly post-synaptic localization [48].

## Physiological roles of mGluRs

### Role of mGluRs in synaptic plasticity
A widely studied form of synaptic plasticity is the phenomenon of LTP in the rat hippocampus. LTP is a sustained increase in the efficiency of a synapse, usually brought about by high-frequency electrical stimulation, which is believed to be the cellular process underlying learning and memory [49].

### Post-synaptic mechanisms
The induction of LTP requires a rise in the intracellular concentration of calcium in the post-synaptic neuron. This can be achieved by either (1) the influx of extracellular calcium via NMDA receptors or voltage-dependent $Ca^{2+}$ channels, or (2) the mobilization of intracellular calcium stores, for example by the production of Ins$P_3$ via the activation of phospholipase C-coupled (group 1) mGluRs [50].

Co-application of the mGluR agonist 1$S$,3$R$-ACPD with sub-threshold tetanic stimulation of the Schaeffer collateral fibres has been reported to generate LTP, an effect which was blocked by NMDA receptor antagonists or protein kinase C inhibitors [51,52]. Furthermore, it has been shown that the specific mGluR antagonist MCPG can completely inhibit the induction of LTP by tetanic stimulation [53]. It is well established that, under most experimental conditions, the synaptic activation of NMDA receptors is required for the induction of LTP in the CA1 region of the hippocampus [49], and more recently it has been suggested that activation of mGluRs is also needed for the induction of LTP [54,55]. However, the role of mGluRs appears to be fundamentally different from that of NMDA receptors. In contrast with NMDA receptors, which need to be triggered each time a tetanus is delivered to induce LTP, it has been proposed that an initial stimulation of mGluRs activates a molecular switch (possibly a phosphorylation event) which then negates the need for subsequent mGluR stimulation during the induction of LTP [56].

### Presynaptic mechanisms
A contribution of presynaptic mechanisms to the induction of LTP has also been proposed. Since the activation of presynaptic protein kinase C with phorbol esters enhances glutamate release from nerve terminals [57], one scenario involves presynaptic phospholipase C-coupled mGluRs [58]. It is interesting that, while activation of a presynaptic receptor has been shown to enhance L-glutamate exocytosis in cerebrocortical nerve terminals, this potentiation was observed only in the presence of arachidonic acid [58]. This is significant because it has been suggested that arachidonic acid is released into the synaptic cleft from the post-

synaptic cell when both ionotropic glutamate receptors and mGluRs are stimulated [59]. Thus arachidonic acid may provide positive feedback on glutamate exocytosis via presynaptic mGluRs.

## mGluRs in the retina

A role for mGluRs has been well established in the retina, a structure that has a very well characterized neuronal circuitry. Briefly, the photoreceptor cells (rods and cones) synapse on retinal bipolar cells, which are essentially interneurons, and these, in turn, synapse on retinal ganglion cells, which exit the retina and carry information to the visual regions of the brain via the optic nerve. Photoreceptor cells release L-glutamate tonically, since they are maintained in a depolarized state in the absence of light by the constitutive activation of a cGMP-gated cation channel. Closing these channels causes hyperpolarization and thus a decrease in L-glutamate release.

Photoreceptor cells synapse on to two major classes of bipolar cells: ON and OFF. Thus the net effect of photon activation of the rods and cones is depolarization of the OFF cells and hyperpolarization of the ON cells in response to the released L-glutamate. The opposite effects of L-glutamate on the bipolar cells can be explained by the glutamate receptors they possess. The main receptor type on OFF cells is an ionotropic AMPA/kainate subtype [60,61], whereas the main receptor type on ON cells is an L-AP4-sensitive mGluR, activation of which inhibits the cGMP-activated current via a pertussis-toxin-sensitive mechanism [62,63].

Screening of a rat retina cDNA library yielded mGluR6, which is potently activated by L-AP4 and L-serine-O-phosphate [61,64]. Furthermore, it is exclusively expressed in the retina, specifically in the inner nuclear layer where the ON cells are located. These data provide compelling evidence that mGluR6 is the receptor that mediates the responses of ON cells, and represent perhaps the first example of a defined physiological process being accounted for by an individual mGluR subtype.

## mGluRs in the accessory olfactory bulb

Another system where an mGluR has been shown to modulate synaptic transmission is in the accessory olfactory bulb of the rat [65]. Mitral cells receive inputs from the vomeronasal nerve and project excitatory outputs. The mitral cells have dentrodentritic synapses with granule cells at which L-glutamate, released from the mitral cells, depolarizes the granule cells, which in turn release γ-aminobutyric acid (GABA) to inhibit the mitral cells. The granule cells synapse on a population of mitral cells, some of which will be depolarized but most of which will not be excited. However, following stimulation by the excited mitral cells, the granule cells release GABA at all synapses, thereby inhibiting the responsiveness of the stimulated and unstimulated mitral cells.

*In situ* hybridization and immunohistochemical evidence has suggested that mGluR2 is highly expressed in the granule cells. Therefore Nakanishi and co-workers utilized 2-(2,3-dicarboxycyclopropyl)glycine, which they identified as a selective and potent mGluR2 agonist, to study the effects of mGluR2 activation of GABA transmission in the accessory olfactory bulb [65]. They found that L-glutamate release from the mitral cell activates presynaptically located mGluR2 on the granule cell, and thereby relieves GABA inhibition. Since the mGluR2 activation is localized to the synapses at which the mitral cell releases L-glutamate, the GABA inhibition of neighbouring mitral cells would be unaffected.

## Pathophysiological roles of mGluRs

L-Glutamate has been proposed as a crucial mediator in a range of neurodegenerative disorders, including both Huntington's disease and Alzheimer's disease [66,67]. While the exact roles of L-glutamate in chronic diseases remain difficult to define, its involvement in acute neuronal damage following trauma and ischaemic insults is becoming increasingly well understood [68,69]. The combination of excitatory and toxic effects that can be evoked by L-glutamate stimulation of neurons gave rise to the term 'excitotoxicity' [70–72].

Elucidation of the precise roles of mGluRs as opposed to ionotropic receptors in excitotoxicity has been severely hampered by the lack of suitable pharmacological probes. However, the relatively recent and continuing development of increasingly selective mGluR agonists and antagonists is beginning to allow progress to made in this area.

## Concluding remarks

There is no doubt that great strides have been made over the last decade in defining the structures, pharmacology and physiological roles of mGluRs. The molecular biology in particular has advanced at an impressive pace, due in large part to the efforts of the group of Professor S. Nakanishi at Kyoto University. It is likely that major future developments will depend heavily on the development of novel subtype-specific agonists and antagonists. As outlined in this chapter, these new tools are now being developed, notably by Jeff Watkins, David Jane and co-workers at the University of Bristol, and as they become widely available the growth of the mGluR field will continue to yield important information regarding the role of these receptors in normal and pathophysiological conditions.

We are grateful to the Medical Research Council and the Wellcome Trust for financial support.

## References


1. Nicoletti, F., Meek, J.L., Iadarola, M., Chuang, D.M., Roth, B.L. and Costa, E. (1986) J. Neurochem. **6**, 40–46
2. Recasens, M., Sassetti, I., Nourigat, A., Sladeczek, F. and Bockaert, J. (1987) Eur. J. Pharmacol. **414**, 87–93
3. Sladeczek, F., Recasens, M. and Bockaert, J. (1988) Trends Neurosci. **11**, 545–549
4. Sladeczek, F., Pin, J.P., Recasens, M., Bockaert, J. and Weiss, S. (1985) Nature (London) **317**, 717–719
5. Sugiyama, H., Ito, I. and Hirono, C. (1987) Nature (London) **325**, 531–533
6. Sugiyama, H., Ito, I. and Watanabe, M. (1989) Neuron **3**, 129–132
7. Schoepp, D.D. and Conn, P.J. (1993) Trends Pharmacol. Sci. **14**, 13–20
8. Watkins, J.C., Krosgaard-Larsen, P. and Honoré, T. (1990) Trends Pharmacol. **11**, 25–33
9. Palmer, E., Monaghan, D.T. and Cotman, C.W. (1988) Mol. Brain Res. **4**, 161–165
10. Recasens, M., Guiramand, J. and Vignes, M. (1991) Neurochem. Res. **16**, 659–668
11. Conn, P.J. and Patel, J. (1994) The Receptors, Humana Press, Totowa, NJ
12. Houamed, K.M., Kuijper, J.L., Gilbert, T.L., Haldeman, B.A., Ohara, P.J., Mulvihill, E.R., Almers, W. and Hagen, F.S. (1991) Science **252**, 1318–1321
13. Masu, M., Tanabe, Y., Tsuchida, K., Shigemoto, R. and Nakanishi, S. (1991) Nature (London) **349**, 760–765
14. Pin, J.P., Waeber, C., Prezeau, L., Bockaert, J. and Heinemann, S.F. (1992) Proc. Natl. Acad. Sci. U.S.A. **89**, 10331–10335
15. Tanabe, Y., Masu, M., Ishii, T., Shigemoto, R. and Nakanishi, S. (1992) Neuron **8**, 169–179
16. Abe, T., Sugihara, H., Nawa, H., Shigemoto, R., Mizuno, N. and Nakanishi, S. (1992) J. Biol. Chem. **267**, 13361–13368
17. Okamoto, N., Hori, S., Akazawa, C., Hayashi, Y., Shigemoto, R., Mizuno, N. and Nakanishi, S. (1994) J. Biol. Chem. **269**, 1231–1236
18. Nakanishi, S. (1994) Neuron **13**, 1031–1037
19. Suzdak, P.D., Thomsen, C., Mulvihill, E. and Kristensen, P. (1994) in The Metabotropic Glutamate Receptors (Conn, P.J. and Patel, J., eds.), pp. 1–30, Humana Press, Totowa, NJ
20. Aramori, I. and Nakanishi, S. (1992) Neuron **8**, 757–765
21. Thomsen, C., Mulvihill, E.R., Haldeman, B., Pickering, D.S., Hampson, D.R. and Suzdak, P.D. (1993) Brain Res. **619**, 22–28
22. Tanabe, Y., Nomura, A., Masu, M., Shigemoto, R., Mizuno, N. and Nakanishi, S. (1993) J. Neurosci. **13**, 1372–1378
23. Nakanishi, S. (1992) Science **258**, 597–603
24. Barnes, J.M. and Henley, J.M. (1992) Prog. Neurobiol. **39**, 113–133
25. O'Hara, P.J., Sheppard, P.O., Thogersen, H., Venezia, D., Haldeman, B.A., Mcgrane, V., Houamed, K.M., Thomsen, C., Gilbert, T.L. and Mulvihill, E.R. (1993) Neuron **11**, 41–52
26. Seal, A.J., Collingridge, G.L. and Henley, J.M. (1994) Neuropharmacology **33**, 1065–1070
27. Conquet, F., Bashir, Z.I., Davies, C.H., Daniel, H., Ferraguti, F., Bordi, F., Franzbacon, K., Reggiani, A., Matarese, V., Conde, F., Collingridge, G.L. and Crepel, F. (1994) Nature (London) **372**, 237–243
28. Aiba, A., Chen, C., Herrup, K., Rosenmund, C., Stevens, C.F. and Tonegawa, S. (1994) Cell **79**, 365–375
29. Aiba, A., Kano, M., Chen, C., Stanton, M.E., Fox, G.D., Herrup, K., Zwingman, T.A. and Tonegawa, S. (1994) Cell **79**, 377–388
30. Schoepp, D.D. and Johnson, B.G. (1993) Neurochem. Int. **22**, 277–283
31. Nicoletti, F., Iaddarola, M.J., Wroblewski, J.T. and Costa, E. (1986) Proc. Natl. Acad. Sci. U.S.A. **83**, 1931–1936
32. Schoepp, D.D. (1993) Biochem. Soc. Trans. **21**, 97–102
33. Winder, D.G. and Conn, P.J. (1992) J. Neurochem. **59**, 375–378
34. Boss, V. and Conn, P.J. (1992) J. Neurochem. **59**, 2340–2343
35. Schoepp, D., Bockaert, J. and Sladeczek, F. (1990) Trends Pharmacol. Sci. **11**, 508–515
36. Watkins, J. and Collingridge, G. (1994) Trends Pharmacol. Sci. **15**, 333–342
37. Hayashi, Y., Sekiyama, N., Nakanishi, S., Jane, D.E., Sunter, D.C., Birse, E.F., Udvarhelyi, P.M. and Watkins, J.C. (1994) J. Neurosci. **14**, 3370–3377

38. Jane, D.E., Jones, P.L.S.J., Pook, P.C.-K., Tse, H.-W. and Watkins, J.C. (1994) Br. J. Pharmacol. **112**, 809–816

39. Nakanishi, S., Masu, M., Bessho, Y., Nakajima, Y., Hayashi, Y., Nomura, A. and Shigemoto, R. (1994) Neuropsychopharmacology **10**, S8–S10

40. Saugstad, J.A., Kinzie, J.M., Mulvihill, E.R., Segerson, T.P. and Westbrook, G.L. (1994) Mol. Pharmacol. **45**, 367–372

41. Shigemoto, R., Nakanishi, S. and Mizuno, N. (1992) J. Comp. Neurol. **322**, 121–135

42. Condorelli, D.F., Dellalbani, P., Amico, C., Casabona, G., Genazzani, A.A., Sortino, M.A. and Nicoletti, F. (1992) Mol. Pharmacol. **41**, 660–664

43. Catania, M.V., Desocarraz, H., Penney, J.B. and Young, A.B. (1994) Mol. Pharmacol. **45**, 626–636

44. Catania, M.V., Hollingsworth, Z., Penney, J.B. and Young, A.B. (1993) Neuroreport **4**, 311–313

45. Wright, R.A., McDonald, J.W. and Schoepp, D.D. (1994) J. Neurochem. **63**, 938–945

46. Baude, A., Nusser, Z., Roberts, J.D.B., Mulvihill, E., McIlhinney, R.A.J. and Somogyi, P. (1993) Neuron **11**, 771–787

47. Nusser, Z., Mulvihill, E., Streit, P. and Somogyi, P. (1994) Neuroscience **61**, 421–427

48. Shigemoto, R., Nomura, S., Ohishi, H., Sugihara, H., Nakanishi, S. and Mizuno, N. (1993) Neurosci. Lett. **163**, 53–57

49. Bliss, T. and Collingridge, G. (1993) Nature (London) **361**, 31–39

50. Gallagher, J.P., Zheng, F. and Shinnick-Gallagher, P. (1994) in The Metabotropic Glutamate Receptors. (Conn, P.J and Patel, J., eds.), pp. 173–194, Humana Press, Totowa, NJ

51. McGuinness, N., Anwyl, R. and Rowan, M. (1991) Eur. J. Pharmacol. **197**, 231–232

52. Benari, Y., Aniksztejn, L. and Bregestovski, P. (1992) Trends Neurosci. **15**, 333–339

53. Bashir, Z.I., Bortolotto, Z.A., Davies, C.H., Berretta, N., Irving, A.J., Seal, A.J., Henley, J.M., Jane, D.E., Watkins, J.C. and Collingridge, G.L. (1993) Nature (London) **363**, 347–350

54. Bortolotto, Z.A. and Collingridge, G.L. (1992) Eur. J. Pharmacol. **214**, 297–298

55. Bashir, Z.I., Jane, D.E., Sunter, D.C., Watkins, J.C. and Collingridge, G.L. (1993) Eur. J. Pharmacol. **239**, 265–266

56. Bortolotto, Z.A., Bashir, Z.I., Davies, C.H. and Collingridge, G.L. (1994) Nature (London) **368**, 740–743

57. Barrie, A.P., Nicholls, D.G., Sanchezprieto, J. and Sihra, T.S. (1991) J. Neurochem. **57**, 1398–1404

58. Herrero, I., Mirasportugal, M.T. and Sanchezprieto, J. (1992) Nature (London) **360**, 163–166

59. Dumuis, A., Oomagari, K., Pin, J.-P., Sebben, M. and Bockaert, J. (1990) Nature (London) **347**, 182–184

60. Bloomfield, S.A. and Dowling, J.E. (1985) J. Neurophysiol. **53**, 699–713

61. Conn, P.J., Winder, D.G. and Gereau, R.W. (1994) in The Metabotropic Glutamate Receptors (Conn, P.J and Patel, J., eds.), pp. 195–229, Humana Press, Totowa, NJ

62. Nawy, S. and Jahr, C.E. (1991) Neuron **7**, 677–683

63. Shiells, R. and Falk, G. (1992) NeuroReport **3**, 845–848

64. Nakajima, Y., Iwakabe, H., Akazawa, C., Nawa, H., Shigemoto, R., Mizuno, N. and Nakanishi, S. (1993) J. Biol. Chem. **268**, 11868–11873

65. Hayashi, Y., Momiyama, A., Takahashi, T., Ohishi, H., Ogawameguro, R., Shigemoto, R., Mizuno, N. and Nakanishi, S. (1993) Nature (London) **366**, 687–690

66. Choi, D.W. (1992) Science **258**, 241–243

67. Zorumski, C.F. and Olney, J.W. (1993) Pharmacol. Ther. **59**, 145–162

68. Meldrum, B.S. (1994) Neurology **44**, 14–23

69. Choi, D.W. (1992) J. Neurobiol. **23**, 1261–1276

70. Schousboe, A., Frandsen, A. and Krogsgaardlarsen, P. (1992) Cell. Biol. Toxicol. **8**, 93–100

71. Frandsen, A., Schousboe, A. and Griffiths, R. (1993) J. Neurosci. Res. **34**, 331–339

72. Frandsen, A. and Schousboe, A. (1993) J. Neurochem. **60**, 1202–1211

# Non-*N*-methyl-D-aspartate (NMDA) glutamate receptors: molecular properties

**Robert J. Wenthold* and Ronald S. Petralia**

Section on Neurotransmitter Receptor Biology, Laboratory of Neurochemistry, NIDCD, NIH, Bethesda, MD 20892, U.S.A.

## Introduction

Glutamate, or a related molecule, is the neurotransmitter at most excitatory synapses in the mammalian brain. There is extensive evidence that glutamate and glutamate receptors are involved in learning, memory, synapse development and other plastic changes in the central nervous system (Mayer and Westbrook, 1987; Monaghan et al., 1989; Collingridge and Singer, 1990; Watkins et al., 1990; Malenka and Nicoll, 1993). Glutamate, and several analogues such as kainic acid, α amino 3 hydroxy-5-methyl-4-isoxazolepropionate (AMPA) and *N*-methyl-D-aspartate (NMDA), are also potent neurotoxins, and abnormalities in the regulation of glutamate and glutamate receptors may underlie some human neurodegenerative diseases such as Huntington's disease and Alzheimer's disease (Choi and Rothman, 1990; Meldrum and Garthwaite, 1990; Choi, 1992; Appel, 1993). There is also promising evidence that blockers of the action of glutamate may be used therapeutically to prevent neuron degeneration due to stroke. The wide range of glutamate's action and its involvement in essentially every neuronal system have made the study of glutamate and glutamate receptors one of the most active and important fields in neurobiology today.

The action of glutamate is mediated by three distinct ionotropic receptor families, the AMPA, kainate and NMDA receptors, and by eight metabotropic receptors. As discussed below, the subunits comprising these receptors, as well as additional 'orphan receptors', have been identified through molecular cloning. While the glutamate receptor field is complex and most questions concerning these receptors are only beginning to be addressed, the status of glutamate receptors is now considerably more settled than it was only a few years ago, when there was legitimate doubt as to whether or not glutamate was actually a neurotransmitter. The first reports of the actions of glutamate on neurons appeared more than 35

**To whom correspondence should be addressed.*

years ago (Hayashi, 1954; Curtis et al., 1959; see Watkins, 1986, for a review). However, the facts that glutamate is ubiquitous in the central nervous system, that it lacks an enzymic inactivation system and that it excites almost every neuron to which it is applied were used as an argument against its role as a neurotransmitter. It is now clear that these early results simply reflected the widespread nature of glutamatergic synapses and the fact that essentially every neuron in the brain expresses at least one type of glutamate receptor.

The status of the glutamate receptor field today is owed largely to the pioneering efforts of a few physiologists who persisted in studies to identify selective agonists and antagonists in order to begin to characterize the glutamate-induced responses of neurons. NMDA, developed in 1962 (Watkins, 1962) as a more potent excitant than glutamate, was the critical compound in defining the NMDA subclass of glutamate receptor. Other key molecules in defining the ionotropic glutamate receptors include ibotenic acid (Johnston et al., 1968), kainic acid (Shinozaki and Konishi, 1970), the NMDA antagonists 2-amino-5-phospho-nopentanoate (AP5) and 2-amino-7-phosphonoheptanoate (AP7) (Davies et al., 1981; Evans et al., 1982; Perkins and Stone, 1992), AMPA (Krogsgaard-Larsen et al., 1980) and the non-NMDA antagonists 6-cyano-7-nitroquinoxaline-2,3-dione and 6,7-dinitroquinoxaline-2,3-dione (Honoré et al., 1988). The identification of metabotropic glutamate receptors was first reported in 1985 by Sladeczek et al., who showed the stimulation of phosphoinositide hydrolysis in cultured neurons by glutamate and quisqualate. Although a repertoire of selective and potent agonists and antagonists of metabotropic receptors remains to be developed, this area has benefitted from molecular cloning studies that have identified eight distinct receptors, many of which are expressed in alternatively spliced forms.

The development of radioactive ligands which bind selectively to glutamate receptors has also been important to our understanding of these receptors. The first binding studies were reported by Roberts (1974) and used [³H]glutamate as a ligand. The lack of selective competitors with which to displace the radioactive ligand, and the likelihood that glutamate is also binding to molecules other than receptors, made the interpretation of these early studies problematic. However, more selective ligands have proven specific for receptor subtypes, including [³H]kainate, first reported by Simon et al. (1976), [³H]AMPA (Honoré et al., 1982) and ligands for the NMDA receptor, including [³H]AP7 (Ferkany and Coyle, 1983), [³H]AP5 (Olverman et al., 1984) and [³H]MK-801 (Wong et al., 1986). [³H]Glutamate has also proven to be a useful ligand when used with selective blockers of receptor subtypes.

# Cloning of glutamate receptors

## Cloning strategies

Essentially all of our knowledge about the molecular properties of glutamate receptors, as well as a great deal of their physiological and pharmacological properties, is based on the molecular cloning of the cDNAs of the receptor subunits. The first three members of the glutamate receptor family were cloned in 1989 using different experimental approaches and reported simultaneously (Gregor et al., 1989; Hollmann et al., 1989; Wada et al., 1989). The cloned subunits were related but sufficiently dissimilar to portend the diversity of this receptor family that would soon be demonstrated as other members were cloned. The first member of the AMPA family (GluR1) was cloned by functional expression in oocytes from cDNA libraries of rat brain (Hollmann et al., 1989). This approach was favoured because of the difficulties in obtaining purified receptor from mammalian brain using biochemical approaches. The other two studies used more traditional biochemical approaches in which the proteins were purified and partially sequenced, and cDNA libraries were probed with oligonucleotides based on these sequences. These studies, however, took advantage of the high levels of expression of kainate binding proteins (KBPs) in the brains of some lower vertebrates, including frogs, fish and birds (London et al., 1980). These proteins are so named because they bind [$^3$H]kainate, in some cases with a pharmacology indistinguishable from that of mammalian brain (Hampson and Wenthold, 1988). One KBP was purified from frog brain using affinity chromatography with immobilized domoic acid, a potent glutamate receptor agonist (Wada et al., 1989). In bird cerebellum a low-affinity KBP is present at very high levels, such that traditional biochemical techniques were sufficient for purification (Gregor et al., 1989). Both KBPs have a size and hydrophobicity pattern similar to those of previously cloned ligand-gated ion channels, including the nicotinic acetylcholine receptor, the γ-aminobutyric acid (GABA)$_A$ receptor and the glycine receptor. Although they bind [$^3$H]kainate, the KBPs do not form functional ion channels when expressed in oocytes. On the other hand, with a calculated $M_r$ of 99800, the AMPA receptor is about twice the size of other ligand-gated ion channels; the larger size of the AMPA receptor is due to a longer N-terminal portion of the molecule. It does, however, have a hydrophobicity pattern like that of other receptors and of the KBPs. While the KBPs are about half the size of the AMPA receptor, the sequence identity for the conserved region is remarkably high considering the differences in species. The amino acid identity is nearly 40% between the KBPs from frog and chick and the AMPA receptor from rat. The sequences of these three molecules served as the basis for the cloning of the additional members of the non-NMDA receptor family (Keinänen et al., 1990). While the KBPs played an important role in the cloning of the first glutamate receptors and continue to add information to the glutamate receptor family as additional molecules are cloned, their function remains unknown. The lack of channel function may simply reflect the requirement for

**Fig. 1.**           **Mammalian excitatory amino acid (EAA) receptors**

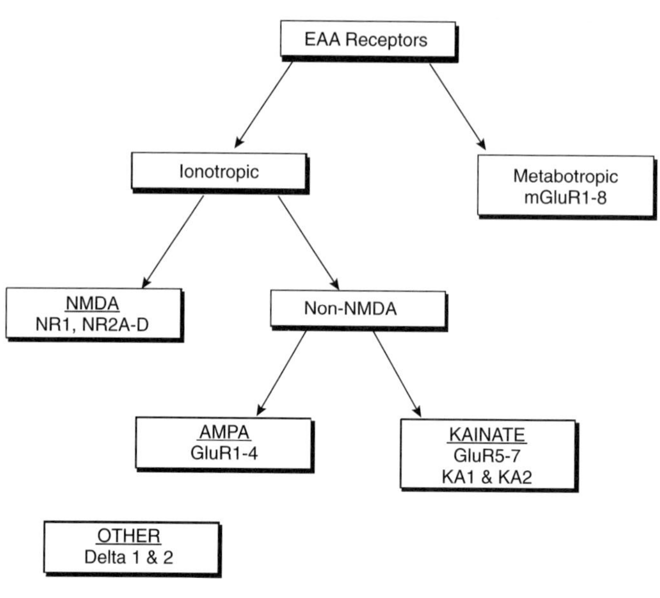



additional subunits, as is the case for some mammalian kainate receptors. Since this chapter addresses mammalian receptors, the KBPs will not be discussed except in reference to mammalian glutamate receptors. Other members of the glutamate receptor family have also been cloned from non-mammals, including two KBPs from goldfish brain (Wo and Oswald, 1994), one KBP from duck brain (Kimura et al., 1993), two glutamate receptors from *Drosophila* (Schuster et al., 1991; Ultsch et al., 1992) and one glutamate receptor from the pond snail (Hutton et al., 1991).

The members of the mammalian glutamate receptor family that have been cloned are shown in Fig. 1. This includes four members of the AMPA receptor family (GluR1–GluR4 or GluRA–GluRD) and five members of the kainate receptor family (GluR5–GluR7 and KA1/KA2). In addition, there are two δ subunits; neither is active, so it is not yet known if they are non-NMDA receptors, NMDA receptors or some other receptor. However, their identity is greatest to the non-NMDA receptors. Although the receptors from rat have been the most thoroughly studied, homologous receptors have been cloned from human and mouse. In the mouse, for which many of the NMDA and non-NMDA receptors have now been cloned, a different nomenclature has been introduced where the following are mouse equivalents: α1, GluR1; α2, GluR2; β2, GluR6; γ2, KA2 (Sakamura et al., 1990, 1992; Morita et al., 1992). The δ subunits cloned from both rat and mouse have the same nomenclature. The amino acid identities among the non-NMDA receptor subunits are shown in Table 1. In addition to displaying

**Table 1.     Amino acid sequence identity of non-NMDA glutamate receptors**

Values are percentage amino acid identity for rat receptors. Values in parentheses in the first column refer to identities of the subunits within the family. Data from the following: Bettler et al. (1990, 1992); Keinänen et al. (1990); Egebjerg et al. (1991); Moriyoshi et al. (1991); Herb et al. (1992); Katsuwada et al. (1992); Sommer and Seeburg (1992); Ishii et al. (1993); Lomeli et al. (1993); Hollmann and Heinemann (1994).

| | Identity (%) | | | | |
| | Kainate | | NMDA | | |
| | GluR5–GluR7 | KA1 and KA2 | NR1 | NR2A–NR2D | $\delta$1 and $\delta$2 |
|---|---|---|---|---|---|
| AMPA GluR1–GluR4 (68–73%) | 39–42 | 35–37 | 25–27 | 21–26 | 28–30 |
| Kainate GluR5–GluR7 (74–81%) | | 42–45 | 27–29 | 22–26 | 29–32 |
| Kainate KA1 and KA2 (68%) | | | 26–27 | 22–27 | 25–28 |
| NMDA NR1 | | | | 25–27 | 25–28 |
| NMDA NR2A–NR2D (42–56%) | | | | | 21–25 |
| $\delta$1 and $\delta$2 (56%) | | | | | |

significant sequence identity, the non-NMDA receptor subunits are similar in size (approx. $M_r$ 100000), each has multiple consensus glycosylation and phosphorylation sites, and all have similar locations of hydrophobic stretches which are putative transmembrane domains. Because of the extensive similarities, much of the data on one subunit, such as structure and subunit organization, can be tentatively applied to the other subunits. In the original models of the glutamate receptor, four transmembrane domains were proposed (TM1–TM4). Thus subsequent literature has used these as landmarks for noting positions on the receptor molecule. While it now appears that there are only three transmembrane domains, we will use the original designation of TM1–TM4 in reference to the AMPA, kainate and $\delta$ receptor subunits.

## Alternative splicing and RNA editing

Several of the non-NMDA receptor subunits exist in alternatively spliced variants. Most notable are the 'flip' and 'flop' forms of the AMPA receptors, which refer to alternatively spliced segments of 38 amino acids on the N-terminal side of TM4 (Sommer et al., 1990). Both forms are widely distributed throughout the brain, but the flip and flop variants are differentially expressed in brain and have different developmental profiles (Sommer et al., 1990; Monyer et al., 1991). The ligand binding properties of the two splice variants are the same, but glutamate is 4–5 times more effective in activating flip channels than flop channels. An additional splice variant of GluR4, which produces a novel C-terminus and shorter molecule, has been described (Gallo et al., 1992). For GluR5, five splice variants, affecting both the C- and N-termini, have been described in the rat (Bettler et al., 1990; Sommer et al., 1992), and additional variants have been described in mouse and human (Gregor et al., 1993). One splice variant of GluR6 has been reported in the mouse (Gregor et al., 1993).

Several glutamate receptor subunits have been found to be modified through RNA editing. This was first discovered for GluR2 (Sommer et al., 1991) when it was found that cDNAs encoding the subunit contained arginine at position 586, while glutamine was encoded in the genes for all four AMPA receptor subunits. This position, which is in TM2, was found to be critical in controlling the calcium permeability of the channel. Editing is complete in the adult rat brain, while low levels of the unedited form are found in fetal brain (Burnashev et al., 1992). In human brain a significant amount of GluR2 may be unedited; for example, in the substantia nigra only 72% is edited (Nutt and Kamboj, 1994). A second site of editing of the AMPA subunits was recently reported (Lomeli et al., 1994). This site is located immediately before the alternatively spliced flip or flop modules and involves conversion of an arginine into a glycine. Editing occurs in all AMPA subunits except GluR1, and the extent of editing is different for the different subunits and the flip and flop forms. Editing is developmentally regulated, with the adult forms more fully edited than those of younger animals. Functionally, the two forms differ in their desensitization rate and time course of recovery from desensitization. GluR5 and GluR6 are also edited at the same site as GluR2, but the extent of editing is only 40% for GluR5 and 80% for GluR6 (Sommer et al., 1991). GluR6 is also edited at two additional sites which are in the TM1 region (Köhler et al., 1993); this editing converts an isoleucine into valine and a tyrosine into cysteine. As is the case for the glutamine/arginine site of GluR5 and GluR6, the sites in the TM1 region are not fully edited. (For a more detailed discussion of editing, see Sommer, 1997.)

# Functional properties of cloned glutamate receptors

## AMPA receptors

Each AMPA receptor subunit forms functional ion channels when expressed alone in oocytes or transfected cells (Hollmann et al., 1989; Keinänen et al., 1990; Boulter et al., 1990). AMPA and kainate are both potent agonists of AMPA receptors, but differ in their desensitization properties, with AMPA causing rapid desensitization and kainate being non-desensitizing. Cyclothiazide selectively blocks desensitization of AMPA receptors (Partin et al., 1993). Expression of combinations of receptor subunits, which presumably form heteromeric receptor complexes, leads to more potent responses. Among the AMPA receptor subunits, GluR2 stands out because of its dominant effect on co-expressed subunits with respect to the calcium permeability of the ion channel and the properties of the current/voltage ($I/V$) relationship (Hollmann et al., 1991). Channels formed without GluR2 are calcium-permeable and have inwardly rectifying $I/V$ curves, while those with GluR2 are not calcium-permeable and have linear or outwardly rectifying $I/V$ curves. This unique property of GluR2 was shown to be due to a single amino acid difference in TM2, which is an arginine in GluR2 and a glutamine in GluR1, GluR3 and GluR4 (Hollmann et al., 1991; Hume et al., 1991; Verdoorn et al., 1991). As noted above, the amino acid difference in GluR2 is not due to a difference in the DNA, but arises through RNA editing (Sommer et al., 1991). The AMPA receptor subunits are abundantly expressed throughout the brain and appear to account for most of the fast glutamate-gated ion channels. The AMPA receptors appear to account for most or all of the [³H]AMPA binding activity found in brain, although this has not been thoroughly explored. For intact brain membranes or detergent-solubilized membranes, [³H]AMPA binds with at least two affinities: a high-affinity site with a $K_d$ of approx. 10 nM, with some values reported to be as high as 75 nM, and one or more low-affinity sites with a $K_d$ of about 500 nM, with reported values ranging from 200 to 3900 nM (Hunter et al., 1990; Hunter and Wenthold, 1992). Homomeric AMPA receptors contain only the high-affinity [³H]AMPA binding site, with a $K_d$ of 11 nM for GluR2 expressed in HEK293 cells and a $K_d$ of 30 nM for α1 (the mouse equivalent of GluR1) in baculovirus-infected Sf21 cells (Kawamoto et al., 1994, 1995). In the baculovirus-infected cells, [³H]AMPA binding is activated by KSCN, similar to that in rat brain.

Using antibodies selective for GluR1–GluR4, [³H]AMPA binding activity can be immunoprecipitated from detergent-solubilized rat brain (Wenthold et al., 1990, 1992), while antibodies to kainate receptor or δ2 receptor subunits fail to immunoprecipitate [³H]AMPA binding (Wenthold et al., 1994; May at et al., 1995). More than half of the AMPA binding activity is immunoprecipitated with an antibody that recognizes both GluR2 and GluR3. Scatchard analysis of the binding activity immunoprecipitated with anti-GluR1 antibodies shows both high- and low-affinity sites (Wenthold et al., 1990), indicating that heteromeric receptor complexes in brain contain both high- and low-affinity binding sites. The lack of the low-affinity site in the transfected or infected cells may be due to the fact that a maximum AMPA concentration of 100 nM was used for the transfected HEK293 cells and a maximum of 200 nM was used for the infected Sf21 cells; under these conditions a low-affinity site may not be detected. Alternatively, these results may suggest that only heteromeric complexes have low-affinity sites, or that additional modifications or proteins, that are not present in 293 or Sf21 cells, are required to produce the low-affinity site that is seen in brain. For example, glycosylation may be different in expression systems and neurons, and it has been shown that glycosylation is essential for [³H]AMPA binding to AMPA receptor subunits (Kawamoto et al., 1995). Furthermore, GluR3 or GluR4, neither of which has been studied extensively with respect to ligand binding properties, may be responsible for the low-affinity binding of native AMPA receptors.

## Kainate receptors

Of the kainate receptor subunits, only GluR5 and GluR6 form functional homomeric channels when expressed *in vitro* (Bettler et al., 1990; Egebjerg et al., 1991). Both are most potently activated by kainate and domoate, and respond only weakly (GluR5), or not at all (GluR6), to AMPA. Kainate-evoked responses are rapidly desensitized, and this desensitization is blocked by concanavalin A (Partin et al., 1993). While the other kainate receptor subunits (GluR7, KA1 and KA2) are not functional when expressed alone, they do combine with GluR5 and GluR6 to form functional ion channels, in some cases with unique properties (Werner et al., 1991; Bettler et al., 1992; Herb et al., 1992; Sakamura et al., 1992). For example, channels comprising GluR6 and KA2 respond to AMPA, whereas homomeric GluR6 channels do not respond to AMPA, and KA2 is not functional (Herb et al., 1992). In brain, kainate receptor complexes are presumably heteromers made up of different combinations of subunits, as discussed below.

Although they are widely distributed, the kainate receptors are less abundant than AMPA receptors in brain. Comparison of the $B_{max}$ values for binding of [³H]AMPA and [³H]kainate to Triton X-100-solubilized rat brain membranes shows more than 50 times more [³H]AMPA binding sites (Hampson et al., 1987; Hunter et al., 1990). Functional kainate receptors have been identified in only a few cases, including the dorsal root ganglia and a population of hippocampal neurons (Huettner, 1990; Lerma et al., 1993). The kainate receptor subunits, GluR5–GluR7, KA1 and KA2, may account for most or all of the [³H]kainate binding activity in brain. Most studies of [³H]kainate binding to intact or detergent-solubilized brain membranes show both high- and low-affinity binding sites, with $K_d$ values of approx. 5 and 50 nM respectively (Hampson et al., 1987). Based on [³H]kainate binding to transfected cell membranes, the kainate receptor subunits can be divided into high-affinity receptors, i.e. KA1 and KA2, with $K_d$ values of 2.6–15 nM, and low-affinity receptors, i.e. GluR5–GluR7, with $K_d$ values of 36–95 nM (Table 2). Therefore, in brain, it is likely that the high- and low-affinity binding sites arise from complexes containing various combinations of high- and low-affinity

**Table 2.** **Ligand binding properties of AMPA and kainate receptor subunits**

All subunits are cloned from rat, except EAA1 and EAA2 which are from human. Binding data for receptor subunits were obtained from cultured cells transfected with the cDNA. Rat brain data are from studies with membrane fractions. $K_d$ values were obtained using [3H]kainate and [3H]AMPA for kainate and AMPA receptors, respectively. Abbreviations: DA, domoic acid; KA, kainic acid; Glu, glutamic acid; QA, quisqualic acid; IP, immunoprecipitated.

| Subunit | $K_d$ (nM) | Displacement order | Reference |
|---|---|---|---|
| Kainate receptors | | | |
| GluR5-2a(R) | 67 | DA > QA = Glu ≫ AMPA | Sommer et al. (1992) |
| GluR5-2b(R) | 73 | DA > QA = Glu ≫ AMPA | Lomeli et al. (1992) |
| GluR6 | 36 | DA > QA > Glu ≫ AMPA | Lomeli et al. (1992) |
| GluR7 | 77 | DA > Glu > QA | Bettler et al. (1992) |
| GluR7 | 63 | DA ≫ Glu > QA ≫ AMPA | Lomeli et al. (1992) |
| KA1 | 4.7 | KA > QA > Glu ≫ AMPA | Werner et al. (1991) |
| KA1 (EAA1) | 2.3 | KA > QA > DA > Glu > AMPA | Kamboj et al. (1994) |
| KA2 | 15.2 | QA > DA = Glu ≫ AMPA | Herb et al. (1992) |
| KA2 (EAA2) | 2.6 | KA > QA > DA > Glu > AMPA | Kamboj et al. (1992) |
| GluR5/KA2 | 90 | | Herb et al. (1992) |
| Rat brain | 5.5, 50 | DA ≫ QA > Glu | Hampson et al. (1987) |
| | | | |
| AMPA receptors | | | |
| GluR1 | 30 | QA > AMPA > Glu > KA | Kawamoto et al (1994) |
| GluR2 | 12 | QA > Glu > KA | Keinänen et al. (1990) |
| GluR1 (IP, rat brain) | 4.6, 323 | QA > Glu > KA | Wenthold et al. (1990) |
| Rat brain | 5.4, 624 | QA > AMPA > Glu > KA | Hunter et al. (1990) |

subunits. Whether or not the high- and low-affinity sites are associated with the same molecular complex remains to be determined. Immunoprecipitation of detergent-solubilized brain membranes using antibodies to GluR6 and KA2 immunoprecipitated [³H]kainate binding, whereas antibodies to the AMPA receptor subunits or to δ2 did not (Wenthold et al., 1994; Mayat et al., 1995).

### δ **receptors**

The δ receptor subunits, δ1 and δ2, do not form functional homomeric channels when expressed *in vitro* (Yamazaki et al., 1992; Araki et al., 1993; Lomeli et al., 1993). Co-expression of the two subunits also fails to produce a functional channel, and co-expression of the δ subunits with other glutamate receptor subunits does not alter the properties of AMPA, kainate or NMDA receptor subunits, suggesting that δ subunits do not form functional complexes with them. Therefore inclusion of the δ subunits in the glutamate receptor family is based only on their sequence similarity with glutamate receptor subunits. The δ1 subunit is expressed at low levels in the adult brain, and is somewhat more abundant in the developing animal (Lomeli et al., 1993). On the other hand, δ2 is heavily expressed in cerebellar Purkinje neurons (Araki et al., 1993; Lomeli et al., 1993; Mayat et al., 1995). The very distinct developmental and distributional properties of δ1 and δ2 suggest they are not normally associated together in the same molecular complex. Antibodies to the δ2 subunit did not co-immunopreciptitate AMPA, kainate, NMDA or mGluR1α receptor subunits from detergent-solubilized cerebellum, indicating that δ subunits are not associated with these receptors (Mayat et al., 1995). Antibodies to δ2 also did not immunoprecipitate [³H]AMPA, [³H]kainate or [³H]glutamate binding activity from detergent-solubilized rat cerebellum, suggesting that the δ subunits do not bind these ligands under conditions in which binding to other glutamate receptors has been demonstrated.

# Structure and topology of glutamate receptors

## Topology

In order to understand the mechanisms regulating functions ranging from agonist interaction to subunit assembly, knowledge of receptor structure is essential. Although the amino acid sequences of the KBPs and glutamate receptor subunits are not similar to those of the previously cloned ligand-gated ion channels, i.e. the nicotinic acetylcholine, $GABA_A$ and glycine receptors, they resemble these proteins in the number and general spacing of the hydrophobic domains that are candidates for transmembrane segments. Thus the first three reports of the cloning of glutamate receptors proposed a model of receptor topology with four transmembrane segments, although there was initial disagreement on the assignment of one of these segments (Gregor et al., 1989; Hollmann et al., 1989; Wada et al., 1989). With such a model, the N- and C-termini are both extracellular (Fig. 2a). However, experimental results from a variety of studies soon challenged such a model. First, immunocytochemistry using antibodies directed to the C-terminus of AMPA receptors showed a reaction product on the cytoplasmic side of the membrane at the post-synaptic density, rather than in the synaptic cleft as would be expected if the C-terminus is extracellular. This type of labelling pattern was consistent for all the AMPA subunits (Petralia and Wenthold, 1992; Eshhar et al., 1993; Molnár et al., 1993), as well as for subunits of kainate (Petralia et al., 1994a), NMDA (Petralia et al., 1994b) and δ (Mayat et al., 1995) receptors, as shown later. A second point arguing against a four transmembrane model was the finding that a splice variant of the NMDA receptor was shown to be phosphorylated at a site near the C-terminus, which would support an intracellular location for the C-terminus (Tingley et al., 1993). Finally it was shown that GluR6 and KA2, of the kainate family, contain consensus glycosylation sites between the third and fourth transmembrane segments, which would be on the cytoplasmic side of the membrane with a four transmembrane model. Mutational analyses verified that this site on GluR6 was glycosylated in the functional receptor (Roche et al., 1994; Taverna et al., 1994), supporting an extracellular location for this part of the molecule.

Recent studies have addressed the topology issue more directly. In the goldfish brain a KBP is expressed which has four consensus glycosylation sites between TM3 and TM4, of which two are glycosylated. Since deleting the second transmembrane region did not change the glycosylation pattern of the receptor, the

## Fig. 2.      Topology of glutamate receptors

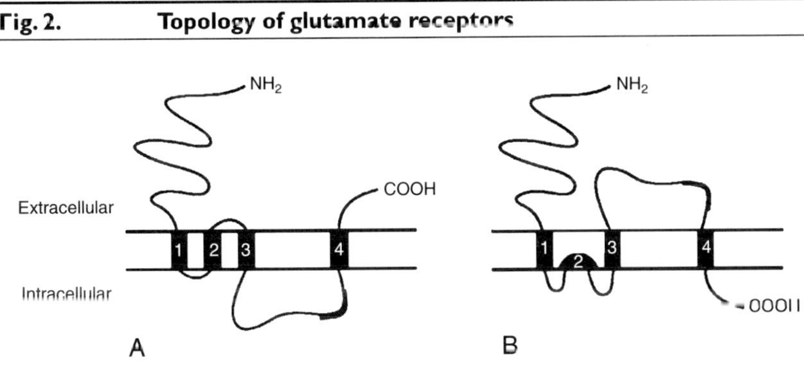


*(a) Original model with four transmembrane segments proposed for AMPA receptors and KBPs (Gregor et al., 1989; Hollmann et al., 1989; Wada et al., 1989; Keinänen et al., 1990). While the number of proposed transmembrane segments was the same, the location of the sites was different in the model proposed by Hollmann et al. (1989). (b) The current model is based on data from Hollmann et al. (1994) and Wo and Oswald (1994). In this model, TM2 does not pass through the membrane but either loops in and out of the membrane or does not enter the membrane at all. The transmembrane regions are numbered 1–4. The bold section of line preceding TM4 indicates the flip/flop alternative splice region.*

authors proposed a three transmembrane model for the KBP (Wo and Oswald, 1994). A more extensive analysis was carried out for the GluR1 subunit of the AMPA receptor by Hollmann et al. (1994), who added and deleted glycosylation sites throughout the molecule. These results also supported a three transmembrane structure in which the second transmembrane segment of the former four transmembrane model does not span the membrane, but either loops into the membrane or does not associate with the membrane at all. In this case the loop between TM3 and TM4 is entirely extracellular. A similar model was proposed based on studies of GluR3 in which an epitope was inserted into the molecule and its position determined with antibodies (Bennett and Dingledine, 1995).

The strongest evidence arguing against the three transmembrane model arises from studies on GluR6 which showed a functional phosphorylation site in the area between TM3 and TM4 (Raymond et al., 1993; Wang et al., 1993), which would imply that this area is intracellular; a three transmembrane model would place this site extracellularly. Given the sequence similarities and the fact that functional chimaeras can be created between the AMPA subunits and GluR6, it is unlikely that these subunits have different topologies. There are several other explanations (Hollmann et al., 1994) for this discrepancy which must be addressed experimentally. In summary, while some additional questions remain, the existing data indicate that glutamate receptors have a topology different from that of other ligand-gated ion channels.

## Ligand binding domain

With a model of the glutamate receptor in which the molecule spans the membrane three times, one or both of the two extracellular regions, the N-terminus to TM1 and the sequence between TM3 and TM4, are expected to make up the ligand binding domain. Nakanishi et al. (1990) first proposed that the area immediately preceding TM1 of the AMPA receptor was involved in ligand binding, based on its sequence similarity with the glutamine binding subunit of glutamine permease of *Escherichia coli* and the conservation of sequence in this region with the KBPs. Uchino et al. (1992) tested the effects of altering individual amino acids using site-directed mutagenesis in a region on the N-terminal side of TM1 (spanning about 125 amino acids) of the mouse $\alpha$1 subunit (analogous to rat GluR1). In the mutated receptors, changes ranging from a total loss of receptor function to slight changes in agonist affinity were obtained. Stern-Bach et al. (1994) created a series of GluR3 and GluR6 chimaeras to study the agonist binding site of the receptor. Since AMPA and kainate receptors differ in their responses to several agonists and have distinctly different binding characteristics for [³H]AMPA and [³H]kainate, the functional contributions of each of the molecules could be determined. This study identified both the region of 150 amino acids preceding TM1 and the extracellular loop between TM3 and TM4 as being involved in agonist binding. The involvement of the loop between TM3 and TM4 in agonist binding, an area which also contains the flip/flop variants, is additional evidence that this region is extracellular. Based on the sequence identities between *E. coli* proteins and regions of the AMPA receptor (O'Hara et al., 1993) (the leucine/isoleucine/valine binding protein is similar to the region on the N-terminal side of TM1, and the lysine/arginine/ornithine binding protein is similar to the region between TM3 and TM4), a model for the ligand binding regions of the AMPA receptor has been proposed (Stern-Bach et al., 1994). Recently, it has been shown that the N-terminus and the TM3–4 loop of GluR4 can be fused together to form a soluble, functional, ligand-binding protein, adding further proof that these areas are responsible for ligand binding (Kuusinen et al., 1995).

## Size of the receptor complex

As discussed below, native glutamate receptors are predominantly heteromeric complexes with variable subunit stoichiometries. Again, by analogy with the other ligand-gated ion channels, it was assumed that functional glutamate receptors are pentamers. Subsequent experiments addressed the size of the AMPA receptor complex. Gel-filtration and sucrose-density centrifugation of detergent-solubilized membranes showed an $M_r$ of 425000–610000 for AMPA receptors (Hunter et al., 1990; Blackstone et al., 1992; Hunter and Wenthold, 1992) and an $M_r$ of 650000 for kainate receptors (Hampson et al., 1987). A more detailed analysis of the size of the AMPA receptor complex was carried out through chemical cross-linking studies. The synaptic membrane fraction of rat brain was cross-linked with dithiobis(succinimidylpropionate) and then analysed by gel electrophoresis, either directly or after immunoaffinity purification (Wenthold et al., 1992). Since the cross-linking was done before detergent solubilization, the loss of loosely associated proteins would be minimized. From the cross-linked membranes, three high-$M_r$ bands, identified with anti-receptor antibodies, were obtained, migrating with $M_r$ values of 330000, 470000 and 590000. With a monomer $M_r$ of 108000, the largest component would most closely correspond to a pentamer, but the limited accuracy of $M_r$ measurements of proteins of this size, as well as the non-linear structure of the cross-linked proteins, makes this conclusion somewhat tentative, and complexes of four or six subunits should also be considered. Immunoaffinity purification of the receptor failed to show additional co-purifying proteins, in addition to GluR1–GluR4, making it unlikely that such proteins would be included in the cross-linked complexes.

## Organization of glutamate receptor subunits

### Distribution

While many of the glutamate receptor subunits are functional when expressed alone, both distribution studies and analyses of the properties of native receptors suggest that most receptor complexes in the brain are made up of two or more subunits. Of particular interest is GluR2, the subunit which controls calcium flux through the AMPA receptor channel. Neurons that have little or no GluR2 are readily permeable to calcium; the best studied examples include putative inhibitory interneurons of the hippocampus (McBain and Dingledine, 1993; Bochet et al., 1994; Koh et al., 1995; Isa et al., 1996) and neocortex (Jonas et al., 1994). Neurons that readily pass calcium through AMPA receptor channels may be especially vulnerable to excitotoxic damage, and models in which GluR2 is selectively down-regulated in neurodegenerative diseases have been proposed (Pellegrini-Giampietro et al., 1992; Pollard et al., 1993). *In situ* hybridization and immunocytochemistry show a wide range of expression of receptor subtypes and their subunits in different neurons (reviewed by Hollmann and Heinemann, 1994; Petralia and Wenthold, 1996). Some neuronal populations, such as the pyramidal neurons of the hippo-campus (Fig. 3), express high levels of most subunits, while others express relatively few. Cells that express several subunits of a glutamate receptor subtype such as the AMPA receptor presumably express them together as heteromeric complexes, but it is possible that the subunit composition of functional receptor complexes differs in different parts of a neuron, as suggested for hippocampus neurons *in vitro* (also dis-cussed for Purkinje cells; Lerma et al., 1994). Based on physiological studies, these authors suggest that both calcium-impermeable AMPA receptors containing GluR2 and calcium-permeable AMPA receptors lacking GluR2 are present in the same neurons. Thus even neurons that express all four AMPA subunits may express them in different heteromeric complexes. The types of complexes formed in a neuron were investigated in detail for the CA1/CA2 pyramidal neurons (Wenthold et al., 1996). These neurons express moderate to high levels of GluR1, GluR2 and GluR3. Using immunoprecipitation with subunit-specific antibodies, it was shown that complexes contained GluR1 and GluR2 or GluR2 and GluR3, but very few complexes contained GluR1 and GluR3. About 10% of the receptor complexes were homomeric GluR1. These results indicate that neurons are capable of producing different receptor complexes from the same pool of subunits and that the receptor complexes may be targeted to different synaptic populations. Several neuron populations appear to express homomeric AMPA receptors. Based on distribution, the best case for possible homomeric receptors would be various neuron populations that are rich in GluR1, but express little of the other AMPA receptors. Examples include subpopulations of neurons of the neocortex, striatum, subthalamic nucleus, pedunculopontine tegmental nucleus, basal forebrain magnocellular complex and spinal cord (Fig. 4) (Petralia and Wenthold, 1992; Martin et al., 1993a,b; Tachibana et al., 1994).

## Subunit interactions

The relationship among glutamate receptor subtypes and subunits has been investigated biochemically in whole brain, in cell lines expressing several glutamate receptors and in cells transfected with receptor subunits using subunit-specific antibodies. From rat brain solubilized with Triton X-100, which does not dissociate the AMPA receptor complex and retains [$^3$H]AMPA binding activity, antibodies

**Fig. 3.** Coronal sections of the CA1 (a–c) and CA3 (d–f) regions of the hippocampus immunolabelled with antibodies to GluR1 (a, d), GluR2/3 (b, e) and GluR4 (c, f)

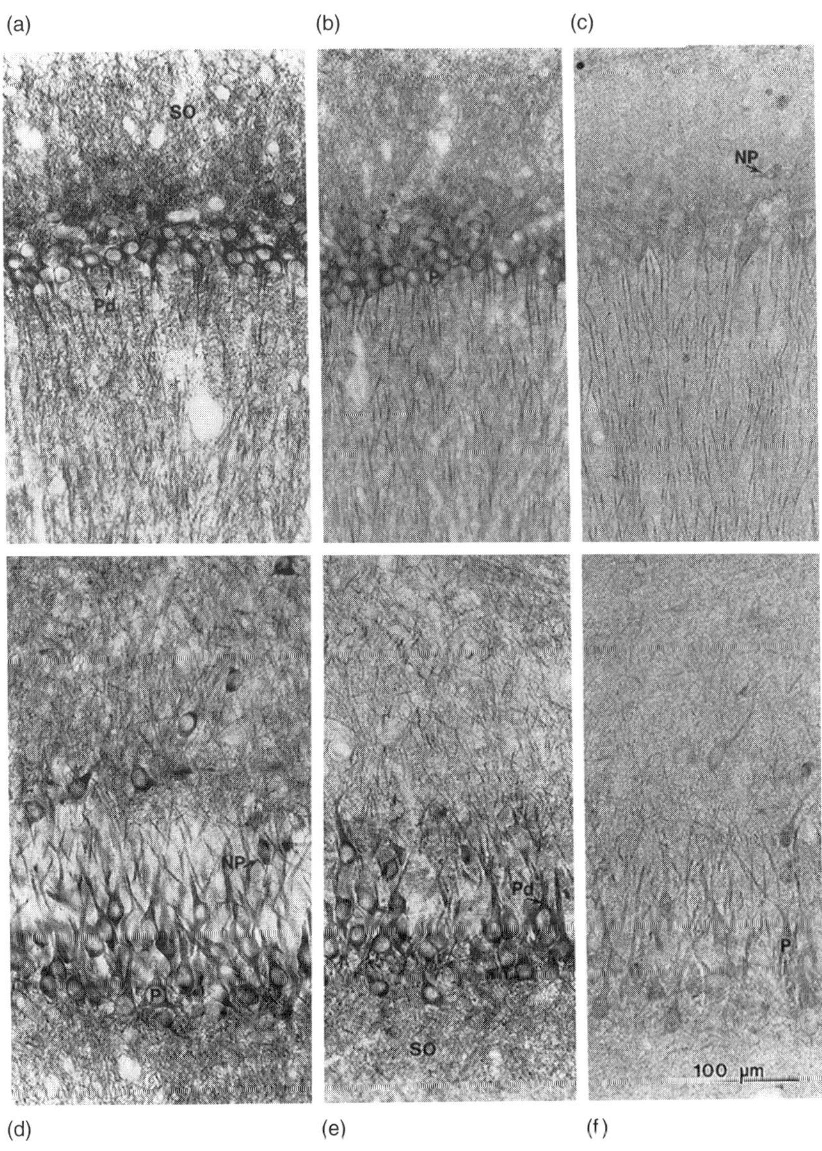



*NP, non-pyramidal cell; P, pyramidal cell; Pd, pyramidal cell apical dendrite; SO, stratum oriens. (From Petralia and Wenthold, 1992.)*

**Fig. 4.**         **Lamina X and surrounding area at the lumbar level of transverse sections of the spinal cord, immunolabelled with antibodies to GluR1 (a, b), GluR2/3 (c) and GluR4 (d)**

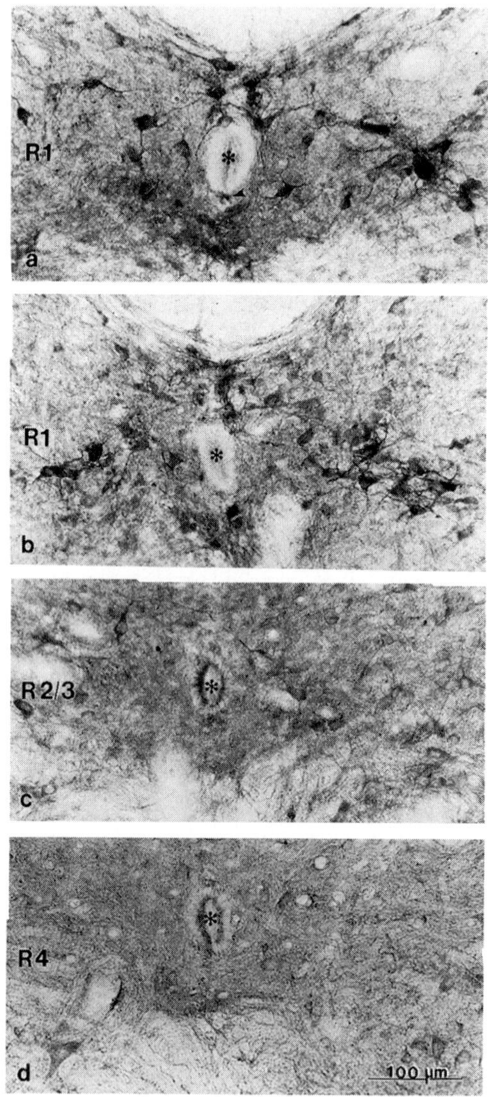



*In (**a**), the arrowhead indicates a distinctive patch of processes and puncta ventral to the central canal; note the numerous 'GluR1 intense' neurons. Asterisks indicate the central canal. (From Tachibana et al., 1994.)*

                                                           

selective for any of the four AMPA receptor subunits immunoprecipitate the subunit to which it was raised as well as the remaining three subunits, showing that most of the receptor complexes are comprised of two or more subunits (Wenthold et al., 1992). A similar result was obtained using the oligodendrocyte progenitor cell line CG4, which expresses GluR2–GluR4, GluR6, GluR7, KA1 and KA2, and has the advantage over the brain in that it is a single cell type (Puchalski et al., 1994). For the kainate receptor subunits in both brain and CG4 cells, immunoprecipitation with anti-GluR6 antibodies also immunoprecipitated KA2, and vice versa. These results support the conclusions that most subunits of the AMPA and kainate families exist as heteromeric complexes.

The question of whether or not subunits of the AMPA subtype assemble with those of the kainate subtype, or other subtypes, has been addressed by immunoprecipitation and functional studies. Using brain or CG4 detergent extracts, significant assembly between AMPA and kainate subunits was not seen, and there was no physiological evidence for hybrid receptors in CG4 cells (Patneau et al., 1994; Puchalski et al., 1994). However, in HEK293 cells co-transfected with GluR1 and GluR6 or GluR2 and GluR6, immunoprecipitation studies showed a small, but reproducible, proportion of GluR1/GluR6 and GluR2/GluR6 complexes (Wenthold et al., 1994). These results may be due to the artificial conditions under which the receptors were synthesized, and such complexes may not be formed in neurons, but it leaves open the possibility that the assembly of AMPA and kainate subunits may occur under some conditions. Since the δ subunits have not been shown to be functional ion channels when expressed *in vitro*, it has been suggested that they may associate with the other receptor subtypes (AMPA, kainate and NMDA). However, immunoprecipitation using detergent-solubilized cerebellum, where δ2 is most abundantly expressed, did not show any complexes with other receptor subtypes (Mayat et al., 1995).


## Distribution of receptors within a neuron

Although mRNA encoding glutamate receptors has been detected in dendrites, most is found in neuronal cell bodies (Craig et al., 1993; Eshhar et al., 1993). Therefore glutamate receptor synthesis occurs in the cell body, and receptor complexes presumably are formed in the endoplasmic reticulum, incorporated into the membrane of vesicles and transported along cytoskeletal pathways to post-synaptic sites, as described for other membrane proteins (e.g. reviews by Rose and Doms, 1988; Hurtley and Helenius, 1989; Atkinson et al., 1992). Ultrastructural analysis of glutamate receptor distribution shows intense post-synaptic labelling (Fig. 5) for all receptor subtypes, including metabotropic receptors (e.g. Martin et al., 1992; Petralia and Wenthold, 1992; reviews by Hollmann and Heinemann, 1994; Petralia and Wenthold, 1996; Petralia, 1997). However, in many cases there is also a surprisingly high level of cytoplasmic staining, indicating a large pool of intracellular receptors (Hampson et al., 1992; Petralia and Wenthold, 1992; Huntley et al., 1993). These analyses were carried out using horseradish peroxidase detection, and an argument could be made for diffusion of the reaction product as the cause of the cytoplasmic staining; however, light microscopy using fluorescent antibodies also showed a large cytoplasmic pool of receptor (Eshhar et al., 1993). Cytoplasmic localizations of glutamate receptors have also been shown with immunogold (Baude et al., 1993; Vidnyánszky et al., 1996). The most reasonable explanation for the cytoplasmic staining is that it represents receptor being transported to or from the synapse, as discussed by others for glutamate receptors (Petralia and Wenthold, 1992; Huntley et al., 1993; Martin et al., 1993a; Petralia, 1997), KBP (Somogyi et al., 1990) and other receptors [e.g. β-adrenergic (Aoki et al., 1987); GABA$_A$/benzodiazepine (Juiz et al., 1989)]. However, in many cases the relative amount of glutamate receptor staining seems to be far greater in some neurons than in others. For example, cytoplasmic staining can be extremely dense in hippocampal neurons stained with anti-GluR1 antibody (Petralia and Wenthold, 1992) compared with GluR4 and δ (Mayat et al., 1995). Examples of low levels of cytoplasmic staining have been noted for other receptors, such as those for GABA (Juiz et al., 1989) and glycine (Wenthold et al., 1988); the glycine receptor has essentially no cytoplasmic staining evident in the somas of ventral cochlear nucleus neurons, although light staining is seen in the cytoplasm of some dendrites. Somogyi et al. (1989) found substantial differences in intracellular staining levels for GABA$_A$/benzodiazepine receptors in different cell types of the cerebellum, and suggested that this reflects differences in the turnover of receptor complexes.

Alternatively, a relatively large pool of receptor in the cytoplasm, especially that adjacent to the synapse, may serve as a reserve, allowing rapid up-regulation of the synaptic receptor pool. If the number of glutamate receptors at a selected population of synapses is indeed subject to up- and down-regulation (Didier et al., 1994; Trevisan et al., 1994), this would be accomplished most efficiently by a local mechanism. In contrast, a mechanism involving changes in mRNA translation (e.g. Sucher et al., 1993) would have several limitations. Since the points of synthesis and utilization can be relatively far apart, significant delays would occur during retrograde transport of a signal from the affected synapse to the soma, and then during the synthesis and transport of the new receptor molecules. Secondly, such a change would be global and affect the receptor concentration throughout the neuron, when changes at a single population of synapses is all that is required. Finally, since a relatively large pool of receptors appears to exist in neurons at steady state, relatively modest changes in receptor concentration may require synthesis of a large number of molecules. Therefore two general mechanisms of receptor regulation may exist in neurons. One would regulate the total number of receptors in the neuron at the level of transcription and translation, and would be used mostly during events affecting the entire neuron, such as development. The second would regulate the number of synaptic receptors using a reserve intracellular receptor pool. Such a reserve pool of receptor molecules has

**Fig. 5.**     **Electron micrograph of cerebral cortex immunolabelled with antibodies to GluR1**

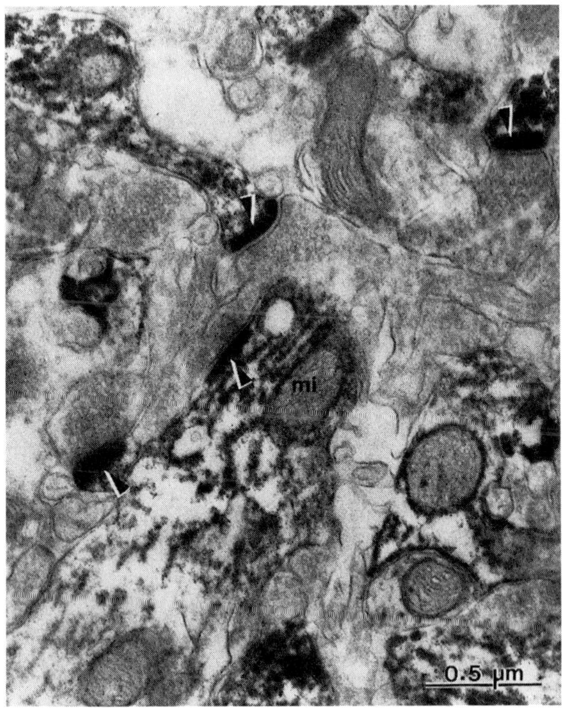

*mi, mitochondrion in labelled dendrite; the arrowheads indicate stained synaptic densities. (From Petralia and Wenthold, 1992.)*

been proposed to explain short-term post-synaptic changes during long-term facilitation in *Aplysia* (Trudeau and Castellucci, 1995).

## Summary

In the past 7 years, molecular cloning studies have formed the basis of significant advances in the glutamate receptor field by identifying the major subunits comprising non-NMDA and NMDA receptors. These studies have led to the functional characterization of the subunits using *in vitro* expression systems, an analysis of their distributions with *in situ* hybridization and immunocytochemistry, and insight into their structures. These studies have also revealed a complex system of regulation and have identified several points where receptor expression is controlled, including transcription, RNA editing, translation, assembly, intracellular targeting and the receptor complex itself, where modifications such as phosphorylation may affect receptor activity. Among the wide range of future research subjects, the mechanisms that control these various points of regulation will be of key importance, as not only will they identify the mechanisms of regulation under normal conditions, but they will also help to elucidate the role of glutamate receptors in the many neurological disorders to which they have been linked.

## References


Aoki, C., Joh, T.H. and Pickel, V.M. (1987) Brain Res. **437**, 264–282

Appel, S.H. (1993) Trends Neurosci. **16**, 3–5

Araki, K., Meguro, H., Kushiya, E., Takayama, C., Inoue, Y. and Mishina, M. (1993) Biochem. Biophys. Res. Commun. **197**, 1267–1276

Atkinson, S.J., Doberstein, S.K. and Pollard, T.D. (1992) Curr. Biol. **2**, 326–328

Baude, A., Nusser, Z., Roberts, J.D.B., Mulvihill, E., McIlhinney, R.A.J. and Somogyi, P. (1993) Neuron **11**, 771–787

Bennett, J.A. and Dingledine, R. (1995) Neuron **14**, 373–384

Bettler, B., Boulter, J., Hermans-Borgmeyer, I., O'Shea-Greenfield, A., Deneris, E.S., Moll, C., Borgmeyer, U., Hollmann, M. and Heinemann, S. (1990) Neuron **5**, 583–595

Bettler, B., Egebjerg, J., Sharma, G., Pecht, G., Hermans-Borgmeyer, I., Moll, C., Stevens, C.F. and Heinemann, S. (1992) Neuron **8**, 257–265

Blackstone, C.D., Moss, S.J., Martin, L.J., Levey, A.I., Price, D.L. and Huganir, R.L. (1992) J. Neurochem. **58**, 1118–1126

Bochet, P., Audinat, E., Lambolez, B., Crépel, F., Rossier, J., Iino, M., Tsuzuki, K. and Ozawa, S. (1994) Neuron **12**, 383–388

Boulter, J., Hollmann, M., O'Shea-Greenfield, A., Hartley, M., Deneris, E., Maron, C. and Heinemann, S. (1990) Science **249**, 1033–1037

Burnashev, N., Monyer, H., Seeburg, P.H. and Sakmann, B. (1992) Neuron **8**, 189–198

Choi, D.W. (1992) Science **258**, 241–243

Choi, D.W. and Rothman, S.M. (1990) Annu. Rev. Neurol. **13**, 171–182

Collingridge, G.L. and Singer, W. (1990) Trends Pharmacol. Sci. **11**, 290–296

Craig, A.M., Blackstone, C.D., Huganir, R.L. and Banker, G. (1993) Neuron **10**, 1055–1068

Curtis, D.R., Phillis, J.W. and Watkins, J.C. (1959) Nature (London) **183**, 161–163

Davies, J., Francis, A.A., Jones, A.W. and Watkins, J.C. (1981) Neurosci. Lett. **21**, 77–81

Didier, M., Mienville, J.-M., Soubrié, P., Bockaert, J., Berman, S., Bursztajn, S. and Pin, J.-P. (1994) Eur. J. Neurosci. **6**, 1536–1543

Egebjerg, J., Bettler, B., Hermans-Borgmeyer, I. and Heinemann, S. (1991) Nature (London) **351**, 745–748

Eshhar, N., Petralia, R.S., Winters, C.A., Niedzielski, A.S. and Wenthold, R.J. (1993) Neuroscience **57**, 943–964

Evans, R.H., Francis, A.A., Jones, A.W., Smith, D.A.S. and Watkins, J.C. (1982) Br. J. Pharmacol. **75**, 65–75

Ferkany, J.W. and Coyle, J.T. (1983) Life Sci. **33**, 1295–1305

Gallo, V., Upson, J.M., Hayes, W.P., Vyklicky, L.J., Winters, C.A. and Buonanno, A. (1992) J. Neurosci. **12**, 1010–1023

Gregor, P., Mano, I., Maoz, I., McKeown, M. and Teichberg, V.I. (1989) Nature (London) **342**, 689–692

Gregor, P., O'Hara, B.F., Yang, X. and Uhl, G.R. (1993) Neuroreport **4**, 1343–1346

Hampson, D.R. and Wenthold, R.J. (1988) J. Biol. Chem. **263**, 2500–2505

Hampson, D.R., Huie, D. and Wenthold, R.J. (1987) J. Neurochem. **49**, 1209–1215

Hampson, R., Huang, X.P., Oberdorfer, M.D., Goh, J.W., Auyeung, A. and Wenthold, R.J. (1992) Neuroscience **50**, 11–22

Hayashi, T. (1954) Keio J. Med. **3**, 183–192

Herb, A., Burnashev, N., Werner, P., Sakmann, B., Wisden, W. and Seeburg, P.H. (1992) Neuron **8**, 775–785

Hollmann, M. and Heinemann, S. (1994) Annu. Rev. Neurosci. **17**, 31–108

Hollmann, M., O'Shea-Greenfield, A., Rogers, S.W. and Heinemann, S. (1989) Nature (London) **342**, 643–648

Hollmann, M., Hartley, M. and Heinemann, S. (1991) Science **252**, 851–853

Hollmann, M., Maron, C. and Heinemann, S. (1994) Neuron **13**, 1331–1343

Honoré, T., Lauridsen, J. and Krogsgaard-Larsen, P (1982) J. Neurochem. **38**, 173–178

Honoré, T., Davies, S.N., Drejer, J., Fletcher, E.J., Jacobson, P., Lodge, D. and Nielsen, F.E. (1988) Science **241**, 701–703

Huettner, J.E. (1990) Neuron **5**, 255–266

Hume, R.I., Dingledine, R. and Heinemann, S.F. (1991) Science **253**, 1028–1031

Hunter, C. and Wenthold, R.J. (1992) J. Neurochem. **58**, 1379–1385

Hunter, C., Wheaton, K.D. and Wenthold, R.J. (1990) J. Neurochem. **54**, 118–125

Huntley, G.W., Rogers, S.W., Moran, T., Janssen, W., Archin, N., Vickers, J.C., Cauley, K., Heinemann, S.F. and Morrison, J.H. (1993) J. Neurosci. **13**, 2965–2981

Hurtley, S.M. and Helenius, A. (1989) Annu. Rev. Cell Biol. **5**, 277–307

Hutton, M.L., Harvey, R.J., Barnard, E.A. and Darlison, M.G. (1991) FEBS Lett. **292**, 111–114

Isa, T., Itazawa, S., Iino, M., Tsuzuki, K. and Ozawa, S. (1996) J. Physiol. **491**, 719–733

Ishii, T., Moriyoshi, K., Sugihara, H., Sukurada, K., Kadotani, H., Yokoi, M., Akazawa, C., Shigemoto, R., Mizuno, N., Masu, M. and Nakanishi, S. (1993) J. Biol. Chem. **268**, 2836–2843

Johnston, G.A.R., Curtis, D.R., Degroat, W.C. and Duggan, A.W. (1968) Biochem. Pharmacol. **17**, 2488–2489

Jonas, P., Racca, C., Sakmann, B., Seeburg, P.H. and Monyer, H. (1994) Neuron **12**, 1281–1289

Juiz, J.M., Helfert, R.H., Wenthold, R.J., De Blas, A.L. and Altschuler, R.A. (1989) Brain Res. **504**, 173–179

Kamboj, R.K., Schoepp, D.D., Nutt, S., Shekter, L., Korczak, B., True, R.A., Zimmerman, D.M. and Wosnick, M.A. (1992) Mol. Pharmacol. **42**, 10–15

Kamboj, R.K., Schoepp, D.D., Nutt, S., Shekter, L., Korczak, B., True, R.A., Rampersad, V., Zimmerman, D.M. and Wosnick, M.A. (1994) J. Neurochem. **62**, 1–9

Katsuwada, T., Kashiwabuchi, N., Mori, H., Sakimura, K., Kushiya, E., Araki, K., Meguro, H., Masaki, H., Kumanishi, T., Arakawa, M. and Mishina, M. (1992) Nature (London) **358**, 36–41

Kawamoto, S., Hattori, S., Oiji, I., Hamajima, K., Mishina, M. and Okuda, K. (1994) Eur. J. Biochem. **223**, 665–673

Kawamoto, S., Hattori, S., Sakimura, K., Mishina, M. and Okuda, K. (1995) J. Neurochem. **64**, 1258–1266

Keinänen, K., Wisden, W., Sommer, B., Werner, P., Herb, A., Verdoorn, T.A., Sakmann, B. and Seeburg, P.H. (1990) Science **249**, 556–560

Kimura, N., Kurosawa, N., Kondo, K. and Tsukada, Y. (1993) Mol. Brain Res. **17**, 351–355

Koh, D.-S., Geiger, J.R.P., Jonas, P. and Sakmann, B. (1995) J. Physiol. **485**, 383–402

Köhler, M., Burnashev, N., Sakmann, B. and Seeburg, P.H. (1993) Neuron **10**, 491–500

Kuusinen, A., Arvola, M. and Keinänen, K. (1995) EMBO J. **14**, 6327–6332

Krogsgaard-Larsen, P., Honoré, T., Hansen, J.J., Curtis, D.R. and Lodge, D. (1980) Nature (London) **284**, 64–66

Lerma, J., Paternain, A.V., Naranjo, J.R. and Mellström, B. (1993) Proc. Natl. Acad. Sci. U.S.A. **90**, 11688–11692

Lerma, J., Morales, M., Ibarz, J.M. and Somohano, F. (1994) Eur. J. Neurosci. **6**, 1080–1088

Lomeli, H., Wisden, W., Kohler, M., Keinanen, K., Sommer, B. and Seeburg, P.H. (1992) FEBS Lett. **307**, 139–143

Lomeli, H., Sprengel, R., Laurie, D.J., Köhr, G., Herb, A., Seeburg, P.H. and Wisden, W. (1993) FEBS Lett. **315**, 318–322

Lomeli, H., Mosbacher, J., Melcher, T., Höger, T., Geiger, J.R.P., Kuner, T., Monyer, H., Higuchi, M., Bach, A. and Seeburg, P.H. (1994) Science **266**, 1709–1713

London, E.D., Klemm, N. and Coyle, J.T. (1980) Brain Res. **192**, 463–476

Malenka, R.C. and Nicoll, R.A. (1993) Trends Neurosci. **16**, 521–527

Martin, L.J., Blackstone, C.D., Huganir, R.L. and Price, D.L. (1992) Neuron **9**, 259–270

Martin, L.J., Blackstone, C.D., Levey, A.I., Huganir, R.L. and Price, D.L. (1993a) Neuroscience **53**, 327–358

Martin, L.J., Blackstone, C.D., Levey, A.I., Huganir, R.L. and Price, D.L. (1993b) J. Neurosci. **13**, 2249–2263

Mayat, E., Petralia, R.S., Wang, Y.-X. and Wenthold, R.J. (1995) J. Neurosci. **15**, 2533–2546

Mayer, M.L. and Westbrook, G.L. (1987) Prog. Neurobiol. **28**, 197–276

McBain, C. and Dingledine, R. (1993) J. Physiol. (London) **462**, 373–392

Meldrum, B. and Garthwaite, J. (1990) Trends Pharmacol. Sci. **11**, 379–387

Molnár, E., Baude, A., Richmond, S.A., Pate., P.B., Somoygi, P. and McIlhinney, R.A.J. (1993) Neuroscience **53**, 307–326

Monaghan, D.T., Bridges, R. and Cotman, C. (1989) Annu. Rev. Pharmacol. Toxicol. **29**, 365–402

Monyer, H., Seeburg, P.H. and Wisden, W. (1991) Neuron **6**, 799–810

Morita, T., Sakamura, K., Kushiya, E., Yamazaki, M., Meguro, H., Araki, K., Abe, T., Mori, K.J. and Mishina, M. (1992) Mol. Brain Res. **14**, 143–146

Moriyoshi, K., Masu, M., Ishii, T., Shigemoto, R., Mizuno, N. and Nakanishi, S. (1991) Nature (London) **354**, 31–37

Nakanishi, N., Shneider, N.A. and Axel, R. (1990) Neuron **5**, 569–581

Nutt, S.L. and Kamboj, R.K. (1994) Neuroreport **5**, 1679–1683

O'Hara, P.J., Sheppard, P.O., Thogersen, H., Venezia, D., Haldeman, B.A., McGrane, V., Houamed, K.M., Thomsen, C., Gilbert, T.L. and Mulvihill, E.R. (1993) Neuron **5**, 41–52

Olverman, J.J., Jones, A.W. and Watkins, J.C. (1984) Nature (London) **307**, 460–462

Partin, K.M., Patneau, D.K., Winters, C.A., Mayer, M.L. and Buonanno, A. (1993) Neuron **11**, 1069–1082

Patneau, D.K., Wright, P.W., Winters, C., Mayer, M.L. and Gallo, V. (1994) Neuron **12**, 357–371

Pellegrini-Giampietro, D.E., Zukin, R.S., Bennett, M.V.L. and Pulsinelli, W.A. (1992) Proc. Natl. Acad. Sci. U.S.A. **89**, 10499–10503

Perkins, N.M. and Stone, T.W. (1982) Brain Res. **247**, 184–187

Petralia, R.S. (1997) in The Ionotropic Glutamate Receptors (Monaghan, D.T. and Wenthold, R.J., eds.), pp. 219–263, Humana Press, Totowa

Petralia, R.S. and Wenthold, R.J. (1992) J. Comp. Neurol. **318**, 329–354

Petralia, R.S. and Wenthold, R.J. (1996) in Excitatory Amino Acids: Their Role in Neuroendocrine Function (Brann, D.W. and Mahesh, V.B., eds.), pp. 55–101, CRC Press, Boca Raton, FL

Petralia, R.S., Wang, Y.-X. and Wenthold, R.J. (1994a) J. Comp. Neurol. **349**, 85–110

Petralia, R.S., Yokotani, N. and Wenthold, R.J. (1994b) J. Neurosci. **14**, 667–696

Pollard, H., Heron, A., Moreau, J., Ben-Ari, Y. and Khristchatisky, M. (1993) Neuroscience **57**, 545–554

Puchalski, R.B., Louis, J.-C., Brose, N., Traynelis, S.F., Egebjerg, J., Kukekov, V., Wenthold, R.J., Rogers, S.W., Lin, F., Moran, T., Morrison, J.H. and Heinemann, S.F. (1994) Neuron **13**, 131–147

Raymond, L.A., Blackstone, C.D. and Huganir, R.L. (1993) Nature (London) **361**, 637–641

Roberts, P.J. (1974) Nature (London) **252**, 399–401

Roche, K.W., Raymond, L.A., Blackstone, C. and Huganir, R.L. (1994) J. Biol. Chem. **269**, 11679–11682

Rose, J.K. and Doms, R.W. (1988) Annu. Rev. Cell Biol. **4**, 257–288

Sakamura, K., Bujo, H., Kushiya, E., Araki, K., Yamazaki, M., Meguro, H., Warashina, A., Numa, S. and Mishina, M. (1990) FEBS Lett. **272**, 73–80

Sakamura, K., Morita, T., Kushiya, E. and Mishina, M. (1992) Neuron **8**, 267–274

Schuster, C.M., Ultsch, A., Schloss, P., Cox, J.A., Schmitt, B. and Betz, H. (1991) Science **254**, 112–114

Shinozaki, H. and Konishi, S. (1970) Brain Res. **24**, 368–371

Simon, J.R., Contrera, J.F. and Kuhar, M.J. (1976) J. Neurochem. **26**, 141–147

Sladeczek, F., Pin, J.-P., Recasens, M., Bochaert, J. and Weiss, S. (1985) Nature (London) **317**, 717–719

Sommer, B. (1997) in The Ionotropic Glutamate Receptors (Monaghan, D.T. and Wenthold, R.J., eds.), pp. 81–98, Humana Press, Totowa

Sommer, B. and Seeburg, P.H. (1992) Trends Pharmacol. Sci. **13**, 291–296

Sommer, B., Keinänen, K., Verdoorn, T.A., Wisden, W., Burnashev, N., Herb, A., Köhler, M., Takagi, T., Sakmann, B. and Seeburg, P.H. (1990) Science **249**, 1580–1585

Sommer, B., Köhler, M., Sprengel, R. and Seeburg, P.H. (1991) Cell **67**, 11–19

Sommer, B., Burnashev, N., Verdoorn, T.A., Keinänen, K., Sakmann, B. and Seeburg, P.H. (1992) EMBO J. **11**, 1651–1656

Somogyi, P., Takagi, H., Richards, J.G. and Mohler, H. (1989) J. Neurosci. **9**, 2197–2209

Somogyi, P., Eshhar, N., Teichberg, V.I. and Roberts, J.D.B. (1990) Neuroscience **35**, 9–30

Stern-Bach, Y., Bettler, B., Hartley, M., Sheppard, P.O., O'Hara, P.J. and Heinemann, S.F. (1994) Neuron **13**, 1345–1357

Sucher, N.J., Brose, N., Deitcher, D.L., Awobuluyi, M., Gasic, G.P., Bading, H., Cepko, C.L., Greenberg, M.E., Jahn, R., Heinemann, S.F. and Lipton, S.A. (1993) J. Biol. Chem. **268**, 22299–22304

Tachibana, M., Wenthold, R.J., Morioka, H. and Petralia, R.S. (1994) J. Comp. Neurol. **344**, 431–454

Taverna, F.A., Wang, L.-Y., Macdonald, J.F. and Hampson, D.R. (1994) J. Biol. Chem. **269**, 14159–14164

Tingley, W.G., Roche, K.W., Thompson, A.K. and Huganir, R.L. (1993) Nature (London) **364**, 70–73

Trevisan, L., Fitzgerald, L.W., Brose, N., Gasic, G.P., Heinemann, S.F., Duman, R.S. and Nestler, E.J. (1994) J. Neurochem. **62**, 1635–1638

Trudeau, L.-E. and Castellucci, V.F. (1995) J. Neurosci. **15**, 1275–1284

Uchino, S., Sakimura, K., Nagahari, K. and Mishina, M. (1992) FEBS Lett. **308**, 253–257

Ultsch, A., Schuster, C.M., Laube, B., Schloss, P., Schmitt, B. and Betz, H. (1992) Proc. Natl. Acad. Sci. U.S.A. **89**, 10484–10488

Verdoorn, T.A., Burnashev, N., Monyer, H., Seeburg, P.H. and Sakmann, B. (1991) Science **252**, 1715–1718

Vidnyánszky, Z., Görcs, T.J., Négyessy, L., Borostyánköi, Z., Kuhn, R., Knöpfel, T. and Hámori, J. (1996) Eur. J. Neurosci. **8**, 1061–1071

Wada, K., Dechesne, C.J., Shimasaki, S., King, R.G., Kusano, K., Buonanno, A., Hampson, D.R., Banner, C., Wenthold, R.J. and Nakatani, Y. (1989) Nature (London) **342**, 684–689

Wang, L.Y., Taverna, F.A., Huang, X.P., MacDonald, J.F. and Hampson, D.R. (1993) Science **259**, 1173–1175

Watkins, J.C. (1962) J. Med. Pharm. Chem. **5**, 1187–1189

Watkins, J.C. (1986) in Excitatory Amino Acids (Roberts, P.J., Storm-Mathisen, J. and Bradford, H.F., eds.), pp. 1–39, MacMillan, London

Watkins, J.C., Krogsgaard-Larsen, P. and Honore, T. (1990) Trends Pharmacol. Sci. **11**, 25–33

Wenthold, R.J., Parakkal, M.H., Oberdorfer, M.D. and Altschuler, R.A. (1988) J. Comp. Neurol. **276**, 423–435

Wenthold, R.J., Hunter, C., Wada, K. and Dechesne, C.J. (1990) FEBS Lett. **276**, 147–150

Wenthold, R.J., Yokotani, N., Doi, K. and Wada, K. (1992) J. Biol. Chem. **267**, 501–507

Wenthold, R.J., Trumpy, V.A., Zhu, W.-S. and Petralia, R.S. (1994) J. Biol. Chem. **269**, 1332–1339

Wenthold, R.J., Petralia, R.S., Blahos, J. and Niedzielski, A.S. (1996) J. Neurosci. **16**, 1982–1989

Werner, P., Voigt, M., Keinänen, K., Wisden, W. and Seeburg, P.H. (1991) Nature (London) **351**, 742–744

Wo, Z.G. and Oswald, R.E. (1994) Proc. Natl. Acad. Sci. U.S.A. **91**, 7154–7158

Wong, E.H.F., Kemp, J.A., Priestley, T., Knight, A.R., Woodruff, G.N. and Iversen, L.L. (1986) Proc. Natl. Acad. Sci. U.S.A. **83**, 7104–7108

Yamazaki, M., Araki, K., Shibata, A. and Mishina, M. (1992) Biochem. Biophys. Res. Commun. **183**, 886–892

# 8

# Molecular biology of *N*-methyl-D-aspartate (NMDA)-type glutamate receptors

**Paul J. Whiting and Tony Priestley**

Neuroscience Research Centre, Merck Sharp and Dohme Research Laboratories, Terlings Park, Eastwick Road, Harlow, Essex CM20 2QR, U.K.

## Introduction

L-Glutamate is the major excitatory neurotransmitter in the vertebrate central nervous system (CNS). It exerts its effect by activating two groups of receptors: ionotropic (ligand-gated ion channels) and metabotropic (G-protein-linked). Ionotropic glutamate receptors can be subdivided using pharmacological tools into three families: so-called kainate receptors, AMPA ($\alpha$-amino-3-hydroxy-5-methylisoxazolepropionate) receptors and NMDA (*N*-methyl-D-aspartate) receptors. It is the last receptor, so called because it is selectively activated by the synthetic agonist NMDA, which is the subject of this chapter.

The NMDA receptor is a complex and intriguing macromolecule. Opening of the intrinsic ion channel requires simultaneous binding of the co-agonists glutamate and glycine to unique recognition sites [1]. The channel is permeable to cations, primarily calcium and sodium (see [2]), which flow down their electrochemical gradients into the cell. The channel has a number of interesting properties, including blockade by magnesium ions [3] and a strongly voltage-dependent gating. The opening of the channel in response to co-agonist binding can be modulated by a variety of agents which are thought to act at distinct sites on the receptor; these include redox modulators, protons, polyamines, neurosteroids, ethanol and zinc (for a review, see [4]).

The physiological role of the NMDA receptor is as complex and intriguing as its molecular structure and function. It probably plays an important role in development of the immature CNS, while in the adult it is thought to be critically involved in synaptic events such as long-term potentiation which are thought to underlie some forms of synaptic plasticity and memory (for reviews, see [5,6]). The NMDA receptor is also thought to be critically involved in certain neuropathological and neurodegenerative conditions. In stroke, anoxia leads in particular to overactivation of NMDA receptors, allowing excess calcium to enter the cell, which leads via a number of different mechanisms to nerve cell death. There is considerable evidence that NMDA receptor antagonists may have utility in treating this condition (see [7]). There is also an increasing body of evidence suggesting that dysfunction of NMDA receptor activity in the mesolimbic pathways may underlie the neuronal dysfunction manifested in schizophrenia [8].

The purpose of the present chapter is not to exhaustively review the NMDA receptor, but to describe the insights into receptor structure, function and pharmacology that have been achieved by application of the tools of molecular biology.

## Cloning of cDNAs encoding subunits of NMDA receptors

The NMDA receptor is, relatively speaking, quite abundant in the CNS. As measured using radioligand binding, it is present at concentrations of pmol/mg of protein [9], whereas neuronal nicotinic receptors, for instance, are present at fmol/mg levels [10]. Despite this abundancy, attempts to purify the NMDA receptor from brain tissue, thereby obtaining protein which could yield some amino acid sequences for the generation of oligonucleotide probes for screening cDNA libraries, proved unsuccessful. Thus a molecular biology approach was needed to finally clone the cDNAs encoding the NMDA receptor subunits. Expression cloning using *Xenopus* oocytes had previously been used successfully to clone a cDNA encoding an AMPA-type glutamate receptor [11]. It was this approach that proved successful in cloning a cDNA encoding the NMDA R1 (NR1) subunit [12], also referred to as ξ1 (mouse cDNA [13]). This was certainly a breakthrough discovery. As has been the case for many other gene families, the sequence information gained from the isolation of one subunit cDNA enabled the isolation of cDNAs encoding other members of the family by using PCR and by low-stringency hybridization screening of cDNA libraries. Using this approach, four additional NMDA receptor subunit cDNAs have been identified, referred to as NR2A–NR2D [14,15], or ε1–ε4 [16–19].

## Structure of NMDA receptor subunits

NMDA receptor subunits are large polypeptides, 2–3 times the length of the subunits of other ligand-gated ion channels such as the nicotinic acetylcholine receptor (Fig. 1). The deduced amino acid sequences of the NMDA receptor subunits reveal that the NR2 subunits show 50–70% sequence identity with each other, but only about 15–18% identity with NR1. There is no significant identity with the polypeptides from the ligand-gated ion channel family comprising the nicotinic, glycine, γ-aminobutyric acid$_A$ and 5-hydroxytryptamine 3 receptors, or

**Fig. 1.**     **Schematic representation of the NMDA receptor family polypeptides**

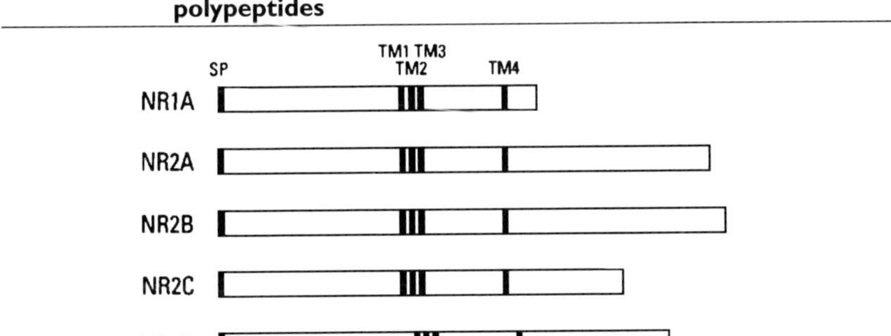


*SP, putative signal peptide; TM1–TM4, hydrophobic putative transmembrane spanning domains.*

with the ATP-gated ion channels. They do, however, show amino acid sequence identity with kainate and AMPA-type glutamate-gated ion channels. NR1 and the NR2 subunits exhibit around 20% and 11–16% sequence identity respectively with kainate/AMPA receptor polypeptides. Thus glutamate-gated channels appear to form a further ligand-gated ion channel gene family. The degree of deduced amino acid sequence conservation between species appears to vary considerably. While the rat and mouse NMDA receptor subunits are very well conserved (greater than 99% deduced amino acid sequence identity), there is greater diversity when the human sequences are examined. For instance, while the NR1 subunit is well conserved (only seven or eight amino acid differences between rat and human [20–22]), the NR2A subunit shows significantly greater diversity [22] (81 amino acid differences between rat and human).

Analysis of the deduced amino acid sequences of the NMDA receptor subunits allows prediction of a number of structural motifs, including a putative signal peptide and four hydrophobic domains, TM1–TM4 (Fig. 1). Interpretation of hydrophobic domains as membrane-spanning regions must be done with caution. It is, however, of interest to note that, when comparing the NR2 subunits, the highest amino acid sequence identity is in the N-terminal half of the polypeptide; the C-terminal sequences beyond the fourth hydrophobic region are quite divergent. The polypeptides also have a number of putative N-glycosylation sites and putative phosphorylation sites. The functional consequences of the latter are discussed elsewhere in this chapter, and may be important in the light of the reported roles of protein kinases in the establishment and maintenance of long-term potentiation (see [5,6]).

# Transmembrane topology of NMDA receptor subunits

As mentioned above, hydropathy analysis of NMDA receptor subunits reveals four major hydrophobic regions (see Fig. 1), which have been proposed as putative transmembrane domains, with a long N-terminal extracellular domain and an extracellularly located C-terminus. Evidence supporting this model is limited. Site-directed mutagenesis studies have identified amino acid residues in the N-terminal domain which contribute to the glycine binding site (see below), supporting the proposal that this domain is extracellular. Similarly, site-directed mutagenesis studies have demonstrated that amino acid residues in the putative TM2 domain influence channel function (see below), suggesting that TM2 may in some way contribute to the ion channel and thus be associated with the cell membrane. There is, however, a growing body of evidence suggesting that this four transmembrane model is incorrect. Tingley et al. [23] have demonstrated that putative phosphorylation sites located in the alternatively spliced C-terminus of NR1 (see below) are indeed phosphorylated in both cortical neuronal cultures and transiently transfected cells. This clearly suggests that the C-terminus is intracellular and not extracellular. Kuryatov et al. [24] and Hirai et al. [25] identified amino acid residues in the domain between TM3 and TM4 which may contribute to the glycine binding site. Further, Hirai et al. used epitope tagging to demonstrate that the N-terminal domain and the domain between TM3 and TM4 are extracellular, whereas the C-terminus is cytoplasmic. This new model of the transmembrane topology of NMDA receptor subunits is in agreement with studies on the related AMPA-type glutamate receptors [26,27].

# Gene structure and splice isoforms of the NR1 subunit

The first indication of the compexity and diversity of the NR1 subunit came from the work of Nakanishi and colleagues, who described cDNAs encoding a number of NR1 subunit isoforms, presumably generated by alternative splicing of RNA [28]. Since then, others have also described NR1 isoforms, in humans [22,29] as well as rodents. Eight splice variants have been identified, and these are shown in Fig. 2, together with two different nomenclatures that have been used. Analysis of the rat NR1 gene indicates that these isoforms are generated by alternative splicing of RNA transcribed from a single gene, allowing the omission or inclusion of additional exons at the N- and/or C-terminal end of the polypeptide [30]. According to the nomenclature of Hollmann et al. [30], NR1 subunits containing the N-terminal 21-amino-acid insert (exon 5) are referred to as 'b-type', whereas those lacking it are 'a-type'. The rat gene, at least the region containing the coding sequences, is approx. 24 kb, and contains 22 exons [30].

**Fig. 2.**  **Schematic representation of the rat NR1 subunit isoforms generated by alternative splicing**

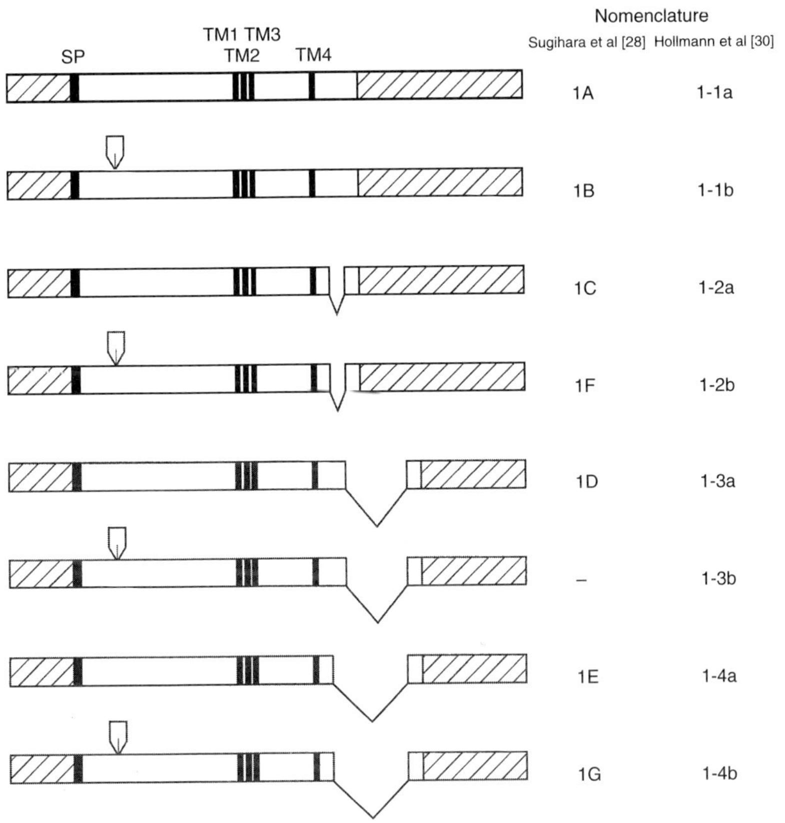

The open boxes represent the coding regions, and the hatched areas represent untranslated regions. SP, putative signal peptide; TM1–TM4, hydrophobic putative transmembrane spanning domains. The 'pen-nib' symbol above the sequence bar represents the 63 bp insert of exon 5. Two alternative nomenclatures for the splice isoforms are indicated on the right.

The functions of the NR1 isoforms are not clear, although, as discussed in more detail below, some data are emerging that suggest that the splice isoforms influence some of the functional properties of the receptor.

## Gene structure and splice isoforms of the NR2 subunits

To date, only the structure of the mouse NR2C gene has been elucidated [31]. It covers 13 kb and contains 12 exons. Comparison with the structure of the NR1 gene indicated a relatively close evolutionary relationship, with several of the introns interrupting the coding region in similar or the same positions.

Currently there is no evidence of alternatively spliced NR2A and NR2B subunits. There is, however, evidence for alternative splicing of the NR2C and NR2D subunits. The mouse NR2C mRNA is alternatively spliced in the 5′ end of the coding region; an 82 bp portion is alternatively included or deleted from the mRNA, resulting in two different C-terminal sequences [15]. Again, the functional consequences of this are unclear.

## Chromosome assignment of NMDA receptor subunit genes

The chromosomal localization of some of the NMDA receptor subunit genes has now been determined. The human NR1, NR2A and NR2C subunit genes have been mapped to 9q34.3, 16q13 and 17q35 [22,32]. Thus, in common with other receptor families, the genes appear to be distributed throughout the genome.

## Expression of NMDA receptor subunits in the CNS

The regional expression of rat NMDA receptor subunit mRNAs has been addressed by *in situ* hybridization [33–39] and by ribonuclease protection assays [40]. One generalization that can be made is that the NR1 subunit is remarkably ubiquitous, being expressed in essentially every neuronal type in the CNS, while the NR2 subunits show a more restricted pattern of expression. The NR1 splice isoforms do, however, show some spatial variations in their patterns of expression [38,39]. NR1 a-type and NR1-2 are most abundant and widespread in the adult brain, while NR1 b-type, NR1-1 and NR1-4 exhibit a more restricted expression. For instance, NR1 b-type and NR1-4 are absent from the caudate–putamen, while NR1-1 is of low abundance in the colliculi. NR1 a-type constitutes the majority of NR1 isoforms in the adult cortex and hippocamus, while in the cerebellum there are relatively similar proportions of NR1 a-type and NR1 b-type isoforms [40]. The NR1-3 isoform is the rarest of all. These clear differences in expression patterns strongly suggest an important functional correlate with the alternative splicing. NR1 mRNA can first be detected at embryonic day 8, and the regional differences in the expression patterns of the splice isoforms become most apparent after birth.

The NR2 subunits show very distinct patterns of expression [33–36,39]. NR2A is perhaps the most ubiquitously expressed in the adult rat brain, being especially abundant in the cortex and hippocampus. This subunit is essentially absent from the embryo. NR2B is also very abundant throughout the adult rat forebrain, but is particularly rare in the adult cerebellum. This subunit is quite widely expressed during development, before becomimg somewhat more restricted in its expression in the adult. NR2C contrasts with the other subunits in being very rare in the adult forebrain, its expression being confined almost exclusively to the cerebellum and spinal cord. It is essentially absent from the embryonic CNS. NR2D is by far the rarest NR2 subunit in the adult rat brain, its expression being confined primarily to some regions of the basal ganglia, such as the globus pallidus, substantia nigra and subthalamic nuclei. It is also expressed in the spinal cord [34]. The NR2D subunit is relatively more abundant in the embryonic rat brain, being found in both the forebrain and the brainstem, before declining significantly in expression after birth.

The above gives, of course, only a very broad description of the expression patterns of the individual subunits. By focusing on and determining precisely expression at the cellular level, it is possible to generate very discriminating data. Such information is already becoming available for anatomical structures such as the hippocampus and the cerebellum. Not only will these data give clues as to possible functional roles of receptors, but correlating the co-expression of subunits will aid in our understanding of the subunit arrangement of NMDA receptor subtypes. Additionally, it is important to be aware that *in situ* hybridization experiments only localize the mRNA primarily in the cell body. They give no indication as to whether the protein is actually being expressed, or where the protein is localized, which may be at a synaptic location some distance from the cell body. Subunit-specific antibodies should prove to begin to answer some of these questions. Western blot analysis of membranes prepared from various brain regions, and imunohistochemical analyses, have confirmed the distinct regional distribution of NMDA receptor subunits and the changes in levels of expression during development [41–43]. Detailed immunoelectron microscopic analyses have been able to demonstrate the presence of NMDA receptor subunits in dendrites and post-synaptic densities [44,45].

## Subunit structure of NMDA receptor subtypes

A key issue in the understanding of the NMDA receptor is determining the subunit composition and arrangement of receptor subtypes. As inferred from the discussion throughout this chapter, the mimimal subunit arrangement of native NMDA receptors is almost certainly a heteromeric structure of NR1 along with one or more NR2 subunits. The large size of the receptor subunits and the receptor macromolecule, and its resistance to solubilization from the membrane in a native form, make the elucidation of the structure of the native receptor a difficult task. Therefore, to date there is a relative paucity of data, and that which exists is somewhat contradictory. The approach of choice is that used to characterize other neurotransmitter receptors, namely the use of subunit-specific antibodies as biochemical tools. Such antibodies can be used in Western blot and immunoprecipitation experiments to determine the subunit arrangement of receptors, and in immunocytochemical approaches to visualize the receptor.

Western blot analysis using membranes prepared from different brain regions has in general confirmed the regional expression of subunit mRNAs suggested by *in situ* hybridization. The NR1 subunit is a glycosylated polypeptide with a molecular mass of around 115–120 kDa [23,46] which is widely expressed in the CNS. The NR2A and NR2B subunit polypeptides have apparent molecular masses of between 165 and 175 kDa respectively [41,42,47–49], and the NR2C and NR2D have apparent molecular masses of between 140 and 160 kDa [41,43,48].

Sheng and colleagues [47] have used immunoprecipitation/Western blot techniques to begin to analyse the subunit composition of native receptor subtypes expressed in the rat brain. They were able to demonstrate the coexistence of NR1 with an NR2A or NR2B subunit in a receptor macromolecule, confirming the conclusions obtained on expressing such subunit combinations in *Xenopus* oocytes or transfected cells (see below). Interestingly, there was also the suggestion that a proportion of native receptors contain both NR2A and NR2B, presumably co-assembled with NR1. Didier et al. [48] have also reported that in rat cerebellum the majority of the NR2B subunit is co-assembled with NR1 and NR2A, while Blahos II and Wenthold found that most NR2A and NR2B in rat forebrain is not co-assembled in the same macromolecule [49]. The reason for these contradictions is not clear, but presumably lies in the nature of the antibodies, the methodology used for receptor solubilization and immunoprecipitation, or the brain region used. Sheng et al. found some evidence that the NR1 splice isoforms may segregate in a specific manner with NR2 subunits [47], while Blahos II and Wenthold found no evidence for this, but did demonstrate that a single receptor molecule can contain at least two NR1 splice isoforms [49].

The photoaffinity label [$^{125}$I-]CGP 55802A, which is a high-affinity antagonist at the glutamate site of the NMDA receptor, has been used as a tool to probe the subunit structure [50]. The data obtained were not straightforward to interpret. When recombinant receptors expressed in transiently transfected cells were photoaffinity labelled, only the NR1/NR2A combination was photoreactive, the photolabel being incorporated into both the NR1 (115 kDa) and NR2A (175 kDa) subunits. This could of course be because the radioligand has too low an affinity at the other (NR1/NR2B and NR1/NR2C) combinations tested. A similar labelling pattern was observed when early postnatal rat brain was labelled, but the pattern changed in the adult, where only a 175 kDa (NR2A?) polypeptide was labelled. In contrast, when cultured cerebellar granule cells (which are thought to express primarily NR1 and NR2A subunits [51]) were photoaffinity labelled, only a 115 kDa polypeptide (NR1?) was photoreactive. This hints at the complexity of native NMDA receptors and their plasticity during development, but it is currently difficult to interpret these data further.

*In vitro* experiments using recombinant NMDA receptors have also addressed the issue of subunit arrangement. The first indication that an NR1 subunit could co-assemble with more than one type of NR2 subunit came from Wafford and colleagues [52]. They were able to demonstrate that when NR1, NR2A and NR2C subunits were co-expressed in *Xenopus* oocytes, they co-assembled to form a receptor molecule which had an affinity for glycine which was distinct from that of NR1/NR2A or NR1/NR2C combinations, suggesting the formation of an NR1/NR2A/NR2C receptor. Subsequent biochemical experiments by Chazot et al. [53], who expressed these subunit combinations in transiently transfected HEK 293 cells, also demonstrated the co-assembly of NR1, NR2A and NR2C in the same receptor molecule, and suggested that this may be the major receptor subtype expressed in adult cerebellar granule cells. However, Didier et al., precipitating receptors from rat cerebellum, found no evidence for co-assembly of NRC2 with NR2A (or indeed NR2B) [48].

There is thus increasing evidence that the 'minimal' subunit arrangement of NR1 plus an NR2 may be simplistic, and that receptor subtypes containing more than one type of NR2 subunit may exist. However, this is clearly a very controversial question generating contradictory data.

The subunit stoichiometry of the NMDA receptor macromolecule is currently unknown. The size of the native receptor is approx. 730 kDa [46], which agrees well with the size of recombinant NR1/NR2A or NR1/NR2A/NR2C receptors [53]. Given the size of the NR1 subunit (115–120 kDa) and NR2 subunits (160–170 kDa), this suggests that the receptor could be a pentamer (like members of the nicotinic receptor gene family), or even a hexamer. Since the glycine binding site is known to be on the NR1 subunit (discussed below), and there are two glycine sites per receptor molecule [54], then one could speculate that a pentameric NMDA receptor consists of two NR1 subunits and three NR2 subunits.

## Association of the NMDA receptor with other synaptic proteins

An important question, which is key not only to the NMDA receptor but many other membrane proteins, is how do they end up at specific locations in the neuron (synaptic, extra synaptic, etc.), and how do they stay there? Recent studies in a number of laboratories have demostrated a direct interaction of NMDA receptor subunits with a family of synapse-associated proteins which includes PSD95 (SAP90), PSD93, SAP97, SAP102 and chapsyn 110 [55–58]. These proteins, all around 90–110 kDa, have a modular structure of three PDZ domains (named after three homologous proteins which contain these domains), an SH3 domain (which may be involved in signal transduction/protein–protein interaction), and a guanylate cyclase homology domain. The proteins are thought to interact with the NMDA receptor subunits through their PDZ domains [55–56,58]. The site of interactions on the NMDA receptor subunit is a sequence motif Ser-Xaa-Val at the C-terminus of NR1-3, NR1-4 and all four NR2 subunits. This interaction almost certainly serves to anchor/cluster the receptor at the synapse. In addition, since other proteins — such as nitric oxide synthase [59] — may also interact with these synapse-associated proteins, it may be that they act as a skeleton for the formation of a multi-protein synaptic signalling complex.

## Functional and pharmacological properties of recombinant NMDA receptors

The molecular cloning of the various subunits comprising an NMDA receptor, and the successful reconstitution of recombinant receptors in *Xenopus* oocytes and mammalian cell expression systems, has in a relatively short time period greatly enhanced our understanding of the function and pharmacology of this receptor.

As discussed above, the first NMDA receptor subunit cDNA to be obtained, encoding the NR1 subunit, was done by expression cloning using *Xenopus* oocytes [12]. It is important to note, however, that the currents generated by 'homomeric' NR1 expressed in *Xenopus* oocytes are very small, and in transfected mammalian cells NR1 alone does not form detectable functional channels [60,61]. Similarly, NR2 subunits expressed alone do not form functional channels [14]. It is only the co-expression of NR1 with one or more NR2 subunits that leads to the formation of robust glutamate/glycine-gated channels. Thus, as discussed above, the current working hypothesis is that NMDA receptors are heteromeric macromolecules consisting of an NR1 subunit and one or more NR2 subunits.

### Glutamate and glycine sites of recombinant NMDA receptors

The pharmacology and functional properties of the glutamate and glycine sites of recombinant NMDA receptors expressed in *Xenopus* oocytes or transfected mammalian cells have been characterized using both electrophysiological and radioligand binding approaches. Studies have investigated the properties of both homomeric NR1 receptors and heteromeric NR1/NR2 receptors. The pharmacology is determined by both the type of NR1 splice isoform and the type of NR2 subunit. Glutamate has been shown to have a slightly higher affinity for homomeric NR1 a-type than b-type receptors [65], whereas glutamate antagonists have higher affinities for the NR1 b-type variants [28,30,62–64]. However, in heteromeric assemblies the affinities of glutamate-site ligands are also influenced by the NR2 subunit. Thus combinations comprising NR1/NR2B have a modest (~2-fold) higher affinity for glutamate compared with NR1/NR2A [66–68]. Interestingly, glutamate antagonists show the reverse profile, having higher affinity for the NR2A-containing receptor over other NR2-containing receptors [66–68].

Glycine affinity does not appear to differ between the NR1 a-type and b-type splice isoforms, but is influenced by the NR2 subunit. Thus NR1/NR2B subunit assemblies have an approx. 10-fold higher affinity for glycine than does

NR1/NR2A [68,69]. Glycine and glutamate affinities at NR1/NR2C appear to be comparable with those at NR1/NR2B (again ~10-fold higher than at NR1/NR2A) [69]. There have been few comparable studies with the NR2D subunit, although kinetic studies of offset decay time constants suggest that the NR1/NR2D receptor may have an exceptionally high affinity for glutamate which far exceeds that seen with the other NR2 subunits [55].

The NR1 subunit appears to contain many of the determinants that constitute the glycine binding site of the NMDA receptor. This is suggested by radioligand binding experiments, which have demonstrated that homomeric NR1 expressed in transfected cells is able to form a binding site for glycine-site antagonists which is essentially indistinguishable in its antagonist pharmacology from that of native NMDA receptors [61]. Site-directed mutagenesis experiments have identified several amino acids in the NR1 sequence that are involved in the putative pharmacophore for glycine. Three regions appear to form composite parts of the overall binding domain. Two of these lie in the distal third of the extracellular N-terminal domain of NR1 [24,70], and the third region is situated between the putative transmembrane regions TM3 and TM4 [24]. Specific amino acid substitutions in these regions produce substantial changes in glycine agonist affinity, but have little or no effect on glutamate affinity. Furthermore, there appears to be considerable sequence identity between these domains and the ligand binding regions of a bacterial amino acid transporter, the so-called lysine/arginine/ornithine binding protein, suggesting that there may be a conserved molecular architecture defining ligand affinity in these two evolutionarily diverse proteins [24,71]. Although these studies imply that the glycine binding site resides on NR1, they do not rule out the possibility that additional components of the binding domain are provided by the NR2 subunits. Indeed, as discussed above, the type of NR2 subunit co-assembled with NR1 affects the affinity of the receptor for glycine.

## Properties and structure of the ion channel

In addition to the co-agonist gating requirement, NMDA receptor channels have a number of other unique features, including a relatively non-selective permeability to cations, a high permeability to $Ca^{2+}$ (as much as 15-fold higher than that to $Na^+$ [72]) and a voltage-dependent block by $Mg^{2+}$ ions, all of which have been the subject of a recent extensive review [4]. The generation and expression of recombinant receptors comprising different subunit combinations together with site-directed mutagenesis manipulations has provided a molecular basis for the $Ca^{2+}$ permeability and the $Mg^{2+}$ block. Monyer and colleagues [35] have shown that $Mg^{2+}$ block is more prominent in recombinant receptors comprising NR1/NR2A or NR1/NR2B compared with that in NR1/NR2C or NR1/NR2D.

The cationic selectivity of the channel appears to be due, at least in part, to the presence of asparagine (N) residues within the channel pore. These residues occupy position 598 in NR1, 595 in NR2A and 593 in NR2C (based on sequence alignment), they are equivalent to the glutamine/arginine (Q/R) site in AMPA channels which also influences ion permeability. Collectively, this motif is often referred to as the Q/R/N site or, more recently, the N-site [73], Q and R being the relevant residues in the case of AMPA receptors and N the equivalent in the case of NMDA receptors. Glutamine confers high permeability to calcium in AMPA receptors but is edited in about 90% of cases to arginine, which significantly reduces calcium permeability [74–75]. However, Q/R mutations in AMPA receptors do not confer $Mg^{2+}$ block and, by analogy, residues additional to those effecting $Ca^{2+}$ permeability are likely be involved in the $Mg^{2+}$ block of NMDA channels. NR1($N^{598} \rightarrow Q$)–NR2A mutations have a similar, decreased $Ca^{2+}$-permeability effect in the case of NMDA receptors [76]. The picture is complicated further because although NR1–NR2A($N^{595} \rightarrow Q$) or NR1–NR2C($N^{593} \rightarrow Q$) mutations do reduce $Ca^{2+}$ permeability, the magnitude (~2-fold) of the change is much less than that seen with the equivalent mutation in NR1 (4- to 10-fold). Furthermore, the NR2A($N^{595} \rightarrow Q$) mutation increases the relative $Mg^{2+}/Ca^{2+}$ permeability, i.e. opposite to that seen with NR1($N^{598} \rightarrow Q$) [76]. Hence, a common site of interaction within the channel for these two divalent cations is unlikely and it may be inferred that the asparagine residues contributed by NR1 and NR2 subunits are unlikely to form a ring of polar residues situated at equal depths within the channel [76–78]. Further evidence for this has been provided by site-directed mutagenesis studies designed to identify the structural determinants contributed by NR1 and NR2A to the selectivity filter of the ion channel pore. Wollmuth et al. [79] used the permeability ratios of a number of permeant cations, of different sizes, to probe the amino acid residues critically involved in determining the narrowest region of the channel. These experiments found that residues of both NR1 and NR2A contribute to the constriction but that these residues were not at equivalent depths within the channel. Thus, in the case of NR1, replacing N-site asparagine for the smaller glycine [NR1($N^{598} \rightarrow G$)–NR2A] produced the largest increase in pore size, whereas the homologous mutation in NR2A had only minor effects on permeability ratios. However, mutation of a residue C-terminal to the N-site (N+1) of NR2A had the most prominent effects on permeability [NR1–NR2A($N^{596} \rightarrow G$)], and this mutation had little effect in the case of NR1. A double mutation, NR1($N^{598} \rightarrow G$)–NR2A($N^{596} \rightarrow G$) increased pore size by an amount equivalent to the sum of the two individual point mutations. Once again, this suggests that the narrowest portion of the channel is determined by asparagine residues located at non-homologous positions according to sequence alignment. Further confirmation of this and evidence to support the currently accepted view that TMII is a re-entrant loop has been put forward by Kuner et al. [80]. Using site-directed mutation of each pore residue in turn to cysteine, they have probed the accessibility of the pore region of TMII by making use of the ability of methanethiosulphonate (MTS) derivatives to form covalent bonds with the sulphydryl moieties. The ability of MTS derivatives to block current flux when applied via the external or the internal (cytoplasmic) surface was examined for a large number of subunit combinations (mutated either NR1 + WT NR2C, or vice versa). Irreversible block was observed to occur at more cysteine-mutated residues when MTS compounds were applied from the cytoplasmic face than from the external face. As this profile included residues which were C-terminal to the N-site, the only possible explanation is to assume a pocket formed by TMII residues with cytoplasmic-directed residues situated both N- and C-terminal to the N-site, i.e. a re-entrant loop. Furthermore, mutation of the N-site asparagine residue itself to cysteine resulted in a consistent irreversible block by MTS reagents from both the cytoplasmic and extracellular route of application, implying that the N-site forms a hinge region and effectively is the selectivity filter. Ascher and colleagues [81] have also confirmed the re-entrant loop model for the pore by examining the effect of several mutations in TMII downstream to the N-site on the extent of channel block by external and internal $Mg^{2+}$. Thus mutations involving residues 5 to 7 positions downstream to the N 'hinge' did not affect the block of current flux by external $Mg^{2+}$, but clearly influenced that produced by internal $Mg^{2+}$. This suggests not only that the C-terminal part of TMII projects back towards the cytoplasmic face of the membrane, but also that there are separate sites for internal and external block by $Mg^{2+}$.

Analysis of recombinant NMDA receptor subtypes has also provided a molecular correlate of the different single-channel conductance states of native NMDA receptors. Heteromeric assemblies comprising NR1 together with NR2A or NR2B have similar single-channel characteristics, with a principal conductance (in 1 mM $Ca^{2+}$) of around 50 pS and which is associated with brief sojourns to a subconductance level of ~38 pS [82]. This profile compares extremely well with that of hippocampal CA1 pyramidal cells [83,84], and also with that of rat cultured cerebellar granule cells [85], which have an NR1/NR2A pharmacology [86] and express principally NR1 and NR2A subunit mRNAs [51]. By comparison, NR1/NR2C assemblies appear quite different, having a main conductance of ~36 pS with brief transitions to ~19 pS [82], a profile which is correlated more closely with that of tissue cultured 'large cerebellar' neurons (presumed Purkinje cells) taken from fetal rats [87,88]. Very little is known about the NR1/NR2D receptor, although there is a suggestion that neonatal (P0–P8) cerebellar Purkinje cells may express this assembly [35], and single-channel recordings of P2–P12 rat Purkinje cells do appear to have a distinct single channel characteristics which may represent an NR1/NR2D receptor [89]. Single-channel studies of recombinant NR1/NR2D would clarify this situation.

A number of non-competitive antagonists, such as MK-801, phency-clidine and ketamine, have been proposed to exert their effects by a use-dependent block of the NMDA receptor channel (see e.g. [4]). MK-801 blocks the function of homomeric NR1 and heteromeric NR1/NR2 receptors expressed in *Xenopus* oocytes, and of heteromeric receptors expressed in transiently transfected cells (e.g. [12,15,22,28,90]). There is disagreement over whether [$^3$H]MK-801 binds with high affinity to homomeric NR1 in transfected cells [61,66,91–93]. However, it is clear that [$^3$H]MK-801 binds with high affinity to heteromeric NR1/NR2A and NR1/NR2B receptors, and with somewhat lower affinity to NR1/NR2C and

NR1/NR2D receptors [53,66,92,94]. A similar profile was obtained by analysing the potency of the functional block at these receptor combinations [90]. Interestingly, NR1/NR2A/NR2C receptors expressed in HEK293 cells had an affinity for [³H]MK-801 of 22 nM, which is identical to that found for the binding of [³H]MK-801 to mouse cerebellum membanes, which is further evidence suggesting that this may be the subunit structure of NMDA receptors of adult cerebellar granule cells [53].

The site of action of these drugs has been confirmed by site-directed mutagenesis. Again, asparagine-598 of TM2 has been implicated; substitution of this residue (in NR1 or NR2A) with a glutamine or arginine leads to a significant decrease in the potency of the block [76,77].

## Modulation by polyamines and ifenprodil

The ability of polyamines such as spermine and spermidine to enhance [³H]MK-801 binding [95,96] has stimulated considerable interest in the effects of these polycations on NMDA receptor function. It has been established that polyamines such as spermine can act at a number of distinct sites on the NMDA receptor to elicit glycine-dependent and -independent potentiation, as well as voltage-dependent inhibition. The net effect of this complex interaction depends upon a number of factors, including polyamine concentration, glycine and glutamate concentrations, membrane potential and receptor subtype. The antagonist actions of polyamines are potentiated at negative membrane potentials, and are therefore described as being voltage-dependent [96]. The effects of the atypical, non-competitive NMDA antagonist, ifenprodil, have been attributed to the polyamine site [97,98], although several studies have consistently argued against this being the case [95,99]. The most convincing evidence in favour of non-identical sites for polyamines and ifenprodil has been provided by recent site-directed mutagenesis experiments. A single arginine residue at position 337 in the $\epsilon$2 (NR2B) subunit is an absolute requirement for high-affinity ifenprodil inhibition in NR1a/$\epsilon$2 subunit combinations, but substitution at this position has little, if any, effect on polyamine potentiation of receptor function [100]. However, some degree of overlap is suggested by other studies in which substitution of aspartate-669 in the NR1a subunit has been found to completely abolish glycine-independent spermine potentiation of heteromeric NR1a/NR2B receptors but also significantly attenuated ifenprodil antagonism [101].

Together, the polyamines and ifenprodil have revealed important aspects of the molecular pharmacology of the NMDA receptor and, arguably, represent the greatest pharmacological potential for subunit-selectivity to date. The expression of homomeric recombinant NR1 assemblies has shown that glycine-insensitive polyamine potentiation only occurs in those NR1 splice variants lacking the N-terminal 21-amino-acid insert encoded by exon 5 (NR1 a-type variants) [64,65]. In contrast, glycine-dependent polyamine potentiation occurs at all NR1 splice variants [102]. The potential importance of a net increase in positive charge conferred by the insertion of the exon 5 amino acid cassette was recognized relatively early on [63]. More recent site-directed mutagenesis experiments have demonstrated that six positively charged amino acids grouped into two clusters of three within this 21-amino-acid insert are responsible for the polyamine insensitivity of those NR1 variants containing the insert [103]. Furthermore, it appears as though this alternatively spliced exon 5 also determines the sensitivity of the receptor to protons [104].

The influence of NR2 subunits on polyamine and ifenprodil pharmacology has been addressed by several groups. Zhang et al. [102] have shown that the modest effect of spermine on glycine affinity (the glycine-dependent polyamine potentiation) is retained when NR1 subunits are combined with NR2A or NR2B but not with NR2C, whereas the more robust glycine-independent spermine potentiation appeared to be critically dependent upon the presence of an NR2B subunit. Dimeric subunit assemblies of an NR1 a-type variant in combination with NR2B, and trimeric assemblies containing (in addition to NR2B) either NR2A or NR2C, all retained glycine-independent spermine potentiation. However, this polyamine effect was lost in dimeric assemblies containing NR1 a-type variants co-expressed together with NR2A, NR2C [102,105] or NR2D [106]. The NR1 subunits continue to play a dominant role in such heteromeric receptors, as the co-assembly of NR2B with NR1 b-type variants (i.e. containing exon 5) does not confer glycine-independent spermine potentiation [105]. In summary, whether one or both forms of polyamine potentiation are expressed is governed by the presence or absence of the N-terminal exon 5 sequence. The polyamine binding site could be located either exclusively on individual NR1 subunits or could involve components from juxtaposed NR1 (homomeric) or NR1/NR2 (heteromeric) subunits [105]. Depending upon which of these scenarios holds true, it is conceivable either that certain structural features of NR2A, NR2C or NR2D allosterically modulate the conformation of NR1 and thereby alter the affinity for polyamines, or that these NR2 subunits lack (or occlude) part of the necessary binding domain formed by assembly with NR1. In contrast, the NR2B subunit is apparently permissive or contributory in dimeric assemblies, and in more complex trimeric assemblies its presence apparently overrides the negative effects of the other NR2 subunits when they are also part of the overall subunit structure.

The antagonist actions of polyamines and ifenprodil have also received considerable attention. At negative membrane potentials the antagonist action of polyamines predominates over either form of potentiation. Thus at −80 to −100 mV, spermine exerts a substantial yet incomplete antagonism of NMDA responses [107]. This may be due to a surface charge screening effect, and is observed in hetero-oligomeric combinations containing NR1 together with NR2A or NR2B, but not NR2C [107]. Interestingly, a number of polyamine spider toxins also show voltage-dependent blockage of NMDA responses [108,109] and, as with spermine, appear to be effective in blocking NR2A- and NR2B- but not NR2C-containing assemblies [110]. Ifenprodil has both voltage-dependent and -

independent antagonist actions; at NR1A/NR2B subunit combinations this compound shows a high-affinity non-competitive antagonism, whereas at NR1A/NR2A its affinity is 300–400-fold lower [68,111] and, at least in part, involves a voltage-dependent component [112]. The substantial difference in the affinity of ifenprodil at NR2A- and NR2B-containing receptors has been exploited to reveal subpopulations of native NMDA receptors [86,113].

## Modulation by histamine

Several reports have indicated that histamine can potentiate native NMDA receptor function [114,115]. Studies using recombinant rat NMDA receptors expressed in *Xenopus* oocytes have shown that this is a direct effect on the NMDA receptor protein and has very precise subunit requirements [116]. Thus histamine potentiation was only seen in heteromeric NR1/NR2 receptors containing NR1 a-type splice variants (NR1-1a or NR1-4a), together with NR2B but not NR2A or NR2C subunits [116]. These subunit requirements have clear parallels with those necessary for polyamine potentiation. Indeed spermine and histamine potentiations are non-additive, but subtle differences in other aspects of their action suggest that the histamine effect may turn out to represent an action at yet another novel site on the NMDA receptor complex [116].

## Redox modulation of receptor function

Chemical reduction of thiol groups on the NMDA receptor ion channel by the reducing agent dithiothreitol elicits a selective potentiation of native NMDA receptor function [117–122], whereas kainate and AMPA-type receptors are relatively unaffected [117]. The potentiation of NMDA responses by thiol reduction appears to be due to an increase in the channel open probability [120,121]. The opposite effect, i.e. attenuation of NMDA receptor function and, in some cases, reversal of the dithiothreitol effect, is imparted by the oxidizing agent 5,5′-dithiobis-2-nitrobenzoic acid [119,122,126–128], and an altered redox potential is one consequence of cerebral ischaemia [129]. Oxidation of NMDA receptors may thus play a part in the pathology of NMDA-receptor-mediated neurodegeneration following such brain insults.

## Modulation by zinc

Zinc has long been known to antagonize native NMDA receptor function in a complex manner which involves both a weak voltage-sensitive component and a voltage-insensitive component (see [4] for a review). Zinc also antagonizes recombinant homomeric NR1 [12,16]. The apparent affinity appears to differ depending on the splice variant [64]. Curiously, Hollmann et al. [30] have reported a zinc-induced potentiation of homomeric NR1 a-type assemblies expressed in *Xenopus* oocytes, an effect which is seen with concentrations of the cation much lower than those required for the blocking effect and which is not seen with NR1 b-type splice variants, or when NR1 a-type is combined with NR2A, NRB or NRC. The functional significance of this observation remains to be established.

## Modulation by protons

NMDA receptor activity is modulated by protons with an affinity in the physiological pH range [130–132]. Studies using recombinant receptors have demonstrated that the degree of proton inhibition is dependent upon the expression of N-terminal amino acids encoded by exon 5 in NR1 subunits. NR1 a-type variants lacking the insert show greater sensitivity to proton inhibition. The presence or absence of C-terminal inserts had no effect on proton sensitivity; nor did co-assembly of NR1 subunits with NR2A or NR2B [133]. Site-directed mutagenesis experiments have identified lysine-211, at the C-terminal end of exon 5, as the residue that mediates the relief of proton inhibition conferred by exon 5 [104]. Relief from proton inhibition could also be conferred by polyamines. The structural similarity between the positively charged polyamines and the positively charged sequence of exon 5 led Traynelis et al. [104] to speculate that exon 5 may act as a pH-sensitive integral modulator of receptor function. Additionally, the potentiation by polyamines at receptors containing NR1 a-type variants lacking exon 5 (as discussed above) may be explained by the relief of proton inhibition.

## Modulation by ethanol

Ethanol is known to inhibit NMDA receptor function in a variety of neuronal preparations and at what may be considered physiological concentrations [134]. Several studies have addressed the effects of ethanol on recombinant NMDA receptor subtypes [135,136]. Ethanol was both more potent and more efficacious in its inhibition at NR1/NR2A and NR1/NR2B that at NR1/NR2C receptors, and this apparent selectivity may explain the variation in the sensitivity to ethanol of the NMDA responses seen in different types of neurons [134].

## Phosphorylation and receptor function

Sequence analysis of the deduced amino acid sequences of NMDA receptor subunits reveals numerous putative phosphorylation sites. Stimulation of protein kinase C (PKC) activity by phorbol esters such as phorbol 12-myristate 13-acetate (PMA) has been shown to potentiate the function of both native [137] and recombinant [64,138–140] NMDA receptors. Attention has, quite justifiably, been focused on the role of potentially phosphorylatable sites within the now presumed intracellular domain C-terminal to TM4 (see above). Indeed, several phospho rylated residues are apparent in phosphopeptide maps of trypsin-digested NR1 subunits expressed in HEK cells following PMA treatment and, furthermore, a similar profile is observed with a bacterial fusion protein containing only the C-terminal 105 amino acids of NR1 [23]. Significantly, of the eight major phosphorylated residues, those involved specifically in the potentiating effect of PMA were suggested to be contained within the amino acid cassette encoded by the alternatively spliced exon 21 (i.e. serine-889, -890, -896 and -897). Mutation of these four serine residues to alanine greatly decreased PMA-induced phosphorylation [23]. However, the fact that these two pairs of C-terminal serine residues are phosphorylated by phorbol esters appears to be inconsequential in terms of the functional modulation of NMDA receptors by PKC activation. For example, PMA has been shown to potentiate homomeric NR1 variants [64,139] which lack the C-terminal sequence encoded by exon 21. Even more enigmatically, Durand et al. [64] found that PMA produced a much more pronounced potentiation of homomeric NR1b (which is isoform NR1-4b in Fig. 2) compared with NR1a (isoform NR1-1a in Fig. 2), suggesting that not only is the C-terminal exon 21 domain redundant but the putatively extracellular N-terminal insert encoded by exon 5 appears to influence the effect of PMA.

A number of studies have begun to investigate the role of the NR2 subunits in the modulation of function by phosphorylation, and there is evidence that NR2 subunits may dictate the extent of modulation by PKC in hetero-oligomeric assemblies. Co-expression of mouse NR1/NR2A or NR1/NR2B results in receptors that are potentiated by PMA, as are homomeric NR1 receptors, but combination of mouse NR2C with NR1 results in heteromeric assemblies which are not sensitive to PMA [13,18]. The role of the putative PKC phosphorylation sites within the NR2 subunits is unknown. Using C-terminal mouse NR2B/NR2C subunit chimaeras co-expressed with NR1, Mori et al. [139] found that the domain between TM3 and TM4 of NR2B conferred PMA-sensitivity to NR2C. On the other hand, Sigel et al. [140] found that rat NR2A co-expressed with NR1C (which lacks both the C-terminal exon 21 containing the four phosphorylatable serine residues identified by Tingley et al. [23] and the N-terminal exon 5 insert) showed greater potentiation by PMA than did NR2A/NR1A. Furthermore, deletion of additional PKC consensus sites from the region separating TM3 and TM4 in both NR1C and NR2A failed to modify PMA potentiation compared with the wild type. The data of Sigel et al. [140] thus suggest that none of the putative PKC phosphorylation sites in the domain between TM3 and TM4 of NR1 or NR2 subunits, or in the alternatively spliced exon 21 of NR1, are essential for PKC modulation of NMDA receptor activity.

More recent work has suggested that native NMDA receptors may be tonically phosphorylated, since receptor activity can be potentiated by serine/threonine phosphatase inhibitors such as okadaic acid, calyculin and FK506 [141,142], although these studies disagree as to which phosphatase (1/2A or 2B) is responsible. There is evidence that tyrosine phosphorylation may also be involved in NMDA receptor modulation [141]. It is clear from these studies that the NMDA receptor complex may be modulated by kinases and phosphatases. It is also clear, however, that there is considerable confusion over the relationship between the biochemical and functional consequences of phosphorylation. Elucidation of the relative contributions of direct (subunit phosphorylation) and indirect (phosphorylation of regulatory proteins or cytoskeletal elements; see [143]) components will help to clarify some of the present confusion.

## NMDA receptor knock-out mice

One of the key questions is the role of the receptor subtypes *in vivo*, and one of the ways that this question is being approached is through transgenic technology. The first generation of such studies have allowed production of knock-out mice, i.e. mice in which a particular gene has been disrupted to prevent production of the cognate RNA and protein. NR1 knock-out mice have been generated [144,145]. Unfortunately, and perhaps not suprisingly, these null mutants died soon after birth and thus were not very informative. NMDA-mediated responses were abolished in NR1 null mice, confirming that NR1 is the key subunit of the NMDA receptor. An NR2A knock-out mouse has also been generated [146]. This mutation was not lethal, and the mice were reported to have reduced long-term potentiation (LTP) in CA1 synapses, and a deficit in spatial learning. The deficit in LTP increased with age [147]. NR2B knock-out mice exhibit a complete loss of NMDA-mediated responses in the hippocampus of the neonate [148], reflecting the preponderance of the NR2 subunit at this stage of development. These NR2B null mice also exhibited a deficit in long-term depression in the hippocampus, and in the formation of certain neuronal structures (the whisker-related neuronal barrelette, and the cluster of sensory afferent terminals in the brainstem trigeminal nucleus), suggesting a role for NR2B in both neuronal pattern formation and synaptic plasticity. NR2D null mice have also generated, but results to date suggest only a very mild phenotype, with reduced spontaneous activity in the open-field behavioural test [149].

Future developments in transgenic technologies will allow development of a new generation of more sophisticated animal models, e.g. (i) which have amino acid point mutations introduced into a subunit, (ii) which allow disruption of the gene in the adult animal (inducible knock-outs) or (iii) which allow disruption of the gene in discrete brain regions. An example of the third possibility has been reported by Tonegawa et al. [150], who have developed a mouse with the NR1 gene deletion restricted to the CA1 hippocampal pyramidal cells. Such mice grow to adulthood without any gross abnormalities, but lack NMDA-mediated synaptic currents and LTP at CA1 synapses, and have deficits in spatial learning. These data suggest that NR2A containing NMDA receptors in the CA1 region mediate the synaptic plasticity underlying the acquisition of spatial memories.

# Future perspectives

Cloning of cDNAs encoding subunits of the NMDA receptor has allowed a huge leap forward in our knowledge of this complex macromolecule. It is obvious, however, that we have only just begun to scratch the surface in terms of our understanding of its structure, function and regulation, and these are very active areas of research. In the future, studies using 'inducible' knock-outs, and subtype-selective drugs, should allow greater insight into the role of NMDA receptors in brain function.

## References


1.  Kleckner, N.W. and Dingledine, R. (1988) Science **238**, 355–358
2.  Fagg, G.E. and Foster, A.C. (1983) Neuroscience **4**, 701–719
3.  Nowak, L., Bregestovski, P., Herbet, A. and Prochiantz, Z. (1984) Nature (London) **307**, 462–465
4.  McBain, C.J and Mayer, M.L (1994) Physiol. Rev. **74**, 723–753
5.  Mayer, M.L. and Westbrook, G.L. (1987) Prog. Neurobiol. **28**, 197–276
6.  Collingridge, G.L. and Lester, R.A. (1989) Pharmacol. Rev. **40**, 143–210
7.  Lodge, D. and Collingridge, G. (eds.) (1990) The Pharmacology of Excitatory Amino Acids: A TIPS Special Report, Elsevier Trends Journals, Cambridge
8.  Bristow, L.J., Hutson, P.H., Thorn, L. and Tricklebank, M.D. (1993) Br. J. Pharmacol. **108**, 1156–1163
9.  Grimwood, S., Moseley, A.M., Carling, R.W., Leeson, P.D. and Foster, A.C. (1992) Mol. Pharmacol. **41**, 923–930
10. Marks, M.J. and Collins, A.C. (1982) Mol. Pharmacol. **22**, 554–564
11. Hollmann, M., O'Shea-Greenfield, A., Rogers, S.W. and Heinemann, S. (1989) Nature (London) **342**, 643–648
12. Moriyoshi, K., Masu, M., Ishii, T., Shigamoto, R., Mizuno, N. and Nakanishi, S. (1991) Nature (London) **354**, 31–37
13. Yamazaki, M., Mori, H., Araki, K., Mori, K.J. and Mishina, M. (1992) FEBS Lett. **300**, 39–45
14. Monyer, H., Sprengel, R., Schoepfer, R., Herb, A., Higuchi, M., Lomeli, H., Burnashev, N., Sakmann, B. and Seeburg, P.H. (1992) Science **256**, 1217–1221
15. Ishii, T., Moriyoshi, K., Sugihara, H., Sakurada, K., Kadotani, H., Yokoi, M., Akazawa, C., Shigemoto, R., Mizuno, N., Masu, M. and Nakanishi, S. (1993) J. Biol. Chem. **268**, 2836–2843
16. Yamazaki, M., Mori, H., Araki, K., Mori, K.J. and Mishina, M. (1992) FEBS Lett. **300**, 39–45
17. Meguro, H., Mori, H., Araki, K., Kushiya, E., Kutsowada, T., Yamazaki, M., Kumanishi, T., Arakawa, M., Sakimura, K. and Mishina, M. (1992) Nature (London) **357**, 70–74
18. Kutsowada, T., Kashiwabuchi, N., Mori, H., Sakimura, K., Kushiya, E., Araki, K., Meguro, H., Masaki, H., Kumanishi, T., Arakawa, M. and Mishina, M. (1992) Nature (London) **358**, 36–41
19. Ikeda, K., Nagasawa, M., Mori, H., Araki, K., Sakimura, K., Watanabe, M., Inoue, Y. and Mishina, M. (1992) FEBS Lett. **313**, 34–38
20. Karp, S.J., Masu, M., Eki, T., Ozawa, K. and Nakanishi, S. (1993) J. Biol. Chem. **268**, 3728–3733
21. Planells-Cases, R., Sun, W., Ferrer-Montiel, A.V. and Montal, M. (1993) Proc. Natl. Acad. Sci. U.S.A. **90**, 5057–5061
22. Le Bourdellès, B., Wafford, K.A., Kemp, J.A., Marshall, G., Bain, C., Wilcox, C., Sikela, J.M. and Whiting, P.J. (1994) J. Neurochem. **62**, 2091–2098
23. Tingley, W.G., Roche, K.W., Thompson, A.K. and Huganir, R.L. (1993) Nature (London) **364**, 70–73
24. Kuryatov, A., Laube, B., Betz, H. and Kuhse, J. (1994) Neuron **12**, 1291–1300
25. Hirai, H., Kirsch, J., Laube, B., Betz, H. and Kuhse, J. (1996) Proc. Natl. Acad. Sci. U.S.A. **93**, 6031–6036
26. Hollmann, M., Maron, C. and Heinemann, S. (1994) Neuron **13**, 1331–1343

27. Bennett, J.A. and Dingledine, R. (1995) Neuron **14**, 373–384
28. Sugihara, H., Moriyoshi, K., Ishii, T., Masu, M. and Nakanishi, S. (1992) Biochem. Biophys. Res. Commun. **185**, 826–832
29. Foldes, R.L., Rampersad, V. and Kamboj, R.K. (1994) Gene **147**, 303–304
30. Hollmann, M., Boulter, J., Maron, C., Beasley, L., Sullivan, J., Pecht, G. and Heinemann, S. (1993) Neuron **10**, 943–954
31. Suchanek, B., Seeburg, P.H. and Sprengel, R. (1995) J. Biol. Chem. **270**, 41–44
32. Takano, H., Onodera, O., Tanaka, H., Mori, H., Sakimura, K., Hori, T., Kobayashi, H., Mishina, M. and Tsuji, S. (1993) Biochem. Biophys. Res Commun. **197**, 922–926
33. Watanabe, M., Inoue, Y., Sakimura, K. and Mishina, M. (1992) NeuroReport **3**, 1138–1140
34. Tölle, T.R., Berthele, A., Zieglgängsberger, W., Seeburg, P.H. and Wisden, W. (1993) J. Neurosci. **13**, 5009–5028
35. Monyer, H., Burnashev, N., Laurie, D.J., Sakmann, B. and Seeburg, P.H. (1994) Neuron **12**, 529–540
36. Akazawa, C., Shigemoto, R., Bessho, Y., Hakanishi, S. and Mizuno, N. (1994) J. Comp. Neurol. **347**, 150–160
37. Böckers, T.M., Zimmer, M., Müller, A., Bergmann, M., Brose, N. and Kreutz, M.R. (1994) NeuroReport **5**, 965–969
38. Laurie, D.J. and Seeburg, P.H. (1994) J. Neurosci. **14**, 3180–3194
39. Standaert, D.G., Testa, C.M., Young, A.B. and Penney, J.B. (1994) J. Comp. Neurol. **343**, 1–16
40. Zhong, J., Carrozza, D.P., Williams, K., Pritchett, D.B. and Mollinoff, P.B. (1995) J. Neurochem. **64**, 531–539
41. Wenzel, A., Scheurer, L., Künzi, R., Fritschy, J.M., Mohler, H. and Benke, D. (1995) NeuroReport **7**, 45–48
42. Prtera-Cailliau, C., Price, D.L. and Martin, L.J. (1996) J. Neurochem. **66**, 692–700
43. Dunah, A.W., Yasuda, R.P., Wang, Y-H, Luo, J., Davila-Garcia, M.I., Gbadegesin, M., Vicini, S. and Wolfe, B. (1996) J. Neurochem. **67**, 2335–2345
44. Petralia, R.S., Yokotani, N. and Wenthold, R.J. (1994) J. Neurosci. **14**, 667–696
45. Petralia, R.S., Wang, Y.X. and Wenthold, R.J. (1994) J. Neurosci. **14**, 6102–6120
46. Brose, N., Gasic, G.P., Vetter, D.E., Sullivan, J.M. and Heinemann, S.F. (1993) J. Biol. Chem. **268**, 22663–22671
47. Sheng, M., Cummings, J., Roldan, L.A., Jan, Y.N. and Yan, L.Y. (1994) Nature (London) **368**, 144–147
48. Didier, M., Xu, M., Berman, S.A. and Bursztajn, S. (1996) NeuroReport **6**, 2255–2259
49. Blahos II, J. and Wenthold, R.J. (1996) J. Biol. Chem. **271**, 15669–15674
50. Marti, T., Benke, D., Mertens, S., Heckendorn, R., Pozza, M., Allgeier, H., Angst, C., Laurie, D., Seeburg, P.H. and Mohler, H. (1993) Proc. Natl. Acad. Sci. U.S.A. **90**, 8434–8438
51. Bessho, Y., Nawa, H. and Nakanishi, S. (1994) Neuron **12**, 87–95
52. Wafford, K.A., Bain, C.J., Le Bourdellès, B., Whiting, P.J. and Kemp, J.A. (1993) NeuroReport **4**, 1347–1349
53. Chazot, P.L., Coleman, S.K., Cik, M. and Stephenson, F.A. (1994) J. Biol. Chem. **269**, 24403–24409
54. Clements, J.D. and Westbrook, G.L. (1991) Neuron **7**, 605–613
55. Kornau, H.C., Schenker, L.T., Kennedy, M.B. and Seeburg, P.H. (1995) Science **269**, 1737–1740
56. Niethammer, M., Kim, E. and Sheng, M. (1996) J. Neurosci. **16**, 2157–2163
57. Kim, E., Cho, K-O, Rothschild, A. and Sheng, M. (1996) Neuron **17**, 103–113
58. Müller, B.M., Kistner, U., Kindler, S., Chung, W.J., Kuhlendahl, S., Fenster, S.D., Lau, L.F., Rüdiger, W.V., Huganir, R.L., Gundelfinger, E.D. and Garner, C.C. (1996) Neuron **17**, 255–265
59. Brenman, J.E., Chao, D.S., Gee, S.H., McGee, A.W., Craven, S.E., Santillano, D.R., Wu, Z., Huang, F., Xia, H., Peters, M.F., Froehner, S.C. and Bredt, D.S. (1996) Cell, **84**, 757–767.
60. Boeckman, F.A and Aizenman, E. (1994) Neurosci. Lett. **173**, 189–192
61. Grimwood, S., Le Bourdellès, B. and Whiting, P.J. (1995) J. Neurochem. **64**, 525–530
62. Nakanishi, N., Axel, R. and Shneider, N.A. (1992) Proc. Natl. Acad. Sci. U.S.A. **89**, 8552–8556
63. Anantharam, V., Panchal, R.G., Wilson, A., Kolchine, V.V., Treistman, S.N. and Bayley, H. (1992) FEBS Lett. **305**, 27–30
64. Durand, G.M., Gregor, P., Zheng, X., Bennett, M.V.L., Uhl, G.R. and Zukin, R.S. (1992) Proc. Natl. Acad. Sci. U.S.A. **89**, 9359–9363

65. Durand, G.M., Bennett, M.V.L. and Zukin, R.S. (1993) Proc. Natl. Acad. Sci. U.S.A. **90**, 6731–6735

66. Laurie, D.J. and Seeburg, P.H. (1994) Eur. J. Pharmacol. Mol. Pharmacol. **268**, 335–345

67. Buller, A.L., Larson, H.C., Schneider, B.J., Beaton, J.A., Morrisett, R.A. and Monaghan, D.T. (1994) J. Neurosci. **14**, 5471–5484

68. Priestley, T., Laughton, P., Myers, J., Le Bourdellès, B., Wingrove, P.B. and Whiting, P.J. (1995) Mol. Pharmacol. **48**, 841–848

69. Kutsuwada, T., Kashiwabuchi, N., Mori, H., Sakimura, K., Kushiya, E., Araki, K., Meguro, H., Masaki, H., Kumnaishi, T., Arakawa, M. and Mishina, M. (1992) Nature (London) **358**, 36–41

70. Wafford, K.A., Kathoria, M., Bain, C., Marshall, G.R., Le Bourdellès, B., Kemp, J.A. and Whiting, P.J. (1995) Mol. Pharmacol. **47**, 374–380

71. Galen Wo, Z. and Oswald, R.E. (1995) Trends Neurosci. **18**, 161–167

72. Jahr, C.E. and Stevens, C.F. (1993) Proc. Natl. Acad. Sci. U.S.A. **90**, 11573–11577

73. Burnashev, N. (1993) Cell. Physiol. Biochem. **3**, 318–331

74. Verdoorn, T.L., Burnashev, N., Monyer, H., Seeburg, P.H. and Sakmann, B. (1991) Science **252**, 1715

75. Hulme, R.I., Dingledine, R. and Heinemann, S.F. (1991) Science **253**, 1028

76. Burnashev, N., Schoepfer, R., Monyer, H., Ruppersberg, J.P., Gunther, W., Seeburg, P.H. and Sakmann, B. (1992) Science **257**, 1415–1419

77. Mori, H., Masaki, H., Yamakura, T. and Mishina, M. (1992) Nature (London) **358**, 673–675

78. Sakurada, K., Masu, M. and Nakanishi, S. (1993) J. Biol. Chem. **268**, 410–415

79. Wollmuth, L.P., Kuner, T., Seeburg, P.H and Sakmann, B. (1996) J. Physiol. **491 (3)**, 779–797

80. Kuner, T., Wollmuth, L.P., Karlin, A., Seeburg, P.H. and Sakmann, B. (1996) Neuron **17**, 343–352

81. Kupper J., Ascher, P. and Neyton, J. (1996) Proc. Natl. Acad. Sci. U.S.A. **93**, 8648–8653

82. Stern, P., Behe, P., Schoepfer, R. and Colquhoun, D. (1992) Proc. R. Soc. London B. **250**, 271–277

83. Gibb, A.J and Colquhoun, D.C. (1991) Proc. R. Soc. London B. **243**, 39–45

84. Gibb, A.J. and Colquhoun, D.C. (1992) J. Physiol. (London) **456**, 143–179

85. Howe, J.R., Cull-Candy, S.G. and Colquhoun, D. (1991) J. Physiol. (London) **432**, 143–202

86. Priestley, T., Ochu, E.O. and Kemp, J.A. (1994) NeuroReport **5**, 1763–1765

87. Cull-Candy, S.G. and Usowicz, M.M. (1987) Nature (London) **325**, 525–528

88. Cull-Candy, S.G. and Usowicz, M.M. (1989) J. Physiol. (London) **415**, 555–582

89. Momiyama, A., Feldmeyer, D. and Cull-Candy, S.G. (1995) J. Physiol. (London) **483**, 162P

90. Yamakura, T., Mori, H., Masaki, H., Shimoji, K. and Mishina, M. (1993) NeuroReport **4**, 687–690

91 Cik, M., Chazot, P.L. and Stephenson, F.A. (1993) Biochem. J. **296**, 877–883

92. Grimwood, S., Le Bourdellès, B., Atack, J.A., Barton, B., Cockett, W., Cook, S.M., Gilbert, E., Hutson, P., Kerby, J., McKernan, R.M., Myers, J., Ragan, C.I. and Whiting, P.J (1996) J. Neurochem. **73**, 439–447

93. Chazot, P.L., Cik, M. and Stephenson, F.A. (1992) J. Neurochem. **59**, 1176–1178

94. Lynch, D.R., Anegawa, N.J., Verdoorn, T. and Pritchett, D.B. (1994) Mol. Pharmacol. **45**, 540–545

95. Ransom, R.W. and Stec, N.L. (1988) J. Neurochem. **51**, 830–834

96. Rock, D.M. and MacDonald, R.L. (1992) Mol. Pharmacol. **42**, 157–164

97. Carter, C., Rivy, J.-P. and Scatton, B. (1989) Eur. J. Pharmacol. **164**, 611–612

98. Marvizon, J.C. and Baudry, M. (1994) J. Neurochem. **63**, 963–971

99. Reynolds, I.J. and Miller, R.J. (1989) Mol. Pharmacol. **36**, 758–762

100 Gallagher, M.J., Huang, H., Pritchett, D.B. and Lynch, D.R. (1996) J. Biol. Chem **271**, 9603–9611

101. Kasiwagi, K., Fukuchi, J., Chao, J., Igarashi, K and Williams, K. (1996) Mol. Pharmacol. **49**, 1131–1141

102. Zhang, L., Zheng, X., Paupard, M.C., Wang, A.P., Santchi, L., Friedman, L.K., Zukin, R.S. and Bennett, M.V.L. (1994) Proc. Natl. Acad. Sci. U.S.A. **91**, 10883–10887

103. Zheng, X., Zhang, L., Durand, G.M., Bennett, M.V.L. and Zukin, R.S. (1994) Neuron **12**, 811–818

104. Traynelis, S.F., Hartley, M. and Heinemann, S.F. (1995) Science **268**, 873–876

105. Williams, K., Zappia, A.M., Pritchett, D.B., Min Shen, Y. and Molinoff, P.B. (1994) Mol. Pharmacol. **45**, 803–809
106. Williams, K. (1995) Neurosci. Lett. **184**, 181–184
107. Igarashi, K. and Williams, K. (1995) J. Pharmacol. Exp. Ther. **272**, 1101–1109
108. Priestley, T.P., Woodruff, G.N. and Kemp, J.A. (1989) Br. J. Pharmacol. **97**, 1315–1323
109. Williams, K. (1993) J. Pharmacol. Exp. Ther. **266**, 231–235
110. Raditsch, M., Rppersberg, P., Kuner, T., Gunther, W., Schoepfer, R., Seeburg, P.H., Janh, W. and Witzemann, V. (1993) FEBS Lett. **324**, 63–66
111. Williams, K. (1993) Mol. Pharmacol. **44**, 851–859
112. Legendre, P. and Westbrook, G.L. (1991) Mol. Pharmacol. **40**, 289–298
113. Williams, K., Russell, S.L., Min Shen, Y and Molinoff, P.B. (1993) Neuron **10**, 267–278
114. Bekkers, J.M. (1993) Science **261**, 104–106
115. Vorobjev, V.S., Sharanova, I.N., Walsh, I.B. and Haas, H.L. (1993) Neuron **11**, 837–844
116. Williams, K. (1994) Mol. Pharmacol. **46**, 531–541
117. Aizenman, E., Lipton, S.A. and Loring, R.H. (1989) Neuron **2**, 1257–1263
118. Sucher, N.J. and Lipton, S.A. (1991) J. Neurosci. Res. **30**, 582–591
119. Lei, S.Z., Pan, Z.H., Aggarwal, S.K., Chen, H.S., Hartman, J., Sucher, N.J. and Lipton, S.A. (1992) Neuron **8**, 1087–1099
120. Tang, L.H. and Aizenman, E. (1993) J. Physiol. (London) **465**, 303–323
121. Tang, L.H. and Aizenman, E. (1993) Mol. Pharmacol. **44**, 473–478
122. Tang, L.H. and Aizenman, E. (1993) Neuron **11**, 857–863
123. Sullivan, J.M., Heinemann, S.F., Sucher, N.J., Chen, V.H.S. and Lipton, S.A. (1993) Soc. Neurosci. Abstr. **19**, 624
124. Omerovic, A., Leonard, J.P. and Kelso, S.R. (1993) Soc. Neurosci. Abstr. **19**, 1359
125. Kohr, G., Eckardt, S., Luddens, H., Monyer, H. and Seeburg, P.H. (1994) Neuron **12**, 1031–1040
126. Zangerle, L., Cuenod, M., Winterhalter, K.H. and Quang Do, K. (1992) J. Neurochem. **59**, 181–189
127. Janaky, R., Varga, V., Saransaari, P. and Oja, S.S. (1993) Neurosci. Lett. **156**, 153–157
128. Aizenman, E., Jensen, F.E., Gallop, P.M., Rosenberg, P.A. and Tang, L.H. (1994) Neurosci. Lett. **168**, 189–192
129. Ginsberg, M.D., Reivich, M., Frinak, S. and Harbig, K. (1976) Stroke **7**, 125–131
130. Tang, C.M., Dichter, M. and Morad, M. (1990) Proc. Natl. Acad. Sci. U.S.A. **87**, 6445–6449
131. Traynelis, S.F. and Cull-Candy, S.G. (1990) Nature (London) **345**, 347–350
132. Vyklicky, L.M., Vlachova, V. and Krusek, J. (1990) J. Physiol. (London) **430**, 497–517
133. Traynelis, S.F. and Heinemann, S.F. (1993) Soc. Neurosci. Abstr. **19**, 558.8
134. Weight, F.F. (1992) Int. Rev. Neurobiol. **33**, 289–348
135. Kuner, T., Schoepfer, R. and Korpi, E.R. (1993) NeuroReport **5**, 297–300
136. Masood, K., Wu, C., Brauneis, U. and Weight, F.F. (1994) Mol. Pharmacol. **45**, 324–329
137. Urushihara, H., Tohda, M. and Nomura, Y. (1992) J. Biol Chem. **267**, 11697–11700
138. Yamakura, T., Mori, H., Shimoji, K. and Mishina, M. (1993) Biochem. Biophys. Res. Commun. **196**, 1537–1544
139. Mori, H., Yamakura, T., Masaki, H. and Mishina, M. (1993) NeuroReport **4**, 519–522
140. Sigel, E., Baur, R. and Malherbe, P. (1994) J. Biol. Chem. **269**, 8204–8208
141. Wang, Y.T. and Salter, M.W. (1994) Nature (London) **369**, 233–235
142. Lieberman, D.N. and Mody, I. (1994) Nature (London) **369**, 235–239
143. Rosenmund, C. and Westbrook, G.L. (1993) Neuron **10**, 805–814
144. Li, Y., Erzurumlu, R.S., Chen, C., Jhaveri, S. and Tonegawa, S. (1994) Cell **76**, 427–437
145. Forrest, D., Yuzaki, M., Soares, H.D., Ng, L., Luk, D.C., Sheng, M., Stewart, C.L., Morgan, J.I., Connor, J.A. and Curran, T. (1994) Neuron **13**, 325–338
146. Sakimura, K., Kutsuwada, T., Ito, I., Manabe, T., Takayama, C., Kushiya, E., Yagi, T., Aizawa, S., Inoue, Y., Sugiyama, H. and Mishina, M. (1995) Nature (London) **373**, 151–155
147. Ito, I., Sakimura, K., Mishina, M. and Sugiyama, H. (1996) Neurosci. Lett. **203**, 69–71
148. Kutsuwada, T., Sakimura, K., Manabe, T., Takayama, C., Katakura, N., Kushiya, E., Natsume, R., Watanabe, M., Inoue, Y., Yagi, T., Aizawa, S., Arakawa, M., Takahashi, T., Nakamura, Y., Mori, H. and Mishina, M. (1996) Neuron **16**, 333–344

149.  Ikeda, K., Araki, K., Takayama, C., Inoue, Y., Yagi, T., Aizawa, S. and Mishina, M. (1995) Mol. Brain Res. **33**, 61–71
150.  Tsien, J.Z., Huerta, P.T. and Tonegawa, S. (1996) Cell **87**, 1327–1338

# 9

# Receptor regulation by phosphorylation

**Lynn A. Raymond**

The Kinsmen Laboratory of Neurological Research, Department of Psychiatry, University of British Columbia, 2255 Wesbrook Mall, Vancouver, British Columbia, Canada V6T 1Z3

## Introduction

The precise regulation of synaptic strength in the mammalian central nervous system is essential to such processes as development, sensorimotor processing, learning and memory. Evidence suggests that a number of different mechanisms may be involved in mediating changes in synaptic strength in response to changes in the frequency of synaptic activation, including modification of neurotransmitter receptor responsiveness [1–3], changes in the numbers of active neurotransmitter receptors [4], alterations in neurotransmitter release [3,5,6], changes in synaptic morphology [7] and activation of immediate-early genes which trigger new protein synthesis [3,8]. Each of these mechanisms may contribute to mediating synaptic plasticity under different conditions, but the functional modulation of neurotransmitter receptors, which mediate signal transduction at the post-synaptic membrane, would be among the most rapid means of altering synaptic responsiveness. In this regard, a large body of evidence has suggested that protein phosphorylation is an important mechanism for the regulation of neurotransmitter receptor function [9–11]. Protein phosphorylation is a post-translational modification in which the γ-phosphate of ATP is transferred to a serine, threonine or tyrosine residue of a substrate protein. The phosphate group confers a high negative charge which may change the protein's conformation and thereby alter its function. This reaction is relatively rapid and reversible, making phosphorylation of neurotransmitter receptors a potentially important mechanism underlying synaptic plasticity.

The neurotransmitter receptor channels gated by glutamate and γ-amino-butyric acid (GABA) mediate most fast excitatory and inhibitory neurotrans-mission respectively in the mammalian central nervous system. Glutamate receptors are also involved in neuronal development, the induction and maintenance of long-term potentiation (LTP) and long-term depression (LTD), which are considered as models for learning and memory, and processes underlying stroke, epilepsy and such neurodegenerative disorders as Huntington disease, Parkinson's disease and

Alzheimer's disease [3,12–15]. On the other hand, regulation of GABA receptor function and/or distribution plays a role in mediating the acute and chronic effects of alcohol use, as well as the pathogenesis and treatment of epilepsy [4,16–18]. The distribution of GABA release in the hippocampus may also play a significant role in the co-ordinated induction of synaptic plasticity [19]. Protein phosphorylation and modulation of the ionotropic receptors for glutamate and GABA may therefore play an important role in both synaptic plasticity and neuropathological conditions.

Recent molecular cloning studies indicate that ionotropic glutamate and $GABA_A$ receptors, members of the ligand-gated neurotransmitter receptor superfamily, are composed of oligomeric complexes of homologous subunits, and the details of the molecular structure and function of these receptors are described elsewhere in this volume. In brief, $GABA_A$ receptors consist of combinations of the $\alpha1$–$\alpha6$, $\beta1$–$\beta4$, $\gamma1$–$\gamma3$, $\delta1$ and $\rho1$–$\rho2$ subunits, which have been classified according to sequence identity [20,21]. Ionotropic glutamate receptors are further divided into the subtypes $N$-methyl-D-aspartate (NMDA), $\alpha$-amino-3-hydroxy-5-methyl-4-isoxazolepropionate (AMPA) and kainate, based upon electrophysiological and pharmacological properties. NMDA receptors contain the NR1 subunit along with one or more of the NR2A, NR2B, NR2C and NR2D subunits (cloned from rat brain; the corresponding mouse clones are $\zeta1$ and $\epsilon1$, $\epsilon2$, $\epsilon3$ and $\epsilon4$ respectively); AMPA receptors are composed of combinations of the subunits GluR1(A), GluR2(B), GluR3(C) and GluR4(D); and combinations of the subunits GluR5, GluR6, GluR7, KA1 and KA2 form kainate receptors [22,23]. The ability to express recombinant receptors of defined subunit composition in abundant quantities and in non-neuronal cell lines has greatly facilitated studies of the structure, function and regulation of these receptors. In this chapter the evidence for direct phosphorylation and functional modulation of ionotropic glutamate and $GABA_A$ receptors will be reviewed, along with data suggesting a role for this process in mediating synaptic plasticity.

## Protein phosphorylation and synaptic plasticity

Protein phosphorylation is catalysed by protein kinases and reversed by protein phosphatases. Protein kinases and phosphatases comprise a large and diverse group of enzymes, possessing both substrate and activator specificity. Although many protein kinases are highly expressed in the brain, a smaller group have been more extensively investigated with regard to their role in regulating synaptic function. These include the serine/threonine kinases cyclic AMP-dependent protein kinase (PKA), calcium/calmodulin-dependent kinase type II (CAMKII) and calcium/phospholipid-dependent protein kinase C (PKC), as well as several of the protein tyrosine kinases. As the names indicate, the three serine/threonine kinases are activated by specific intracellular second messengers. Activation of protein tyrosine kinases, on the other hand, is generally mediated by the binding of extracellular ligands. Each different protein kinase recognizes and binds to relatively specific sequences of amino acids surrounding the serine, threonine or tyrosine residue that is targeted for phosphorylation. These consensus sequences, listed in Table 1, provide a starting point for predicting candidate sites for phosphorylation of glutamate and $GABA_A$ receptors, based upon the amino acid sequence of the cloned subunits.

**Table 1.    Protein kinase consensus sequences**

Residues that are targets for phosphorylation are emboldened and indicated by asterisks. X is any polar residue. For PKA and PKC, a rank order of preferred consensus sequences is given.

| Protein kinase | Consensus sequence(s) |
| --- | --- |
| PKA | $\text{R-R/K-X-}\mathbf{S}^*\text{/}\mathbf{T}^* > \text{R-X}_2\text{-}\mathbf{S}^*\text{/}\mathbf{T}^* = \text{R-X-}\mathbf{S}^*\text{/}\mathbf{T}^*$† |
| PKC | $\text{R/K}_{1-3}\text{-X}_{0-2}\text{-}\mathbf{S}^*\text{/}\mathbf{T}^*\text{-X}_{0-2}\text{-R/K}_{1-3} > \mathbf{S}^*\text{/}\mathbf{T}^*\text{-X}_{0-2}\text{-R/K}_{1-3} \geqslant$ $\text{R/K}_{1-3}\text{-X}_{0-2}\text{-}\mathbf{S}^*\text{/}\mathbf{T}^*$† |
| CAMKII | $\text{R-X-X-}\mathbf{S}^*\text{/}\mathbf{T}^*$† |
| Protein tyrosine kinases | $\text{N/Q}_{0\text{ }1}\text{-E/D}_{0\text{ }2}\text{-}\mathbf{Y}^*$‡ |

†Adapted from [115].

‡Adapted from [10].

Protein phosphorylation is known to regulate a wide range of biological functions, but the evidence that it plays a critical role in mediating synaptic plasticity in the central nervous system is particularly compelling. The best studied forms of synaptic plasticity in the mammalian central nervous system are hippocampal LTP (reviewed elsewhere in this volume) and its counterpart, LTD, as well as cerebellar LTD at the parallel fibre–Purkinje cell synapse. In each case, a specific pattern of afferent stimulation induces a change in synaptic strength. Hippocampal LTP at the Schaffer collateral–CA1 synapse is induced by 1–2 s of tetanic stimulation at 100–200 Hz; induction requires the activation of NMDA receptors, and the potentiation is expressed as an increase in amplitude of the excitatory post-synaptic current mediated by AMPA receptors (reviewed in [3]). LTD at this synapse is also triggered by activation of NMDA receptors, but induction occurs after several minutes of low-frequency stimulation (~1 Hz) [24,25]. Cerebellar LTD, on the other hand, is induced by coincident firing of parallel fibres and climbing fibres, requires activation of both AMPA and metabotropic glutamate receptors, and is expressed as a decrease in the synaptic current mediated by AMPA receptors at the parallel fibre–Purkinje cell synapse [26]. As outlined below, in each of these paradigms a variety of data suggest that protein kinases and phosphatases are involved in regulating synaptic strength.

A variety of studies have shown that induction of hippocampal LTP in the CA1 region involves activation of PKC, CAMKII and protein tyrosine kinases (for reviews, see [3,27,28]). Direct injection of inhibitors of PKC or CAMKII into the post-synaptic CA1 neuron has been shown to block induction and, in some cases, expression of LTP [29–31]. Moreover, expression of a constitutively active form of CAMKII in neurons of hippocampal slices resulted in enhanced synaptic transmission and occlusion of further LTP by usual induction protocols [32]. Tyrosine kinase inhibitors applied in the bath or injected into CA1 neurons also inhibit induction of LTP [33]. In other experiments, activation of PKC by phorbol esters or direct intracellular injection of the enzyme was shown to result in a long-lasting potentiation similar to LTP [34–36]. However, the fact that the phorbol ester-evoked potentiation did not occlude the subsequent development of LTP suggested that activation of PKC is necessary but not sufficient for inducing LTP [37,38]. Furthermore, two recent biochemical studies support a role for PKC in both the induction and expression of LTP. One employed a specific PKC substrate to show that this enzyme remains activated during the maintenance phase of LTP [39], and the other demonstrated high levels of a constitutively active form of the $\zeta$ isoform of PKC during expression of LTP [40]. Finally, in transgenic mice lacking one of $\alpha$CAMKII, the tyrosine kinase Fyn or the $\gamma$ isoform of PKC, induction of LTP by standard protocols is severely impaired, providing further support for the hypothesis that activation of these protein kinases is required for long-lasting increases in synaptic strength at the Schaffer collateral–CA1 synapse [41–43].

In contrast to LTP, hippocampal LTD appears to depend upon the activation of protein phosphatases. Bath application or intracellular injection of protein phosphatase inhibitors impairs or eliminates the development of NMDA-receptor-dependent LTD [44]. In addition, another recent study of voltage-pulse-induced (NMDA-receptor-independent) potentiation in CA1 pyramidal neurons, in which protein kinase and phosphatase inhibitors were bath applied or injected intracellularly, demonstrated that activation of a calcium-sensitive kinase (most likely CAMKII) resulted in enhanced AMPA-receptor-mediated post-synaptic currents, and that phosphatase activation antagonized this effect [45]. Both LTP and LTD at the Schaffer collateral–CA1 synapse are triggered by calcium influx through NMDA receptors [24,25,46], and voltage-pulse-induced potentiation in CA1 neurons also depends on the influx of calcium [47]. Taken together, these studies suggest that the spatio-temporal characteristics of post-synaptic intracellular calcium accumulation, in response to the frequency of afferent stimulation, regulate, at least in part, synaptic strength via the opposing activation of protein kinases and phosphatases (Fig. 1).

Just as PKC activation is important in mediating hippocampal LTP, this protein kinase also appears to play a major role in mediating cerebellar LTD. Studies with neuronal cultures and with slices have shown that application of phorbol ester, an activator of PKC, results in long-lasting depression of AMPA-mediated post-synaptic currents, and that inhibitors of PKC abolish LTD in cultured neurons [26].

In each of these paradigms, it is tempting to speculate that activation of protein kinases or phosphatases in the post-synaptic neuron directly affects synaptic strength by regulating the phosphorylation of post-synaptic glutamate

**Fig. 1.    Model of glutamate receptor modulation by phosphorylation**

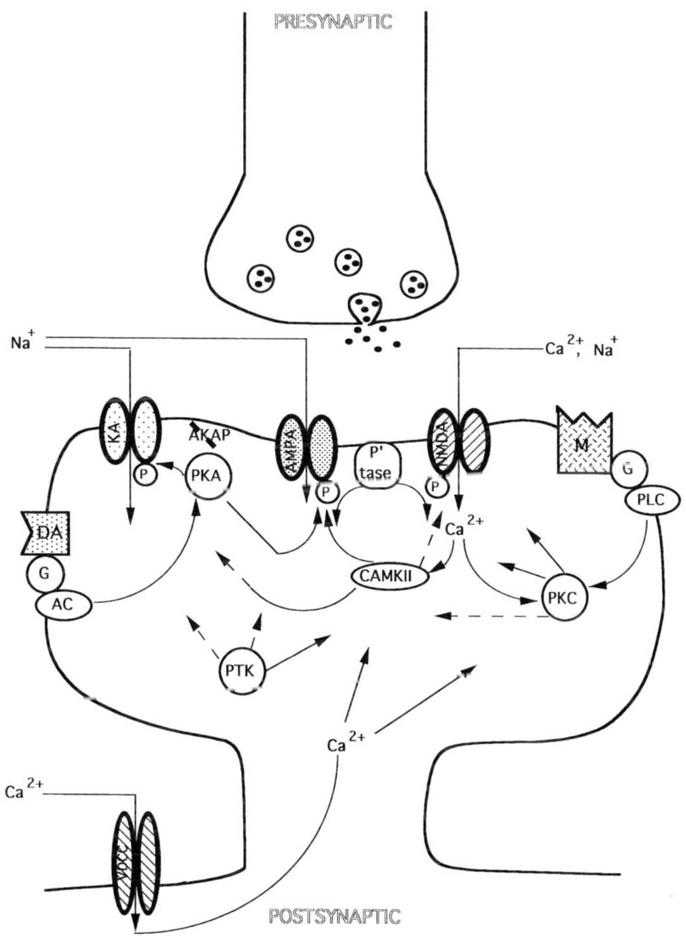



This schematic representation incorporates the results of several different studies which have demonstrated regulation of glutamate receptor function by activation of various protein kinases and phosphatases (see the text for details). The activity of these protein kinases and/or phosphatases can be stimulated as a result of activation of G-protein-coupled receptors such as the dopamine, muscarinic acetylcholine or metabotropic glutamate receptors, or via increases in intracellular calcium mediated by NMDA-receptor- or voltage-dependent calcium channel activation. The solid arrows indicate signalling pathways that have been demonstrated in published work, whereas the broken arrows represent potential protein kinase–receptor interactions. Abbreviations: DA, dopamine receptor; G, G-protein; AC, adenylate cyclase; M, muscarinic acetylcholine receptor or metabotropic glutamate receptor; PLC, phospholipase C; PTK, protein tyrosine kinase; P'tase, phosphatase (PP-1, PP-2A, PP-2B or a protein tyrosine phosphatase); KA, kainate receptor; AKAP, PKA anchoring protein; VDCC, voltage-dependent calcium channel.

receptors, thus modulating the responsiveness of the post-synaptic membrane. Although this question is difficult to approach by applying currently available biochemical or molecular biological techniques in the brain slice, there is increasing evidence from a variety of *in vitro* and *in situ* studies to indicate that phosphorylation of both $GABA_A$ and ionotropic glutamate receptors is a major mechanism for regulating their function. This evidence will be reviewed in the remainder of this chapter.

## Phosphorylation of non-NMDA glutamate receptors

Substantial evidence has accumulated in recent years suggesting that AMPA and kainate receptor function can be modulated by protein phosphorylation. Most of the work has centred on the regulation of these receptors by PKA activation. Early work by Dowling and colleagues showed that kainate-evoked currents in white perch horizontal retinal cells could be potentiated by treatment with dopamine. This effect could be mimicked by activation of PKA with forskolin or cAMP analogues, or by direct perfusion of the cells with PKA [48,49]. Activation of PKA by dopamine was also found to potentiate post-synaptic glutamate-receptor-mediated currents at the goldfish club-ending–Mauthner-cell synapse [50]. Similarly, AMPA-receptor-mediated currents in cultured mammalian hippocampal neurons were increased (and run-down prevented) by cAMP-dependent phosphorylation [51,52]; in both systems, increases in channel opening frequency and open time were found to underlie this potentiation [52,53]. In addition to the work done in cultured neurons, recombinant GluR1/GluR3 AMPA receptors expressed in *Xenopus* oocytes have been shown to be potentiated by bath application of cAMP analogues [54].

The studies cited above suggest that AMPA/kainate receptors may be directly phosphorylated and modulated by PKA. In support of this theory, a recent study by Blackstone et al. [55] provided biochemical evidence for direct PKA phosphorylation of the AMPA receptor subunit GluR1. In this study, native GluR1 subunits from cultured cortical neurons, as well as recombinant GluR1 expressed in HEK293 cells, were shown to be highly phosphorylated at a single site in the basal state, and forskolin activation of PKA resulted in phosphorylation of a second site. Interestingly, another study showed no evidence of PKA-mediated phosphorylation of GluR1 *in vitro* [56]. These seemingly contradictory results might be explained by a recent finding of Rosenmund et al. [57]. This group concluded that modulation of AMPA/kainate-receptor-mediated currents by PKA in cultured hippocampal neurons requires the interaction of the regulatory subunit of PKA with PKA anchoring proteins, since intracellular perfusion of synthetic peptides which specifically inhibit this interaction also blocked the potentiation of AMPA/kainate currents by PKA activation.

Studies combining biochemical and electrophysiological analysis with site-directed mutagenesis in order to analyse recombinant AMPA and kainate receptors expressed in non-neuronal cells have provided the most direct evidence for protein phosphorylation and functional modulation of these receptors. The kainate receptor subunit GluR6 has been shown to be directly phosphorylated by PKA using biochemical techniques, and most of the PKA-mediated phosphorylation was eliminated by using site-directed mutagenesis to substitute an alanine residue for serine-684 (a consensus site for PKA phosphorylation; see Table 1) [58]. Moreover, in whole-cell patch–clamp recordings from GluR6-transfected HEK293 cells, intracellular perfusion of the catalytic subunit of PKA resulted in significant peak current potentiation, whereas identical experiments with GluR6(S684A)-transfected cells showed little or no PKA-mediated increase in current [58,59]. In addition, recent work has demonstrated phosphorylation by PKA of the AMPA subunit GluR1 on serine-845 [59a]. Again, by mutagenesis and patch–clamp recording studies, PKA-dependent phosphorylation at this site was shown to result in potentiation of the peak GluR1-mediated current [59a].

There is also strong evidence for protein phosphorylation and functional modulation of AMPA receptors by CAMKII and PKC. In whole-cell patch–clamp recordings from cultured hippocampal neurons, intracellular perfusion of the constitutively active fragment of PKC resulted in the potentiation of currents elicited by saturating concentrations of AMPA or kainate; potentiation was blocked by co-perfusion of the inhibitory peptide PKCI-(19–36) [60]. In another study, a phosphorylated-form-specific antibody raised against a PKC phosphorylation site located just C-terminal to TM3 in GluR2/3 was used to demonstrate basal phosphorylation of this site in the post-synaptic density fraction from rat cerebellum [61]. Interestingly, the immunoreactivity of the phosphorylated form of GluR2/3 was markedly enhanced after application of glutamate. *In vitro* biochemical studies have demonstrated phosphorylation of the AMPA receptor subunit GluR1 by both PKC and CAMKII, and kainate-gated currents recorded from cultured hippocampal neurons were shown to be potentiated by the intracellular application of CAMKII [56]. The site of CAMKII phosphorylation of GluR1 was suggested to be serine 627 [56], and this was later confirmed [61a] Furthermore, a biochemical analysis of AMPA receptors immunoprecipitated from intact cultured hippocampal neurons showed increases in $^{32}$P incorporation stimulated by CAMKII, PKC and/or NMDA receptor activation [62]. These experiments also demonstrated the phosphorylation and activation of CAMKII in response to activation of NMDA receptors, consistent with the hypothesis that NMDA receptor activation results in stimulation of CAMKII and consequent phosphorylation and modulation of AMPA receptors [62]. The demonstration of direct phosphorylation of an AMPA receptor subunit by CAMKII, a kinase whose activation is required for induction of hippocampal LTP (see above), provides strong support for the theory that direct protein phosphorylation of post-synaptic glutamate receptors is a major mechanism underlying synaptic plasticity (see Fig. 1).

**Fig. 2.** **Two models of ionotropic amino acid receptor transmembrane topology**

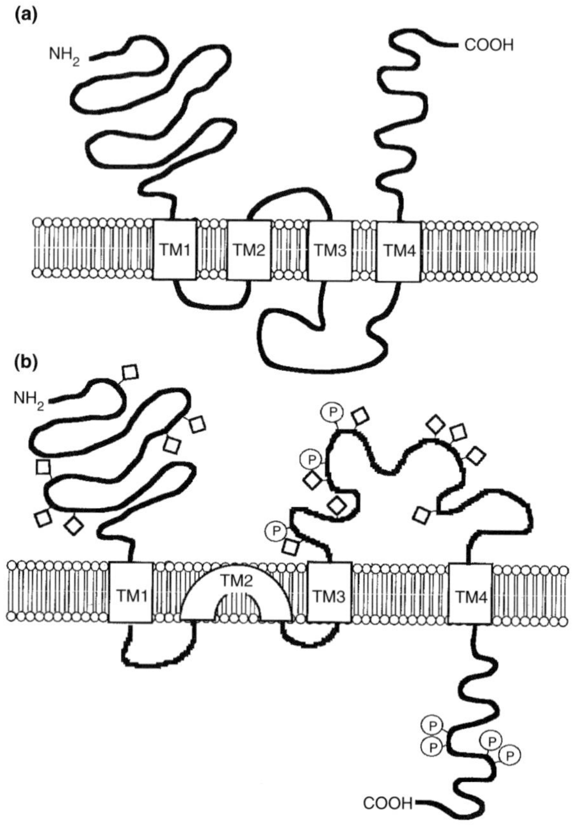



(*a*) This model was originally proposed for the nicotinic acetylcholine receptor cloned from muscle. Initially, the model was based upon hydrophobicity analysis of the receptor's amino acid sequence, but subsequent experimental evidence has supported the assignment of extracellular N- and C-terminal domains, four membrane-spanning regions (TM 1–TM4) and a large cytoplasmic region between TM3 and TM4 containing multiple consensus sites for phosphorylation by a variety of protein kinases. As other members of the ligand-gated ion channel superfamily were cloned and sequenced, including the glycine, $GABA_A$ and ionotropic glutamate receptors, this same model was applied. So far, experimental evidence has supported the model for glycine and $GABA_A$ receptors, but results of recent studies on glutamate receptors are in conflict with some aspects of this original model. (**b**) Revised transmembrane topology model for ionotropic glutamate receptors. This new model is based upon recent evidence from phosphorylation and glycosylation studies (see the text for details). Native and genetically engineered sites that have been shown to be either phosphorylated (P) or glycosylated (◊) are indicated. Although a variety of evidence suggests that this is the correct model for the receptor channel in its closed state, the phosphorylation data suggest that at least part of the region linking TM3 and TM4 may be translocated through the membrane to the cytoplasmic side upon channel opening (not shown).

The studies described above, identifying specific sites of phosphorylation by PKA and CAMKII that are responsible for the modulation of GluR6- and GluR1-mediated currents respectively, as well as the study by Ito and colleagues demonstrating phosphorylation of GluR2/3 at a PKC consensus site [61], all support the originally proposed transmembrane topology model for glutamate receptors. The model, based upon hydropathy plots and analogy with other ligand-gated ion channels, proposed the occurrence of four membrane-spanning segments (TM1–TM4), extracellular N- and C-termini and a large cytoplasmic region betwen TM3 and TM4 (Fig. 2a). Nearly all of the phosphorylation sites that have been identified so far in GluR1, GluR2/3 and GluR6 are located within the N-terminal half of the major loop between TM3 and TM4. However, recent evidence from a variety of studies (reviewed elsewhere in this text) has significantly revised this transmembrane topology model for glutamate receptors. The first challenge to the model came when the C-terminal region of NR1 was shown to be phosphorylated (see below) [63]. Later, several studies suggested that part [64–66] or all [67,68] of the region between the originally proposed TM3 and TM4 segments is extracellular and that TM2 forms a loop in the membrane rather than a fully spanning segment [66,67,69] (Fig. 2b). There are perhaps several explanations for the discrepancy between the phosphorylation data and the new model for glutamate receptors. One is that the conservative substitution of alanine for serine at sites in the region between TM3 and TM4 of GluR1 and GluR6 affects phosphorylation of distant sites yet to be identified, although it seems unlikely that results from four different groups could have been tainted by such a low-probability event. Another very interesting possibility is that the open and closed conformations of glutamate receptor channels are different, such that the entire stretch between TM3 and TM4 lies outside the cell in the closed state, but the N-terminal portion of this region (containing the phosphorylation sites) is translocated to the intracellular side in the open state. The results of Nakazawa et al. [61], showing that phosphorylation of a peptide in the N-terminal half of the TM3–TM4 loop region of GluR2/3 is largely dependent on extracellular glutamate (see above), support this hypothesis. Although these and other hypotheses remain to be tested, precedence for the membrane translocation of large portions of the channel protein during gating between the open and closed states has been established for the voltage-gated bacterial protein channels colicins IA and E1 [70–73].

## Phosphorylation of NMDA receptors

A variety of studies on native NMDA receptors in brain slices and cultured neurons provide strong evidence that these receptors are functionally modulated by protein phosphorylation. In spinal trigeminal neurons, intracellular application of PKC during whole-cell patch–clamp recordings resulted in potentiation of NMDA-receptor-mediated currents [74]. This increase in current appeared to be due, at least in part, to relief of the receptors' voltage-dependent magnesium block [75]. In spinal dorsal horn neurons, bath application of phorbol esters (activators of PKC) also enhanced NMDA-receptor-mediated currents [76]; this and other studies demonstrated effects of phorbol ester application on presynaptic glutamate release as well [28,76,77]. On the other hand, studies of the effects of phorbol esters on NMDA-evoked currents in CA1 pyramidal neurons of the hippocampus have yielded apparently conflicting results, with one group reporting potentiation [78] and another finding inhibition [79]. Activation of G-protein-coupled receptors which are known to activate PKC, such as the μ-opioid and metabotropic glutamate receptors, has also been shown to modulate NMDA receptor currents [74,77,80]. Furthermore, two recent electrophysiological studies using the perforated, cell-attached and inside-out patch techniques indicated that serine/threonine protein phosphatases may play a key role in modulating NMDA receptor currents. Inhibitors of protein phosphatases 1 and 2A, as well as of calcineurin, were shown to prolong the NMDA receptor single-channel open time and the duration of bursts and clusters, whereas application of purified protein phosphatases to inside-out patches inhibited channel opening [81,82].

Recent studies have suggested that tyrosine phosphorylation also plays a major role in regulating NMDA receptor function. Wang and Salter [83] showed that inhibitors of protein tyrosine kinases suppressed, whereas intracellular application of the protein tyrosine kinase pp60[c-src] enhanced, NMDA-receptor-mediated currents in whole-cell recordings from spinal dorsal horn neurons. A biochemical analysis by Kennedy and colleagues [84] demonstrated that the NMDA receptor subunit NR2B purified from rat forebrain is the major tyrosine-phosphorylated protein in the post-synaptic density fraction. Moreover, Lau and Huganir [85] have confirmed that NR2B is tyrosine-phosphorylated when immunoprecipitated from the rat cerebral cortex, and also showed that NR2A could be tyrosine-phosphorylated when synaptic plasma membranes were incubated with ATP and tyrosine phosphatase inhibitors.

Studies of recombinant NMDA receptors expressed in *Xenopus* oocytes that were injected with whole rat brain mRNA, or else with the NR1 subunit alone or in combination with NR2A or NR2B, have uniformly demonstrated potentiation of NMDA-evoked currents following bath application of phorbol esters [80,86–89]. However, in some studies this phorbol-ester-induced potentiation was transient, and in some cases was followed by depression of currents below baseline levels [88,90]. Interestingly, the presence or absence of three different alternatively spliced exons in the NR1 subunit appears to play a role in determining the degree of NMDA receptor current potentiation by phorbol esters, suggesting that these regions may contain phosphorylation sites for PKC [88].

Biochemical analysis of native NMDA receptors in cultured cortical neurons, as well as of recombinant receptors comprising NR1 and expressed in HEK293 cells, has revealed that at least four serine residues in the NR1 subunit are directly phosphorylated by PKC [63] (Fig. 3). These four serines, at positions 889,

## Fig. 3.                    Diagram of the NR1 subunit of the NMDA receptor

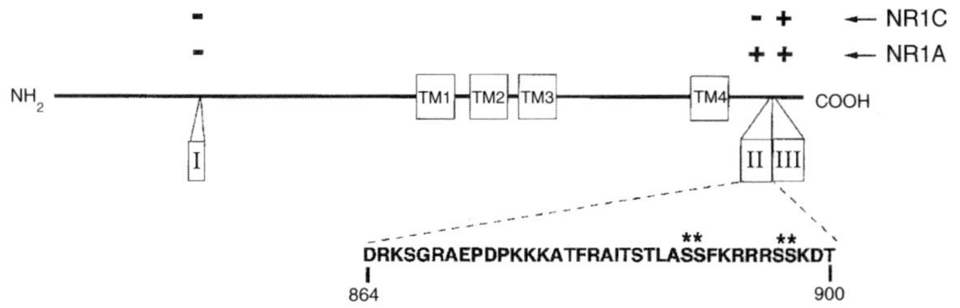

The locations of the three alternatively spliced regions of the NR1 subunit are shown here as I, II and III. From combinations of these three alternatively spliced regions, eight different splice variants of NR1 can be generated; two of these, called NR1A and NR1C (see [116]), exclude the first alternatively spliced region (I) and include the third region (III), but differ by the second alternatively spliced cassette (II). This latter region contains four serines that have been identified as PKC phosphorylation sites (indicated by *; see [63]), and appears to be involved in mediating interactions with cytoskeletal proteins that regulate receptor clustering (see the text).

890, 896 and 897, are all located on the second alternatively spliced exon, which is in the C-terminal region of the NR1 protein. This study provided the first data to suggest the revised view of the transmembrane topology model (see Fig. 2b) and definitive evidence that RNA splicing plays a role in regulating the phosphorylation of glutamate receptors. However, the effect on NMDA channel function of phosphorylation by PKC at these C-terminal sites remains unclear. One study, in which the second alternatively spliced exon was deleted and all of the remaining serine and threonine residues within the NR1 C-terminal domain were mutated to alanines, demonstrated no differences in phorbol-ester-induced potentiation of wild-type compared with mutant NR1-receptor-mediated currents in *Xenopus* oocytes [91]. In addition, two other studies demonstrated a decrease in the phorbol-ester-evoked potentiation of NR1- or NR1/NR2A-mediated currents in *Xenopus* oocytes when the second alternatively spliced exon was present in NR1 [88,90] Taken together, these studies suggest that there are additional PKC phosphorylation sites within NR1 that are responsible for potentiation of NMDA channel function.

Although the second alternatively spliced exon of NR1, with its PKC phosphorylation sites, appears not to be required for potentiation of NMDA channel function, it may play a critical role in the interaction of this receptor with cytoskeletal proteins. A recent study, using confocal microscopy and an anti-peptide antibody specific for NR1, demonstrated clustering of the NR1A splice variant at the plasma membrane of transfected QT cells [91a]. Clustering of NR1A required the presence of the second alternatively spliced exon, since the NR1C splice variant, which lacks this exon, showed a homogeneous membrane distribution. Interestingly, the clustering pattern of NR1A immunostaining in the plasma membranes of QT cells could be converted to a homogeneous distribution within several minutes following extracellular application of phorbol esters. The involvement of the PKC sites located within the second alternatively spliced exon in regulating receptor clustering was demonstrated when QT cells were transfected with mutant NR1A in which all four serine residues (positions 889, 890, 896 and 897) were mutated to alanines. These mutant NR1A receptors showed robust clustering that was resistant to phorbol ester treatment [91a]. In another study, the cytoskeletal protein actin was shown to play a key role in the desensitization/inactivation of NMDA-receptor-mediated currents in cultured neurons [92], and recent work indicates that both $\alpha$-actinin and calmodulin interact with the C-terminal region of NR1 [92a,92b]. It will be interesting to find out whether these interactions are also regulated by PKC-dependent phosphorylation.

## GABA$_A$ receptor phosphorylation

GABA$_A$ receptors, which are generally localized on proximal dendrites and the cell body of neurons, powerfully inhibit neuronal firing by gating a Cl$^-$-selective channel to maintain a hyperpolarized membrane potential. Clearly, alterations in the efficacy of GABA$_A$ receptor function would significantly affect neuronal signalling and thus may be a key mechanism for mediating neuronal plasticity.

Several *in vitro* biochemical studies indicate that GABA$_A$ receptors are substrates for phosphorylation by at least two protein kinases, PKA and PKC. Purified GABA$_A$ receptors from pig brain were phosphorylated by PKA on a 55 kDa polypeptide [93], and those from bovine brain by both PKC and PKA on 56 kDa and 58 kDa polypeptides [94]. All three polypeptides were determined to be $\beta$ subunits, based on muscimol binding. In a third study, fusion proteins of the proposed major intracellular domain (see Fig. 2a) of the GABA$_A$ receptor subunits $\beta$1 and $\gamma$2 were generated and tested for *in vitro* phosphorylation [95]. The $\beta$1 fusion protein was found to be phosphorylated by both PKA and PKC on serine-409. Two different fusion proteins of the $\gamma$2 subunit were produced to reflect the two alternatively spliced forms of this subunit: $\gamma$2$_S$ and $\gamma$2$_L$, which differ due to the addition of eight amino acids to the major intracellular domain of $\gamma$2$_L$. Both $\gamma$2 fusion proteins were phosphorylated by PKC on serine-327, and an additional PKC site was found on serine-343 within the alternatively spliced region in $\gamma$2$_L$. The results of these *in vitro* phosphorylation experiments on fusion proteins of the GABA$_A$ $\beta$ and $\gamma$ subunits have subsequently been confirmed by biochemical analysis of recombinant GABA$_A$ receptors in intact cells [96,97]. Together, these studies indicate that neuronal GABA$_A$ receptors are potential targets for direct protein phosphorylation by at least two protein kinases.

Although several lines of evidence suggest that GABA$_A$ receptors are functionally modulated by protein phosphorylation in neurons, the results of these studies are in some cases conflicting [10,98]. For example, $GABA_A$-receptor-mediated current responses in whole-cell voltage-clamp recordings from dissociated hippocampal neurons or cultured spinal cord neurons exhibited 'run-down' over several minutes unless the intracellular recording solution contained MgATP and substantial calcium buffering capacity [99–101]. Furthermore, the intracellular application of alkaline phosphatase, a dephosphorylating agent, hastened $GABA_A$ receptor current run-down [101]. These results suggested that ATP-dependent processes (perhaps direct receptor phosphorylation) were required to maintain $GABA_A$ receptor currents, whereas calcium-dependent processes (e.g. activation of protein phosphatases and/or calcium-sensitive protein kinases) could inhibit these currents. However, the results of a subsequent study, using whole-cell patch–clamp recording to examine run-down of $GABA_A$ receptor currents in dissociated nucleus tractus solitarius neurons, indicated that ATP may help to maintain these currents via a direct effect on receptor binding affinity which is not mediated by protein kinases [102].

In most cases, the effects of activation of at least two protein kinases, PKA and PKC, have been found to be inhibitory to $GABA_A$ receptor currents in neurons. Studies examining the effects of PKA activation on receptor function have yielded some seemingly contradictory results, however. One group showed that bath application of cAMP analogues to cultured hippocampal neurons resulted in a decrease in GABA-evoked peak current responses recorded under whole-cell voltage-clamp conditions; treatment with these agents also effected an increase in spontaneous neuronal firing [103]. A similar study with cultured chick cortical neurons showed a decrease in the peak current and an increase in the rate of $GABA_A$ receptor desensitization in response to both forskolin, an activator of adenylate cyclase, and cAMP analogues [104]. However, subsequent studies suggested that at least some of the effects of membrane-permeable cAMP analogues, as well as of forskolin, on $GABA_A$ receptor currents may not be mediated by cAMP-dependent protein phosphorylation. In these experiments, the effects of the application of membrane-permeable cAMP analogues were not blocked by the kinase inhibitor H-8 or by intracellular perfusion of cAMP, and bath application of membrane-impermeant cAMP was able to inhibit $GABA_A$ currents as well, suggesting an extracellular site of action [105,106]. The strongest evidence for PKA-mediated phosphorylation and modulation of $GABA_A$ receptor function comes from two studies in which the purified catalytic subunit of PKA was perfused intracellularly during whole-cell patch–clamp recordings in cultured mouse spinal cord neurons [107] or rat superior cervical ganglion neurons [96]. In both studies, the intracellularly applied PKA resulted in a decrease in the amplitude of the GABA-evoked current, due to a decrease in the channel opening frequency [107], and these effects were blocked or reversed by intracellular perfusion of the specific PKA peptide inhibitor PKI. Finally, in contrast with the studies described above, there is also evidence for potentiation of $GABA_A$ receptor function by increases in intracellular levels of cAMP, mediated by activation of β-adrenergic receptors in cerebellar Purkinje cell neurons [108].

As with the studies described above using activators of PKA, some of the results of studies examining the effects of PKC activation on GABA$_A$ receptor function have been in conflict. Three groups have shown a decrease in GABA-evoked currents after application of phorbol esters during voltage-clamp recordings in *Xenopus* oocytes injected with rat, mouse or chick brain mRNA; the application of phorbol esters that do not activate PKC was without effect on GABA-evoked responses [109–111]. In studies of neuronal GABA$_A$ receptors, $^{36}$Cl$^-$ influx in mouse brain cerebellar microsacs was found to be reduced by application of PKC activators [111], whereas a similar technique employed with cultured mouse spinal cord neurons showed no effect of the PKC-activating agent phorbol 12,13-dibutyrate on GABA-evoked responses [106]. In a third neuronal study, intracellular application of phorbol dibutyrate during whole-cell voltage-clamp recordings in sympathetic ganglion neurons resulted in a decrease in GABA$_A$ receptor peak currents; the inactive phorbol 4α-phorbol 12,13-didecanoate had no effect [97].

Some of the variability in the above results from studies on the PKC-, PKA- and/or cAMP-mediated modulation of GABA$_A$ receptor currents may be due to differences in GABA$_A$ receptor subunit composition, which is known to vary among neuronal cell types (for a review, see [112]). Another possible source of variability is diversity in the expression of specific protein kinases and their various isoforms among different mammalian species and neuronal cell types. Several studies have addressed this issue directly by analysing PKA- and PKC-mediated effects on the function of recombinant GABA$_A$ receptors expressed in non-neuronal cell lines or *Xenopus* oocytes. In a study by Moss et al. [96], GABA$_A$ receptors comprising α1 and β1 or α1, β1 and γ2 subunits were transiently expressed in HEK293 cells and analysed using biochemical, electrophysiological and site-directed mutagenesis techniques. In this system, activation of PKA by the extracellular application of forskolin resulted in a marked increase in phosphorylation of the β1 subunit at a single site, as determined by phosphopeptide mapping. Furthermore, phosphorylation of this site by PKA was eliminated by converting serine-409 (located in the proposed major intracellular loop) of β1 to an alanine residue. Moreover, the intracellular application of cAMP or the purified catalytic subunit of PKA during whole-cell patch–clamp recordings in cells transfected with wild-type receptors resulted in an approx. 40% decrease in the GABA-evoked peak current, as well as a decrease in the extent of rapid receptor desensitization for the α1β1 receptor combination. Both of these effects were lacking in receptors containing mutant β1(S409A) subunits. This study provides direct evidence for PKA phosphorylation of GABA$_A$ receptors and, in agreement with many of the previous studies discussed above, demonstrates an inhibitory effect on GABA-evoked currents following activation of PKA for receptors comprising α1β1 or α1β1γ2 subunits. A similar study by Krishek et al. [97] compared the PKC-mediated phosphorylation and modulation of receptors comprising $\alpha1\beta1\gamma2_S$ or $\alpha1\beta1\gamma2_L$ subunits expressed in HEK293 cells and *Xenopus* oocytes. Biochemical analysis revealed phosphorylation by PKC of three different sites (serine-409 of $\beta1$, serine-327 of $\gamma2_S$ and $\gamma2_L$, and serine-343 of $\gamma2_L$). A comparison of voltage-clamp recordings from cells expressing wild-type or mutant receptors (lacking all three sites) demonstrated that phorbol-ester-induced PKC phosphorylation of these three sites resulted in a decrease in the amplitude of the GABA-evoked current without affecting desensitization. Similarly, serine-410 of $\beta2$ and serine-327 of $\gamma2_S$ were found to be involved in mediating decreases in GABA-evoked currents following phorbol ester application in recordings from *Xenopus* oocytes expressing the recombinant subunit combination $\alpha1\beta2\gamma2_S$ [113]. In contrast, the results of a study by Lin et al. [114], in which the subunit combination $\alpha1\beta1\gamma2_L$ was expressed in cultured mouse fibroblasts and $GABA_A$ receptor function was analysed by whole-cell voltage-clamp recording, indicated that the intracellular application of constitutively active PKC mediates an increase in GABA-evoked current amplitudes. The specificity of this effect for PKC phosphorylation of the known sites in $\beta1$ and $\gamma2_L$ (see above) was not tested; however, the differences in experimental results may be due, in part, to differences in signal transduction pathways found in fibroblasts compared with HEK293 cells or *Xenopus* oocytes.


## Summary and conclusion

Evidence is accumulating rapidly to indicate that protein phosphorylation is a major mechanism for regulating amino acid receptor function. Studies in which recombinant glutamate and $GABA_A$ receptors have been expressed in heterologous systems have been particularly useful in determining the functional effects of receptor phosphorylation. These studies indicate that phosphorylation can regulate receptor ion channel function, as well as protein–protein interactions between receptors and cytoskeletal proteins that mediate clustering of surface receptors. In some cases, the amino acid residues phosphorylated by specific protein kinases have been identified within the receptor subunit sequence; a few of these sites have been within alternatively spliced sequences, implicating RNA splicing as a mechanism for regulating receptor phosphorylation. A role for glutamate receptor phosphorylation in hippocampal LTP and LTD has been suggested by the fact that these processes appear to be mediated, at least in part, by modulation of ionotropic glutamate receptor function, as well as the opposing activation of specific protein kinases and phosphatases. Further investigation will be required, however, in order to elucidate the role of protein phosphorylation of amino acid receptors in paradigms of synaptic plasticity *in situ* and *in vivo*. Several recently developed techniques may prove useful in addressing this issue, including the use of phosphoform-specific antibodies to those $GABA_A$ and glutamate receptor subunits with identified phosphorylation sites, and new techniques for modifying protein expression in neurons. It is clear, however, that with the wide variety of neuronal protein kinases, protein phosphatases and G-protein-coupled receptors which activate (or inhibit) these enzymes, coupled with the large number of different $GABA_A$ and glutamate receptor subunits with their alternatively spliced variants, protein phosphorylation of amino acid receptors has the potential to be a powerfully diverse mechanism for regulating synaptic transmission.

## References


1.  Davies, S.N., Lester, R.A.J., Reymann, K.G. and Collingridge, G.L. (1989) Nature (London) **338**, 500–503
2.  Manabe, T., Renner, P. and Nicoll, R.A. (1992) Nature (London) **355**, 50–55
3.  Bliss, T.V.P. and Collingridge, G.L. (1993) Nature (London) **361**, 31–39
4.  Mody, I., De Koninck, Y., Otis, T.S. and Soltesz, I. (1994) Trends Neurosci. **17**, 517–525
5.  Bekkers, J.M. and Stevens, C.F. (1990) Nature (London) **346**, 724–729
6.  Malinow, R. and Tsien, R.W. (1990) Nature (London) **346**, 177–180
7.  Lisman, J.E. and Harris, K.M. (1993) Trends Neurosci. **16**, 141–147
8.  Frank, D.A. and Greenberg, M.E. (1994) Cell **79**, 5–8
9.  Huganir, R.L. and Greengard, P. (1990) Neuron **5**, 555–567
10. Swope, S.L., Moss, S.J., Blackstone, C.D. and Huganir, R.L. (1992) FASEB J. **6**, 2514–2523
11. Raymond, L.A., Blackstone, C.D. and Huganir, R.L. (1993) Trends Neurosci. **16**, 147–153
12. Collingridge, G.L. and Singer, W. (1990) Trends Pharmacol. Sci. **11**, 290–296
13. Choi, D.W. (1988) Neuron **1**, 623–634
14. Albin, R.L. and Greenamyre, J.T. (1992) Neurology **42**, 733–738
15. Coyle, J.T. and Puttfarcken, P. (1993) Science **262**, 689–695
16. Lüddens, H., Pritchett, D.B., Köhler, M., Killisch, I., Keinänen, K., Monyer, H., Sprengel, R. and Seeburg, P.H. (1990) Nature (London) **346**, 648–651
17. Korpi, E.R., Kleingoor, C., Kettenmann, H. and Seeburg, P.H. (1993) Nature (London) **361**, 356–359
18. Bekenstein, J.W. and Lothman, E.W. (1993) Science **259**, 97–100
19. Sik, A., Ylinen, A., Penttonen, M. and Buzsaki, G. (1994) Science **265**, 1722–1724
20. Schöfield, P.R., Darlison, M.G., Fujita, N., Burt, D.R., Stephenson, F.A., Rodriguez, H., Rhee, L.M., Ramachandran, J., Reale, V., Glencorse, T.A., Seeburg, P.H. and Barnard, E.A. (1987) Nature (London) **328**, 221–227
21. Burt, D.R. and Kamatchi, G.L. (1991) FASEB J. **5**, 2916–2923
22. Nakanishi, S. (1992) Science **258**, 597–603
23. Seeburg, P.H. (1993) Trends Neurosci. **16**, 359–365
24. Dudek, S.M. and Bear, M.F. (1992) Proc. Natl. Acad. Sci. U.S.A. **89**, 4363–4367
25. Mulkey, R.M. and Malenka, R.C. (1992) Neuron **9**, 967–975
26. Linden, D.J. (1994) Neuron **12**, 457–472
27. Madison, D.V., Malenka. R.C. and Nicoll, R.A. (1991) Annu. Rev. Neurosci. **14**, 379–397
28. Ben-Ari, Y., Aniksztejn, L. and Bregestovski, P. (1992) Trends Neurosci. **15**, 333–339
29. Malenka, R.C., Kauer, J.A., Perkel, D.J., Mauk, M.D., Kelly, P.T., Nicoll, R.A. and Waxham, M.N. (1989) Nature (London) **340**, 554–557
30. Malinow, R., Schulman, H. and Tsien, R.W. (1989) Science **245**, 862–866
31. Wang, J.-H. and Feng, D.-P. (1992) Proc. Natl. Acad. Sci. U.S.A. **89**, 2576–2580
32. Pettit, D.L., Perlman, S. and Malinow, R. (1994) Science **266**, 1881–1885
33. O'Dell, T.J., Kandel, E.R. and Grant, S.G.N. (1991) Nature (London) **353**, 558–560
34. Akers, R.F., Lovinger, D.M., Colley, P.A., Linden, D.J. and Routtenberg, A. (1986) Science **231**, 587–589
35. Malenka, R.C., Madison, D.V. and Nicoll, R.A. (1986) Nature (London) **321**, 175–177
36. Hu, G.-Y., Hvalby, F., Walaas, S.I., Albert, K.A., Skjeflo, P., Andersen, P. and Greengard, P. (1987) Nature (London) **328**, 426–429
37. Muller, D., Turnbull, J., Baudry, M. and Lynch, G. (1988) Proc. Natl. Acad. Sci. U.S.A. **85**, 6997–7000

38. Gustafsson, B., Huang, Y.-Y. and Wigström, H. (1988) Neurosci. Lett. **85**, 77–81
39. Klann, E., Chen, S.-J. and Sweatt, J.D. (1993) Proc. Natl. Acad. Sci. U.S.A. **90**, 8337–8341
40. Sacktor, T.C., Osten, P., Valsamis, H., Jiang, X., Naik, M.U. and Sublette, E. (1993) Proc. Natl. Acad. Sci. U.S.A. **90**, 8342–8346
41. Silva, A.J., Stevens, C.F., Tonegawa, S. and Wang, Y. (1992) Science **257**, 201–206
42. Grant, S.G.N., O'Dell, T.J., Karl, K.A., Stein, P.L., Soriano, P. and Kandel E.R. (1992) Science **258**, 1903–1910
43. Abeliovich, A., Chen, C., Goda, Y., Silva, A.J., Stevens, C.F. and Tonegawa, S. (1993) Cell **75**, 1253–1262
44. Mulkey, R.M., Herron, C.E. and Malenka, R.C. (1993) Science **261**, 1051–1055
45. Wyllie, D.J.A. and Nicoll, R.A. (1994) Neuron **13**, 635–643
46. Collingridge, G.L., Kehl, S.J. and McLennan, H. (1983) J. Physiol. (London) **334**, 33–46
47. Kullmann, D.M., Perkel, D.J., Manabe, T. and Nicoll, R.A. (1992) Neuron **9**, 1175–1183
48. Knapp, A.G. and Dowling, J.E. (1987) Nature (London) **325**, 437–439
49. Liman, E.R., Knapp, A.G. and Dowling, J.E. (1989) Brain Res. **481**, 399–402
50. Pereda, A.E., Nairn, A.C., Wolszon, L.R. and Faber, D.S. (1994) J. Neurosci. **14**, 3704–3712
51. Wang, L.-Y., Salter, M.W. and MacDonald, J.F. (1991) Science **253**, 1132–1135
52. Greengard, P., Jen, J., Nairn, A.C. and Stevens, C.F. (1991) Science **253**, 1135–1138
53. Knapp, A.G., Schmidt, K.F. and Dowling, J.E. (1990) Proc. Natl. Acad. Sci. U.S.A. **87**, 767–771
54. Keller, B.U., Hollmann, M., Heinemann, S. and Konnerth, A. (1992) EMBO J. **11**, 891–896
55. Blackstone, C., Murphy, T.H., Moss, S.J., Baraban, J.M. and Huganir, R.L. (1994) J. Neurosci. **14**, 7585–7593
56. McGlade-McCulloh, E., Yamamoto, H., Tan, S.-E., Brickey, D.A. and Soderling, T.R. (1993) Nature (London) **362**, 640–642
57. Rosenmund, C., Carr, D.W., Bergeson, S.E., Nilaver, G., Scott, J.D. and Westbrook, G.L. (1994) Nature (London) **368**, 853–856
58. Raymond, L.A., Blackstone, C.D. and Huganir, R.L. (1993) Nature (London) **361**, 637–641
59. Wang, L.-Y., Taverna, F.A., Huang, X.-P., MacDonald, J.F. and Hampson, D.R. (1993) Science **259**, 1173–1175
59a. Roche, K.W., O'Brien, R.J., Mammen, A.L., Bernhardt, J. and Huganir, R.L. (1996) Neuron **16**, 1179–1188
60. Wang, L.-Y., Dudek, E.M., Browning, M.D. and MacDonald, J.F. (1994) J. Physiol. (London) **475**, 431–437
61. Nakazawa, K., Tadakuma, T., Yano, R., Nokihara, K. and Ito, M. (1994) Soc. Neurosci. Abstr. **20**, 741
61a. Yakel, J.L., Vissavajjhala, P., Derkach, V.A., Brickey, D.A. and Soderling, T.R. (1995) Proc. Natl. Acad. Sci. U.S.A. **92**, 1376–1380
62. Tan, S.-E., Wenthold, R.J. and Soderling, T.R. (1994) J. Neurosci. **14**, 1123–1129
63. Tingley, W.G,. Roche, K.W., Thompson, A.K. and Huganir, R.L. (1993) Nature (London) **364**, 70–73
64. Roche, K.W., Raymond, L.A., Blackstone, C. and Huganir, R.L. (1994) J. Biol. Chem. **269**, 11679–11682
65. Taverna, F.A., Wang, L.-Y., MacDonald, J.F. and Hampson, D.R. (1994) J. Biol. Chem. **269**, 14159–14164
66. Wo, Z.G. and Oswald, R.E. (1994) Proc. Natl. Acad. Sci. U.S.A. **91**, 7154–7158
67. Hollmann, M., Maron, C. and Heinemann, S. (1994) Neuron **13**, 1331–1343
68. Stern-Bach, Y., Bettler, B., Hartley, M., Sheppard, P.O., O'Hara, P.J. and Heinemann, S.F. (1994) Neuron **13**, 1345–1357
69. Bennett, J. and Dingledine, R. (1995) Neuron **14**, 167–175
70. Raymond, L., Slatin, S.L., Finkelstein, A., Liu, Q.-R. and Levinthal, C. (1986) J. Membr. Biol. **92**, 255–268
71. Cramer, W.A., Cohen, F.S., Merrill, A.R. and Song, H.Y. (1990) Mol. Microbiol. **4**, 519–526
72. Qiu, X.-Q., Jakes, K.S., Finkelstein, A. and Slatin, S.L. (1994) J. Biol. Chem. **269**, 7483–7488
73. Slatin, S.L., Qiu, X.-Q., Jakes, K.S. and Finkelstein, A. (1994) Nature (London) **371**, 158–161
74. Chen, L. and Huang, L.M. (1991) Neuron **7**, 319–326
75. Chen, L. and Huang, L.-Y.M. (1992) Nature (London) **356**, 521–523
76. Gerber, G., Kangrga, I., Ryu, P.D. and Larew, J.S.A. (1989) J. Neurosci. **9**, 3606–3617

77. Malenka, R.C., Ayoub, G.S. and Nicoll, R.A. (1987) Brain Res. **403**, 198–203
78. Aniksztejn, L., Otani, S. and Ben-Ari, Y. (1992) Eur. J. Neurosci. **4**, 500–505
79. Markram, H. and Segal, M. (1992) J. Physiol. (London) **457**, 491–501
80. Kelso, S.R., Nelson, T.E. and Leonard, J.P. (1992) J. Physiol. (London) **449**, 705–718
81. Wang, L.-Y., Orser, B.A., Brautigan, D.L. and MacDonald, J.F. (1994) Nature (London) **369**, 230–232
82. Lieberman, D.N. and Mody, I. (1994) Nature (London) **369**, 235–239
83. Wang, Y.T. and Salter, M.W. (1994) Nature (London) **369**, 233–235
84. Moon, I.S., Apperson, M.L. and Kennedy, M.B. (1994) Proc. Natl. Acad. Sci. U.S.A. **91**, 3954–3958
85. Lau, L.-F. and Huganir, R.L. (1996) J. Biol. Chem. **270**, 20036–20041
86. Urushihara, H., Tohda, M. and Nomura, Y. (1992) J. Biol. Chem. **267**, 11697–11700
87. Yamazaki, M., Mori, H., Araki, K., Mori, K.J. and Mishina, M. (1992) FEBS Lett. **300**, 39–45
88. Durand, G.M., Bennett, M.V.L. and Zukin, R.S. (1993) Proc. Natl. Acad. Sci. U.S.A. **90**, 6731–6735
89. Kutsuwada, T., Kashiwabuchi, N., Mori, H., Sakimura, K., Kushiya, E., Araki, K., Meguro, H., Masaki, H., Kumanishi, T., Arakawa, M. and Mishina, M. (1992) Nature (London) **358**, 36–41
90. Sigel, E., Baur, R. and Malherbe, P. (1994) J. Biol. Chem. **269**, 8204–8208
91. Yamakura, T., Mori, H., Shimoji, K. and Mishina, M. (1993) Biochem. Biophys. Res. Commun. **196**, 1537–1544
91a. Ehlers, M.D., Tingley, W.G. and Huganir, R.L. (1995) Science **269**, 1734–1737
92. Rosenmund, C. and Westbrook, G.L. (1993) Neuron **10**, 805–814
92a. Ehlers, M.D., Zhang, S., Bernhardt, J.P. and Huganir, R.L. (1996) Cell **84**, 745–755
92b. Wyszynski, M., Lin, J., Rao, A., Wigh, E., Beggs, A.H., Craig, A.M. and Seng, M. (1996) Nature (London) **385**, 439–442
93. Kirkness, E.F., Bovenkerk, C.F., Ueda, T. and Turner, A.J. (1989) Biochem. J. **259**, 613–616
94. Browning, M.D., Bureau, M., Dudek, E.M. and Olsen, R.W. (1990) Proc. Natl. Acad. Sci. U.S.A. **87**, 1315–1318
95. Moss, S.J., Doherty, C.A. and Huganir, R.L. (1992) J. Biol. Chem. **267**, 14470–14476
96. Moss, S.J., Smart, T.G., Blackstone, C.D. and Huganir, R.L. (1992) Science **257**, 661–665
97. Krishek, B.J., Xie, X., Blackstone, C., Huganir, R.L., Moss, S.J. and Smart, T.G. (1994) Neuron **12**, 1081–1095
98. Leidenheimer, N.J., Browning, M.D. and Harris, R.A. (1991) Trends Pharmacol. Sci. **12**, 84–87
99. Stelzer, A., Kay, A.R. and Wong, R.K.S. (1988) Science **241**, 339–341
100. Gyenes, M., Farrant, M. and Farb, D.H. (1988) Mol. Pharmacol. **34**, 719–723
101. Chen, Q.X., Stelzer, A., Kay, A.R. and Wong, R.K.S. (1990) J. Physiol. (London) **420**, 207–221
102. Shirasaki, T., Aibara, K. and Akaike, N. (1992) J. Physiol. (London) **449**, 551–572
103. Harrison, N.L. and Lambert, N.A. (1989) Neurosci. Lett. **105**, 137–142
104. Tehrani, M.H.J., Hablitz, J.J. and Barnes, Jr., E.M. (1989) Synapse **4**, 126–131
105. Lambert, N.A. and Harrison, N.L. (1990) J. Pharmacol. Exp. Ther. **255**, 90–94
106. Ticku, M.J. and Mehta, A.K. (1990) Mol. Pharmacol. **38**, 719–724
107. Porter, N.M., Twyman, R.E., Uhler, M.D. and Macdonald, R.L. (1990) Neuron **5**, 789–796
108. Sessler, F.M., Mouradian, R.D., Cheng, J.-T., Yeh, H.H., Liu, W. and Waterhouse, B.D. (1989) Brain Res. **499**, 27–38
109. Sigel, E. and Baur, R. (1988) Proc. Natl. Acad. Sci. U.S.A. **85**, 6192–6196
110. Moran, O. and Dascal, N. (1989) Mol. Brain Res. **5**, 193–202
111. Leidenheimer, N.J., McQuilkin, S.J., Hahner, L.D., Whiting, P. and Harris, R.A. (1992) Mol. Pharmacol. **41**, 1116–1123
112. Vicini, S. (1991) Neuropsychopharmacology **4**, 9–15
113. Kellenberger, S., Malherbe, P. and Sigel, E. (1992) J. Biol. Chem. **267**, 25660–25663
114. Lin, Y.-F., Browning, M.D., Dudek, E.M. and Macdonald, R.L. (1994) Neuron **13**, 1421–1431
115. Kennelly, P.J. and Krebs, E.G. (1991) J. Biol. Chem. **266**, 15555–15558
116. Sugihara, H., Moriyoshi, K., Ishii, T., Masu, M. and Nakanishi, S. (1992) Biochem. Biophys. Res. Commun. **185**, 826–832

# 10

# Excitatory amino acids and neurodegeneration

**Alan M. Palmer**

Department of Neurochemistry, Cerebrus Ltd, Silwood Park,
Buckhurst Road, Ascot, U.K.

## Introduction

The dicarboxylic amino acids L-aspartate and L-glutamate play a critical role in the mammalian central nervous system, since they represent the major excitatory neurotransmitters [1–4]. This excitatory action was first demonstrated more than 40 years ago following intracarotid injection in humans by workers investigating epileptic phenomena [5,6]. The direct demonstration of neuronal excitation came later with the iontophoretic studies of Watkins and colleagues [7]. This, together with subsequent and more detailed electrophysiological studies [8], led to the recognition that aspartate and glutamate are neurotransmitters in the brain. Neurotransmitter status was amply confirmed by subsequent studies, which further linked excitatory amino acids (EAAs) with a number of neuronal types, including pyramidal cells of the cerebral cortex, mossy parallel and climbing fibres of the cerebellum and thalamocortical neurons [9].

Dysfunction of EAA neurotransmission has been linked to brain dysfunction [10,11], and (under certain circumstances) EAAs themselves have been shown to cause cell death. This toxic action was first described in 1957 by Lucas and Newhouse, who found that, in immature mice, high systemic doses of glutamate induced degeneration of retinal neurons [12]. However, it was the seminal work of John Olney and colleagues that clearly established the toxic actions of EAAs. His histological studies with L-glutamate, and other EAAs, confirmed the work of Lucas and Newhouse and further established that certain brain regions not protected by the blood–brain barrier (notably the arcuate nucleus of the hypothalamus) were damaged by the systemic administration of EAAs [13-17]. These areas had a characteristic cytopathology in which post-synaptic structures (perikarya and dendrites) were destroyed, yet axons, presynaptic terminals and glial cells survived [15,16,18]. This same pattern of damage was evident when L-glutamate or its analogues were administered by local injection into different brain regions of adult animals [19]. In addition, a good correlation was established between the neurotoxic [18] and neuro-excitatory [20] potencies of a range of related amino acids, suggesting that there was a causal relationship between cellular activation and cell death. Thus the term 'excitotoxicity' was introduced. However, it took more than a decade to appreciate the full significance of these observations for the aetiology of disorders associated with both acute and chronic neurodegeneration. This chapter will review the role of excitotoxic mechanisms in a number of neurodegenerative disorders, and approaches to therapy. But first it is necessary to consider EAA neurotransmission and the mechanisms underlying excitotoxin-mediated brain injury in greater detail.

## EAA neurotransmission

### Neurons that release EAAs

L-Aspartate and L-glutamate satisfy most of the criteria for neurotransmitter status [4]. Neurons that release EAAs can be broadly divided, on the basis of the localization of their cell bodies, into cerebrocortical, thalamic and cerebellar. Evidence in support of EAAs being the transmitters in cerebellar and thalamocortical neurons has been reviewed in [9]. EAAs are the transmitters of many (if not all) pyramidal cells, the most abundant neuron in the mammalian cortex [21]. The evidence is greatest for the pyramidal neurons of the corticostriatal pathway (the massive projection system that links virtually the entire neocortex with the neostriatum) and is based on a variety of data. First, selective retrograde labelling of cortical pyramidal neurons occurs after injection of D-[$^3$H]aspartate (a non-metabolizable EAA analogue) into rat neostriatum [22] Secondly, the electrophysiological responsiveness of the neostriatum to electrical stimulation of the neocortex can be mimicked by direct application of EAA receptor agonists to the neostriatum and blocked by EAA receptor antagonists [23,24]. Thirdly, EAA transmission is diminished after ablation or undercutting of the neocortex [25]. Finally, the interstitial concentration of EAAs ([EAA]$_i$) in the neostriatum are increased after stimulation of the frontal cortex [26]. The cerebral cortex innervates a number of other subcortical areas, and there is good neuroanatomical and neurochemical evidence that these projections also use EAAs as transmitters [2,9,27]. Evidence to indicate a transmitter role for EAAs in corticocortical neurons is less advanced, but has been derived from retrograde transport studies in conjunction with D-[$^3$H]aspartate autoradiography and aspartate and glutamate immunocytochemistry, as well as studies of EAA release from areas of the cortex following electrical stimulation of known input pathways [28,29].

### Metabolic and transmitter pools of EAAs

L-Aspartate and L-glutamate play a key role in intermediary metabolism in all cells, in addition to their transmitter role in EAAergic neurons. These distinct roles are uniquely coupled in EAA nerve terminals. The biosynthetic pathways for metabolic and transmitter pools are the same, with concentrations of aspartate and

**Fig. 1.**            **Pathways for the synthesis of glutamate and aspartate**

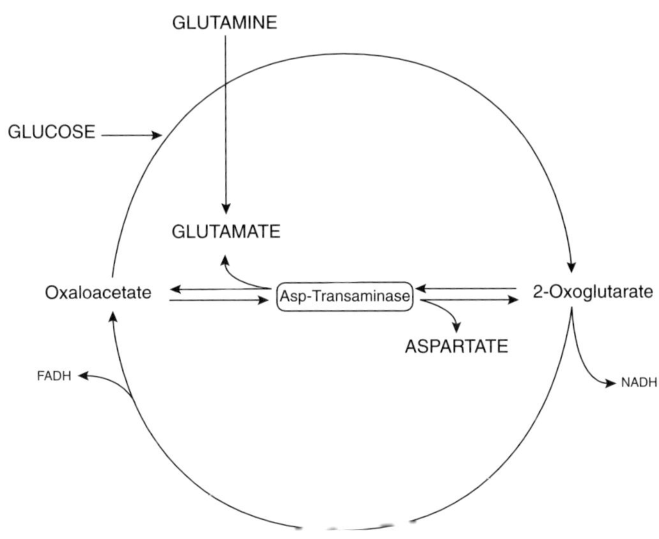


*Glutamate can be produced from glutamine, which is normally abundant in the interstitial space (approx. 0.5 mM) and is deaminated in neurons by the enzyme glutaminase to form glutamate. Glutamate is also produced from 2-oxoglutarate, an intermediate in the tricarboxylic acid (TCA) energy metabolic cycle. The cytosolic enzyme aspartate transaminase (Asp-Transaminase) catalyses the reversible reaction that interconverts glutamate and aspartate, and is thus critically involved in regulating the entry and exit of carbon from the TCA cycle. The equilibrium concentrations of glutamate and aspartate therefore depend on the relative concentrations of glutamine and glucose. Decreasing the availability of glucose should reverse glutamate synthesis from 2-oxoglutarate, causing instead glutamate metabolism in the TCA cycle and increased production of aspartate.*

glutamate being tightly linked by the cytosolic enzyme aspartate transaminase (also known as aspartate aminotransferase). Transmitter glutamate is synthesized primarily from the precursor molecules glutamine and 2-oxoglutarate (Fig. 1). Synthesis by the first route involves deamination of glutamine by the enzyme glutaminase, and that by the second route involves carboxylation of 2-oxoglutarate by the enzyme glutamic acid decarboxylase. The law of mass action determines the net direction of glutamate flow into or out of the tricarboxylic acid cycle; when the glucose concentration and glycolysis are reduced, acetyl-CoA formation is limited, less 2-oxoglutarate is derived, and so the equilibrium changes to favour glutamate catabolism (via the tricarboxylic acid cycle) over glutamate synthesis. Thus the concentrations of aspartate and glutamate are determined by the relative concentrations of glucose and glutamine [30].

## Role of glial cells

Astrocytes are intimately linked to EAA neurotransmission, since not only do they contain a high-affinity EAA transporter, they also serve an important anapleurotic role, providing intermediates required for both metabolism and the formation of transmitter EAAs. The anatomy of brain astrocytes is well suited to such a role, since astrocytic end feet surround intraparenchymal capillaries, the major source of glucose [31]. Thus astrocytes form the first cellular barrier encountered by glucose entering the brain, and glutamine derived from astrocytes serves as a precursor for glutamate formation. In addition, astrocytes transport 2-oxoglutarate to neurons, and glutamate has recently been shown to stimulate glucose utilization and the formation of lactate in astrocytes [32]. This latter observation, coupled with the demonstration of a specific uptake system for lactate in neurons, suggests the existence of a signalling mechanism that allows neuronal activation to be tightly coupled to glucose utilization in brain [32].

## EAA release

$Ca^{2+}$-dependent release of endogenous L-aspartate and L-glutamate from mammalian brain has been demonstrated in numerous studies, both *in vitro* and *in vivo* [2,3,26,33]. The transmitter status of L-aspartate has been questioned, on the basis that neither L- nor D-aspartate competes with L-glutamate for uptake into isolated synaptic vesicles [34]. However, there is good evidence to support a transmitter role for L-aspartate, including a demonstration that both L-aspartate and L-glutamate contribute to post-synaptic responses [4,35]. In addition, the accumulation of radiolabelled L- and D-aspartate and L-glutamate into synaptosomes derived from whole rat brain has recently been demonstrated by pre-accumulating radioactivity into synaptosomes and then isolating vesicles, rather than first isolating synaptic vesicles and then assaying for the uptake of amino acids (M.W. Fleck, G. Barrionuevo and A.M. Palmer, unpublished observations). This approach avoids problems of interpretation associated with incubating isolated synaptic vesicles in non-physiological medium.

Inhibition of EAA release may provide an approach to the treatment of excitotoxic injury. Factors controlling EAA (particularly glutamate) release are reviewed by Nicholls and Sanchez-Priéto in Chapter 1 of this volume.

## EAA uptake

The actions of EAAs released into the synaptic cleft are rapidly terminated by avid high-affinity uptake systems located both on EAA nerve terminals and on astrocytes surrounding the synaptic cleft. Recent molecular advances have led to the cloning of three separate cDNAs encoding $Na^+$-dependent EAA transporters. Of these, immunohistochemical evidence has revealed that GLT-1 is present on astroglia, GLAST is located to astroglia and some neurons, and EAAC1 is selectively localized to neurons (see the review by Kanner in Chapter 2 of this volume). In astrocytes, EAA transport is an electogenic process by which one molecule of L-aspartate or L-glutamate is co-transported with three Na⁺ (or two Na⁺ and one H⁺) ions in exchange for one K⁺ and one OH⁻ (or one HCO₃⁻) [36]. From this stoichiometry, it is clear that EAA uptake into astocytes can be expected to increase the Na⁺ concentration in the cytosol of the astrocyte, and to result in intracellular acidification and extracellular alkalinization.

## EAA receptors

The excitatory actions of EAAs are transduced by both metabotropic and ionotropic receptors. Various metabotropic glutamate receptors (mGluRs) are positively coupled to phosphoinositide hydrolysis (mGluR1 and mGluR5) or negatively coupled to adenylate cyclase activity (mGluRs 3, 4 and 6) (see Chapter 6 of this volume). It is the phosphoinositide-linked receptor that is most relevant to excitotoxic injury, since it is linked (via G-proteins) to phospholipase C, which triggers the hydrolysis of phosphatidylinositol 4,5-bisphosphate to liberate the dual messengers inositol 1,4,5-trisphosphate and diacylglycerol; the latter in turn mobilizes intracellular stores of Ca²⁺. (Glutamate metabotropic receptors may be a misnomer, since most of these receptors also bind aspartate; EAA metabotropic receptors may be a more correct term.) The role of metabotropic receptors in excitotoxic injury has not yet been completely elucidated, although there are reports of neuroprotection by metabotropic receptor antagonists such as 2-amino-3-phosphonopropionic acid [37].

There are two general subtypes of ionotropic EAA receptors: α-amino-3-hydroxy-5-methyl-4-isoxazolepropionate (AMPA)/kainate receptors and *N*-methyl-D-aspartate (NMDA) receptors (reviewed in Chapters 7 and 8 respectively of this volume). Both are widely distributed throughout the mammalian brain [1,38] and are largely co-localized at EAAergic synapses [39,40]. The activity of these ionotropic receptors is central to the role of EAAs in both normal synaptic transmission and neurodegeneration. AMPA/kainate receptors are coupled to a Na⁺/K⁺ ionophore and have fast kinetics, whereas NMDA receptors are coupled to a Na⁺/K⁺/Ca²⁺ ionophore and have relatively slower kinetics [41,42]. However, it has become clear that some AMPA/kainate receptors also gate Ca²⁺, much like the NMDA receptor gated channel. Such AMPA/kainate receptor gated channels have been reported in both hippocampal and cerebellar neurons [43,44], and can be created by transfecting fibroblasts with AMPA receptor subunit combinations that do not include the GluR2/GluRB subunit [45,46]. Since Ca²⁺ ions play a central part in excitotoxic injury, this suggests that cells with this type of receptor are selectively vulnerable to excitotoxic damage. This does appear to be the case [47,48]. The activity of the NMDA receptor is regulated by a number of subsites that interact with a variety of endogenous substances [49]. These sites include a transmitter binding site that binds L-aspartate or L-glutamate, a co-agonist site that binds glycine, a voltage dependent Mg²⁺ binding site (which regulates current flow through NMDA receptors in a voltage-dependent manner and greatly reduces NMDA receptor involvement in fast synaptic transmission), an inhibitory bivalent cation modulatory site that binds $Zn^{2+}$ (Zn appears to be present in nerve terminals and is released during synaptic transmission), and a positive modulatory site that binds polyamines (such as spermine and spermidine), which may also serve a neuromodulatory role. Still further modulation is provided by pH [50] and by oxidizing and reducing agents acting at a thiol redox subsite [51]. These last two modulatory sites may serve to attenuate excitotoxic injury, since acidification (associated with the absence of oxygen) and increased pro-oxidant activity (associated with oxidative stress) in the interstitial space both attenuate the activity of NMDA receptors.

## Excitotoxic hypothesis of brain injury

The excitotoxic hypothesis of brain injury posits that elevated $[EAA]_i$ causes cell death by overactivation of EAA receptors (Fig. 2). A major consequence of this is depolarization of the post-synaptic membrane, which removes the depolarization block to the NMDA receptor channel complex, thus permitting a large influx of $Ca^{2+}$ ions over the period that it is in the open state. $Ca^{2+}$ can also enter through voltage-activated $Ca^{2+}$ channels and, in GABAergic interneurons (where GABA is γ-aminobutyric acid), via $Ca^{2+}$ gating AMPA/kainate receptors [44,48]. A final mechanism by which EAAs increase the cytosolic $Ca^{2+}$ concentration is by activation of $Ca^{2+}$-mobilizing EAA metabotropic receptors. However, it is through the NMDA receptor channel complex that most $Ca^{2+}$ enters the post-synaptic cell during excitotoxic injury [52,53]. Toxicity is then determined by both the magnitude and duration of the rise in the cytosolic $Ca^{2+}$ concentration. Brief exposure of cultured hippocampal neurons to toxic concentrations of glutamate causes an immediate and sustained (up to 1 h) elevation in cytosolic $Ca^{2+}$ [53], which is sufficient to cause cell death [54,55]. The mechanism underlying $Ca^{2+}$-mediated toxicity is not yet clear, but it seems likely to be mediated by some of the many $Ca^{2+}$-dependent enzymes, such as proteases, protein kinases, phospolipases and endonucleases [56–58].

Excitotoxicity is characterized by cellular swelling, membrane blebbing and necrosis [59]. This contrasts with the chromatin condensation and cell shrinkage associated with cell death caused by growth factor deprivation [60]. The changes caused by deprivation of growth factors resemble apoptosis, a naturally occurring process of cell death [61]. Apoptotic cell death is associated with activation of endonucleases, which cleave internucleosomal DNA into fragments of 180–200 bp in length, leaving a characteristic 'fingerprint' of DNA fragmentation. Whether or not excitotoxicity is associated with DNA fragmentation is somewhat equivocal at present [62–64]. However, it is very plausible that the elevations in free intracellular $Ca^{2+}$ associated with excitotoxic injury are sufficient to activate endonucleases. This is supported by evidence indicating that an inhibitor of endonucleases (aurintricarboxylic acid) protects hippocampal neurons (both *in*

**Fig. 2.**     **Schematic representation of the excitotoxic hypothesis of brain injury**

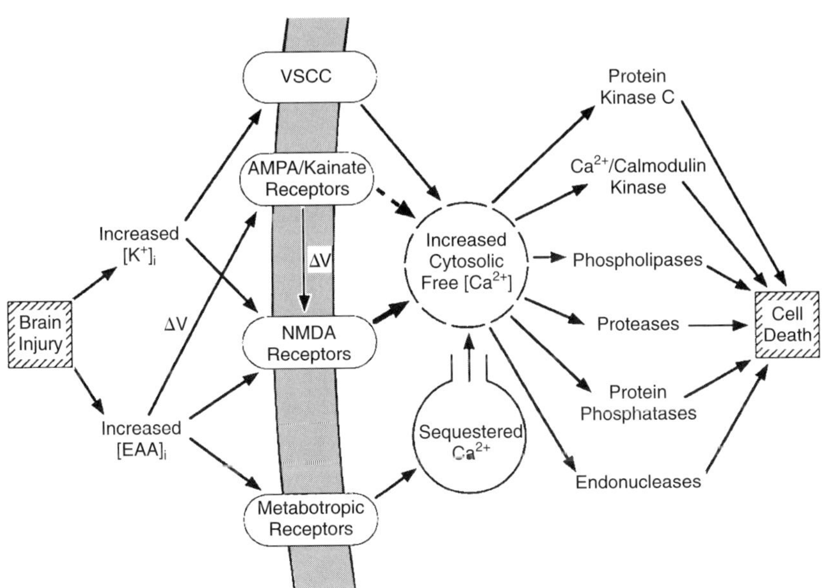


*Brain injury causes an increase in [EAA]ᵢ. This activates AMPA/kainate receptors which, together with elevated interstitial concentrations of K⁺ ([K⁺]ᵢ), depolarizes the post-synaptic membrane and permits Ca²⁺ entry into the post-synaptic cell by activation of voltage-sensitive calcium channels (VSCC); Ca²⁺ can also enter through certain types of AMPA/kainate receptors. However, most Ca²⁺ enters the cell through the NMDA receptor channel complex, which is activated by the combination of increased [EAA]ᵢ and membrane depolarization (ΔV). The concentration of cytosolic free Ca²⁺ is also elevated by activation of EAA metabotropic receptors, which mobilize sequestered intracellular stores of Ca²⁺. Thus EAAs increase the concentration of cytosolic free Ca²⁺ by a variety of mechanisms. This activates a number of Ca²⁺- and calmodulin-dependent enzymes, which can then lead to cell damage and ultimately to cell death.*

*vitro* and *in vivo*) from excitotoxic injury [65,66]. In addition, EAAs are potent enhancers of expression of the c-*fos* gene (a member of the immediate-early gene family and a homologue of the oncogene v-*fos*), which has been implicated in apoptosis [67,68].

## Role of energy metabolism

An impairment of energy metabolism can cause cell death by excitotoxic mechanisms by making cells more vulnerable to normal interstitial concentrations of L-aspartate and L-glutamate. Thus inhibitors of oxidative phosphorylation potentiate the toxicity of glutamate in cultured cerebellar neurons [69]. This probably occurs via ATP depletion. Membrane potential is maintained by the high activity of ion pumps, particularly $Na^+/K^+$ ATPase. A diminution in the steady state concentration of ATP will, therefore, reduce the activity of $Na^+/K^+$ ATPase, leading to a concomitant change in membrane potential. This energy impairment can cause membrane depolarization and relief of the voltage-dependent $Mg^{2+}$ block of the NMDA receptor, leading to persistent activation. Consistent with this, Zeevalk and Nicklas [70] have shown that inhibitors of either glycolysis or oxidative phosphorylation produce NMDA activation and excitotoxicity in the chick retina, without any increase in the interstitial concentration of glutamate.

An example of a metabolic impairment contributing to excitotoxic injury is the toxicity associated with the stress-induced secretion of glucocorticoids from the adrenal gland [71]. The hippocampus is the predominant target for glucocorticoids in the brain, and prolonged exposure causes cell death in this region, which can be prevented by decreasing glucocorticoid exposure [72]. In addition, exposure of the hippocampus to elevated (but not neurotoxic) concentrations of glucocorticoids increases the toxicity of a subsequent excitotoxic insult [73,74]. This increased vulnerability to excitotoxic injury is associated with a greater injury-induced increase in $[EAA]_i$ [74] and is blocked by NMDA receptor antagonists [75]. Glucocorticoid-induced neurotoxicity is attenuated, both *in vivo* and *in vitro*, by the addition of energy (in the form of glucose, mannose and the ketone β-hydroxybutyrate), without affecting the damage caused by the toxin alone [76], and appears to be mediated by inhibition of glucose entry into cells by decreasing the number of glucose transporters [77]. This impairment of energy metabolism in turn diminishes the energy-dependent uptake of EAA, which can then not only mediate hippocampal cell loss, but also determine the severity of neurodegeneration that occurs following both acute and chronic brain injury.

## Role of oxidative stress

Activation of ionotropic EAA receptors increases the rate of oxidative metabolism and thus perturbs the equilibrium that exists between pro-oxidant and antioxidant activities. This can lead to a state of oxidative stress, a condition where pro-oxidant activity becomes greater than the opposing antioxidant forces [78]. This EAA-induced oxidative stress is mediated by the formation of reactive oxygen species (ROS), such as superoxide ions ($O_2^{-\bullet}$) and hydrogen peroxide ($H_2O_2$), which occurs with 'leakage' of high-energy electrons from the mitochondrial electron transport chain. In addition, EAA-evoked elevations in the concentration of cytosolic $Ca^{2+}$ will, in turn, activate a number of enzyme systems that lead to the formation of ROS. Thus the NMDA-receptor-dependent initiation of the arachidonic acid cascade system by phospholipase $A_2$ [79] leads to the excess generation of ROS [80]. Similarly, elevated cytosolic $Ca^{2+}$ concentrations activate peptidases such as calpain I, which can catalyse the enzymic conversion of xanthine dehydrogenase into xanthine oxidase, an enzyme involved in the catabolism of purine bases which concomitantly leads to the formation of $O_2^{-\bullet}$ [81]. Finally, $Ca^{2+}$ influx through the NMDA receptor is associated with activation of the enzyme nitric oxide synthase [82]. The resultant $NO^{\bullet}$ can interact with $O_2^{-\bullet}$ to form the peroxynitrite anion ($ONOO^-$), which decomposes to $^{\bullet}OH$ [83].

Evidence in support of a role for ROS in excitotoxic injury is emerging rapidly, and falls into two categories: (1) studies showing that EAAs can increase the formation of ROS, and (2) studies showing that application of antioxidant molecules or antioxidant-generating systems diminishes the neurodegeneration caused by excitotoxic injury.

One of the first studies to demonstrate the link between EAAs and oxidative stress was that of Murphy and colleagues [84], who attributed glutamate toxicity in a neuronal cell line to the inhibition of cystine uptake, which (because it is a precursor) leads to loss of the important antioxidant glutathione. This was extended by the demonstration that NMDA-induced toxicity is increased in cortical culture that is deficient in glutathione [85]. In addition, AMPA, kainate and NMDA have been found to increase the formation of ROS, detected by assessment of a ROS-sensitive fluorescent probe, in brain synaptoneurosomes [86], and systemic administration of kainate to living animals increases lipid peroxidation in the brain [87,88].

There is a variety of evidence to indicate that ROS formation induced by ionotropic EAA receptor activation is involved in cell death. The catabolism of arachidonic acid is associated with ROS formation, and glutamate toxicity in a neuronal cell line is blocked by inhibition of either phospholipase $A_2$ (which blocks the formation of arachidonic acid) or the subsequent metabolism of arachidonic acid [89]. Glutamate toxicity has also been manipulated by changing the activity of the antioxidant enzyme NAD(P)H:quinone reductase. Thus induction of this enzyme diminishes the vulnerability to glutamate toxicity, whereas inhibition increases vulnerability [89]. Further support comes from the observations that NMDA-mediated cell death of neurons in culture can be attenuated by both spin-trap agents [90] and the 21-aminosteroid U74500A, an inhibitor of lipid peroxidation [91]. In addition, antioxidants block both kainate-induced lipid peroxidation [88] and EAA-induced cell death [92,93]. Finally, transgenic mice overexpressing human Cu–Zn superoxide dismutase demonstrate less ischaemia-induced cell death than control animals, and primary cultures of cortical neurons from these mice are significantly less vulnerable to glutamate-induced swelling and death [94].

There is evidence to suggest that, in addition to being an accomplice in EAA-mediated toxicity, ROS may also interact with EAA nerve terminals to increase EAA efflux. Thus ROS generated from xanthine and xanthine oxidase caused a large ($Ca^{2+}$-independent) increase in the efflux of EAA from hippocampal slices that was not (apparently) attributable to cell lysis [95]. The mechanism underlying this effect is not clear. Other interactions between ROS and EAAs include a marked sensitivity of glutamine synthetase to ROS [96] and a marked inhibition of EAA uptake by arachidonic acid [97], the synthesis of which is associated with ROS production. Moreover, in an *in vitro* model of ischaemia, the amount of oedema caused by middle cerebral artery occlusion was reduced by about 50% by a variety of ROS quenchers (mannitol, superoxide dismutase plus catalase, indomethacin and corticosterone) that prevented the 'ischaemia'-induced increase in EAA concentration. Finally, combined treatment with a non-competitive NMDA receptor antagonist (dizocilpine; MK-801) and 1,3-dimethyl-2-thiourea (a $H_2O_2$ and ${}^{\bullet}OH$ scavenger) provided no significant additive effect over that resulting from treatment with either agent alone, suggesting that EAA and ROS may damage the brain by a common pathway [98].

All of the above interactions point to a scenario whereby excessive activation of ionotropic EAA receptors increases the formation of both ROS and arachidonic acid, which interact with EAA nerve terminals and glial cells to further increase the efflux of EAA or to diminish the uptake of EAA (or both), thus setting up a vicious cycle of mutual co-operation between EAAs and ROS that leads to cell damage and, ulimately, cell death [95].

## Role of the inflammatory response

The brain is regarded as 'immunologically privileged' because it is nearly devoid of a lymphatic system and is protected from circulating factors by the blood–brain barrier. However, brain injury is accompanied by the invasion of peripheral immune cells (e.g. macrophages or neutrophils), which can contribute to the process of neurodegeneration or brain recovery [99]. Both acute and chronic neuro-inflammatory responses are stimulated by a number of cytokines, such as interleukin-1, tumour necrosis factor, interferon-γ, transforming growth factor β and platelet-derived growth factor. For example, interleukin-1 induces the expression of acute-phase proteins such as β-amyloid precursor protein, α-chymotrypsin and cell adhesion molecules. Increased concentrations of a variety of cytokines have been observed in the brain and cerebrospinal fluid (CSF) after acute brain injury and in a number of neurodegenerative diseases [100]. An important role in the process of neurodegeneration is suggested by the observations that interleukin receptor antagonist protein (an endogenous antagonist protein that binds to interleukin-1 receptors) has been shown to attenuate the damage caused by both ischaemic and traumatic brain injury [101,102]. This neuroprotective action of interleukin receptor antagonist protein appears to be mediated largely by blockade of the excitotoxic cascade, since this protein attenuated the tissue damage caused by intrastriatal administration of an NMDA receptor agonist [101]. The precise mechanism by which this occurs remains to be established. Paradoxically, interleukin-1 may also have a neuroprotective action following brain injury, since it induces the formation of nerve growth factor in neurons. While the trophic functions of nerve growth factor are well documented and accepted, this cytokine may also have another very important role: detoxification of ROS. Thus nerve growth factor up-regulates the activity of a number of antioxidant enzymes, and thereby confers protection from oxidative damage [103,104].

## Acute neurodegeneration

### Hypoxia/ischaemia

Cerebrovascular disease, the third most common cause of death in the Western world today, often has neurological consequences. Interruption of the flow of blood to the brain caused by thrombosis and/or detachment of a previously formed clot (embolus) can lead to cerebral ischaemia of varying duration. Transient cerebral ischaemia causes a delayed loss of neurons in both experimental animals and humans [105,106]. A compelling possible mechanism to account for this secondary injury is that cerebral ischaemia causes substantial elevations in the $[EAA]_i$, which in turn overactivate EAA-receptor-operated channels, leading (either directly or indirectly) to a massive increase in the cytosolic concentration of free $Ca^{2+}$ and, ultimately cell death. Evidence for a significant role for EAAs in ischaemia-induced cell death emerged with the demonstration that transient cerebral ischaemia elicits a large increase in $[EAA]_i$ in the hippocampus [107]; in addition, EAA receptor antagonists prevent anoxia-induced neuronal degeneration in cultured hippocampal neurons [108]. The important role of EAAs in ischaemic injury was subsequently amply confirmed by a variety of other data. First, the injuries caused by transient global ischaemia and exogenously applied EAAs both display 'organelle degeneration in post-synaptic elements' and sparing of axon terminals [18,109]. Secondly, both types of injury are associated with a large amount of $Ca^{2+}$ influx [53,110,111]. Thirdly, the severity of the two types of injury increases with either the length of the ischaemic period or the dose of EAA administered, and the resultant injury develops after a delay that is inversely proportional to the severity of the injury [52,112]. Fourthly, all areas of the brain that are selectively vulnerable to an ischaemic insult receive dense EAAergic innervation, e.g. the vulnerable CA1 area of the hippocampus receives excitatory input from the CA3 zone (Schaffer collaterals), the contralateral hippocampus (the commissural fibres) and the entorhinal cortex (perforant pathway). Fifthly, ischaemic neuronal damage in the hippocampus is ameliorated by lesion of EAA innervation to this structure [113]. Finally, EAA receptor antagonists decrease neuronal damage in some models of cerebral ischaemia [114,115].

Two recent studies suggested that the ischaemia-induced elevations in $[EAA]_i$ is biophasic, with initial exocytosis rapidly superseded by non-vesicular efflux, probably occurring as a consequence of energy deficiency [116,117]. This leads to collapse of the $Na^+$ gradient across the plasma membrane and reversal of the $EAA-Na^+$ co-transport system. These changes occur initially in neurons (which have no glycogen stores) and then in astrocytes (which have significant glycogen reserves), since ischaemia-induced elevations in [EAA]$_i$ were either abolished (5 min of ischaemia) or attenuated (20 min of ischaemia) in astrocyte-rich tissue [117].

## Hypoglycaemia

Glucose is the major substrate for brain energy metabolism under normal circumstances, so it is therefore not surprising that a diminished glucose supply can lead to brain damage. Such damage has been reported to occur in humans following insulin shock therapy in psychotic patients, insulin-induced coma in diabetic patients, idiopathic hypoglycaemia in infants and the development of islet cell tumours of the pancreas [109]. There are a variety of data to support the notion that hypoglycaemic brain damage is excitotoxin-mediated. No injury is evident unless the electroencephalographic pattern becomes isoelectric, which is associated with a large elevation in [EAA]$_i$, particularly the concentration of aspartate [118,119]. In addition, hypoglycaemia produces the dendro-somatic axon-sparing lesion characteristic of excitotoxicity, and this injury is prevented by both NMDA and AMPA/kainate receptor antagonists [120–122].

## Epilepsy

Epilepsy is a chronic disorder (affecting 1% of the population) that is characterized by periodic and unpredictable spontaneous recurrent paroxysmal cerebral discharges. These seizures arise from the excessively synchronous and sustained discharge of a group of neurons, and can be categorized on the basis of their origin, severity and nature [123]. Severe forms of epilepsy are often associated with neuronal loss accompanied by a glial cell reaction. This neuronal loss is selective and displays the fine structural characteristics of excitotoxic injury [124]. The first evidence for a role for EAAs in the production of seizure activity came in a study of humans over 40 years ago [6]. Subsequent human studies have shown increased concentrations of EAAs (particularly aspartate) in both the CSF [125] and the interstitial fluid [126,127], and evidence of overactivation of NMDA receptors [128]. These observations are strongly supported by the demonstration that EAA receptor antagonists are powerful anticonvulsants in both genetic and non-genetic animal models of epilepsy [123,129].

## Brain trauma

Traumatic brain injury (TBI) is a common cause of death and disablement, particularly in the young [130]. Like cerebral ischaemia, TBI is associated with secondary or delayed injury processes that develop over a period of minutes to hours to days and contribute to irreversible tissue damage. As discussed above, EAAs play a critical role in the development of the secondary injury associated with cerebral ischaemia. Various data indicate that EAAs play a similar role following TBI. First, elevated brain [EAA]$_i$ has been shown in experimental animals following TBI caused by fluid percussion [131,132], free weight drop [133] and controlled cortical impact [134] (Fig. 3). Secondly, the post-traumatic elevations in [EAA]$_i$ are

**Fig. 3.    TBI-induced elevations in dialysate concentrations of aspartate (A, B) and glutamate (C, D) in the rat neocortex**

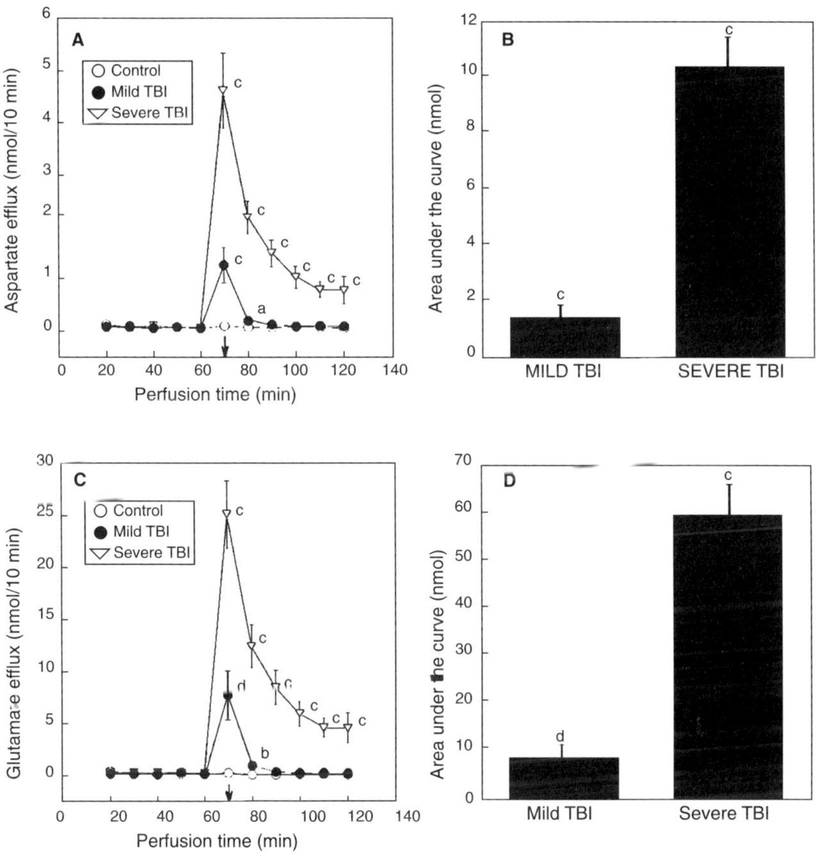



*Either mild (1.5 mm depth of deformation) or severe (3.5 mm depth of deformation) injury was applied, and the area under the peak was calculated for aspartate (B) and glutamate (D) as the sum of the amount of amino acid in each post-TBI sample minus the amount in the baseline samples ($t = 20$–$60$ min). Using a microdialysis probe placed in the cerebral cortex, samples were collected in 10 min fractions for 60 min prior to injury to establish a stable baseline. The probe was then removed from the brain to allow for the impactor to induce brain injury, and then replaced in the same position within 40 s of the impact. As illustrated by the control rats (probe removed and reinserted without contusion injury), probe removal and replacement did not influence dialysate concentrations of aspartate or glutamate. The arrow indicates the point of cortical impact. Values represent means $\pm$ S.E.M for eight sham-treated animals, five with mild TBI and five with severe TBI. Significance of differences compared with control animals: $^{a}P < 0.05$, $^{b}P < 0.01$, $^{c}P < 0.005$ (Kruskal–Wallis ANOVA followed by Mann–Whitney $U$ test). Reproduced, with the publisher's permission, from Palmer et al. [134].*

**Fig. 4.**     **Sections of perfused rat brain 11 days after moderate (A) and severe (B) TBI**

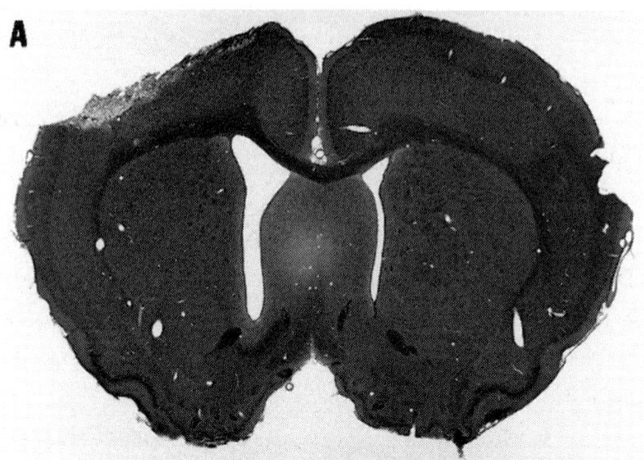
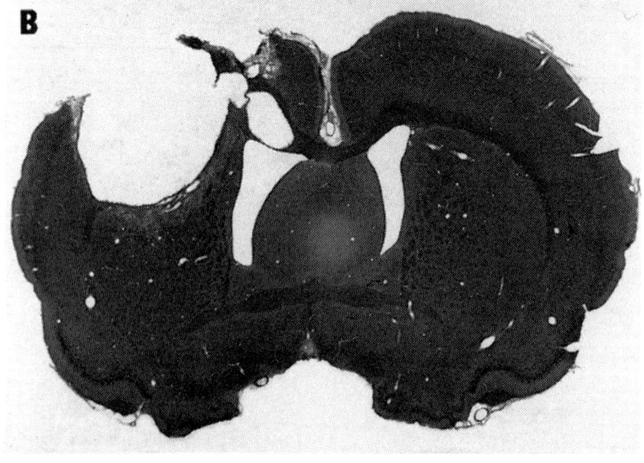

*Frozen sections (40 μm) were stained with haematoxylin and eosin. Reproduced, with the publisher's permission, from Palmer et al. [134].*

correlated with the severity of the injury [131,134] (Figs. 3 and 4). Thirdly, post-traumatic elevations in [EAA]$_i$ are probably sufficient to cause cell death (Table 1). Fourthly, antagonists at both NMDA (Table 2) and AMPA/kainate receptors reduce the severity of TBI [135]. Finally, both AMPA and NMDA produce a cavitary lesion with similar anatomical extent and degree of reactive gliosis to that caused by severe TBI (Fig. 5).

**Table 1.** **Interstitial EAAs rise to toxic concentrations with the collapse of the plasma membrane concentration gradient following TBI**

Values are means ± S.E.M. Interstitial concentrations and cytosolic/interstitial ratios were calculated as described by Palmer and colleagues [134]. It is difficult to establish the toxic concentrations of EAAs from *in vivo* studies. Determination of the toxic concentration of glutamate from studies of neuron cultures or brain slices is complicated by its avid high-affinity uptake into parent nerve terminals and astrocytes; however, the toxic concentration of glutamate has been estimated to be approx. 5 μM [206–208]. The TBI-induced elevations in EAA concentrations are therefore more than sufficient to cause neuronal death. In addition, the massive TBI-induced decline in the cytosolic/interstitial EAA ratio is consistent with complete collapse of ion gradients across the plasma membrane [36].

| | Concentration (μM) | | Cytosolic/interstitial ratio | |
| --- | --- | --- | --- | --- |
| | **Aspartate** | **Glutamate** | **Aspartate** | **Glutamate** |
| Control | 1.7 ± 0.2 | 2.9 ± 0.4 | 397 | 536 |
| Moderate TBI | 63 ± 22 | 231 ± 76 | 10.7 | 6.7 |
| Severe TBI | 123 ± 20 | 414 ± 66 | 5.5 | 3.8 |

**Table 2.** **NMDA receptor antagonists are neuroprotective and improve behavioural outcome in models of TBI**

| Model | Drug | Outcome | Reference |
| --- | --- | --- | --- |
| Vertical fluid percussion | Phencyclidine | Improved behavioural outcome | Hayes et al. [209] |
| Lateral fluid percussion | Dextrophan | Improved behavioural outcome | Faden et al. [131] |
| Lateral fluid percussion | Dizocilpine | Reduced oedema | McIntosh et al. [210] |
| Free weight drop on to skull | Dizocilpine | Reduced oedema | Shapira et al. [211] |
| Lateral fluid percussion | Ketamine | Attenuated post-traumatic memory dysfunction | Smith et al. [212] |

Direct evidence for excitotoxic mechanisms playing a role in TBI in humans is scarce. We recently demonstrated elevated concentrations of aspartate and glutamate in ventricular CSF from patients with severe head injury [136] (Fig. 6). In addition, glutamate concentrations increased linearly for 3 days post-trauma (Fig. 7) (which confirms and extends a previous report in a single patient [137])

**Fig. 5.**          **Glial fibrillary acidic protein immunohistochemistry of rat cortex**

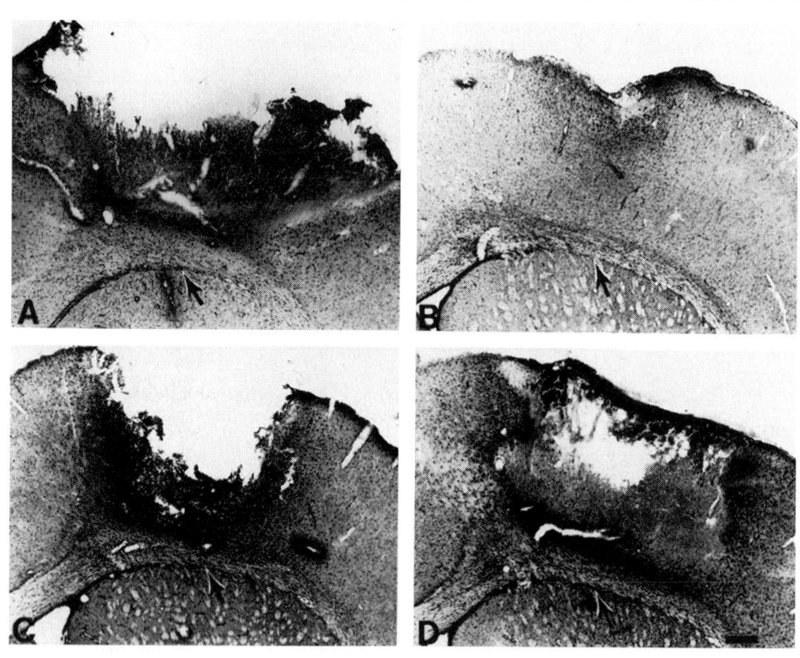

*Immunohistochemistry was carried out 3 days after severe TBI (**A**) or cortical injection of 3 μl of PBS (**B**), 3 μl of PBS containing 30 nmol of NMDA (**C**) or 3 μl of PBS containing 30 nmol of AMPA (**D**). Widespread reactive astrocytes are evident throughout the cortex of all sections except the saline-injected control (**B**), where immunoreactivity is restricted to a nidus associated with the injection site. Lesion depth in (**A**), (**C**) and (**D**) extends to the superior aspect of the corpus callosum, but is essentially absent from (**B**). Arrows indicate the corpus callosum; calibration bar = 0.5 mm. Each figure represents a single animal; similar results were obtained with a duplicate series of animals. Reproduced, with the publisher's permission, from Palmer et al. [134].*

from a concentration of about 3 μM to reach 7 μM at 3 days post-trauma; the mean concentration in controls was 1.4 μM. *In vitro* studies have estimated the toxic concentration of L-glutamate to be about 5 μM when the influence of high-affinity uptake is minimized (Table 1). Thus it seems likely that the concentration of L-glutamate in ventricular CSF post-trauma is sufficient not only to tonically activate both NMDA and AMPA/kainate receptors, but also to cause additional, continuous and more widespread neuronal death. Prolonged increases in the concentrations of aspartate and GABA (relative to neurosurgical controls) were also observed (Fig. 6). Although only transient elevations in $[EAA]_i$ have been

**Fig. 6.** **Concentrations of amino acid transmitters in ventricular CSF from control and severely head-injured humans**

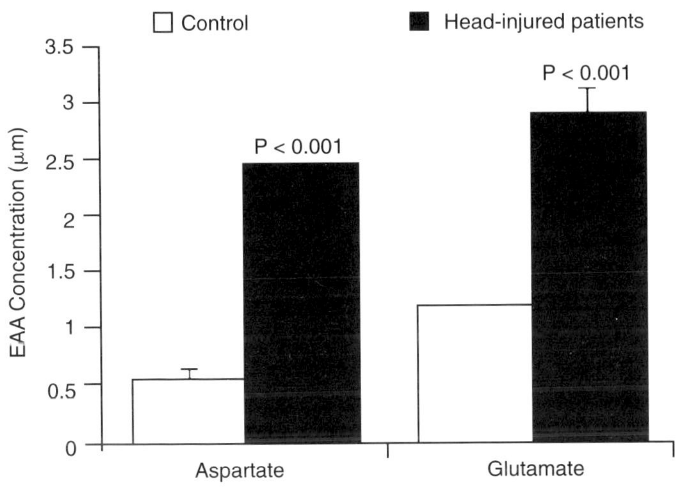


demonstrated following experimental TBI (Fig. 2), it is possible that (at least in the patients examined here) sustained elevations may characterize severe TBI in humans. Such a progressive increase has not been observed following experimental TBI, but a sustained (2 h) elevation of the interstitial concentrations of both glutamate and aspartate is apparent following severe TBI in the rat (Fig. 3). A remarkable observation is the very large post-trauma increase in the CSF concentration of GABA. It has previously been noted that NMDA selectively increases GABA efflux from the embryonic retina in a $Ca^{2+}$- and tetrodotoxin-independent fashion, and this follows, rather than precedes, the associated histological swelling [70,138,139]. This suggests that efflux of GABA is related to osmotic changes rather than release from synaptic pools, and has led to its use as a sensitive index of acute excitotoxicity [139]. Therefore the pronounced increase in the GABA concentration post-trauma may occur as a result of the trauma-induced oedema, which itself is probably mediated by increased activation of EAA receptors. The massive elevation in the concentration of GABA (an inhibitory transmitter) would also be expected to diminish the excitatory action associated with increased concentrations of aspartate and glutamate.

The source of the TBI-induced elevations in $[EAA]_i$ is not clear. Tissue microdialysis studies of interstitial glutamate indicate that ischaemia-induced elevations are abolished by lowering the intra-ischaemic brain temperature to 33°C [140,141]. By contrast, pre-injury hypothermia (32°C) did not affect the TBI-

**Fig. 7.**          **Glutamate concentration in serially sampled ventricular CSF from five patients with severe closed head injury**

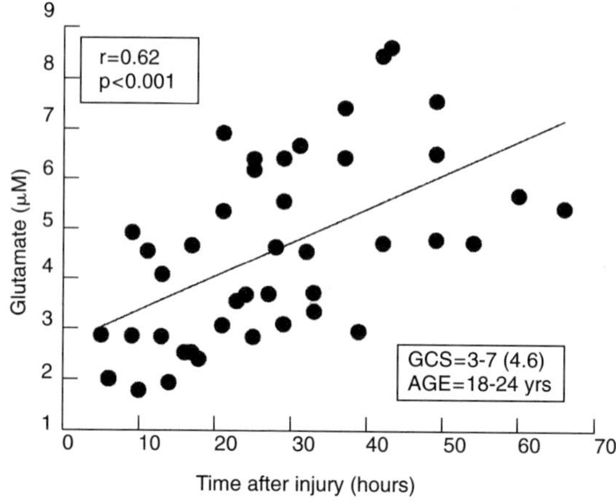


*Samples were collected every 4 h for the first 24 h, and every 6 h thereafter for 2 days. GCS, Glasgow Coma Scale Spearman correlation coefficient = 0.62; P < 0.00 1; n = 42. Modified from Palmer et al. [136].*

induced elevations in [EAA]$_i$, suggesting that these TBI-induced increases occur by a mechanism distinct from that associated with cerebral ischaemia [142].

## Dietary excitotoxins

There seems to be very little deleterious effect of a relatively high dietary intake of aspartate and glutamate. However, there are a number of EAA receptor agonists found in certain foods that are associated with marked toxicity (Table 3). One of the most compelling examples of such a compound causing an acute neurological disorder is the case of domoic acid poisoning on Prince Edward Island in Canada. Domoate is a potent kainic acid receptor agonist that is synthesized by a marine diatom (*Nitschia pungens*) which serves as food for blue mussels. During October 1987, the untypically warm waters around Prince Edward Island led to a rapid multiplication of diatoms, large quantities of which were then taken up by the mussels. These were, in turn, ingested by a number of unfortunate individuals. Those affected rapidly developed a confusional state and many had complex partial seizures and memory impairment. Older individuals seemed particularly vulnerable, since four elderly men died. Post-mortem examination revealed a pattern of pathology that was characteristic of limbic seizures: a predilection for cell loss in the CA1 and CA3 areas of the hippocampus, dentate gyrus, amygdala and claustrum [143,144]. Similar pathological changes were observed in rats administered domoic acid [144].

**Table 3.** **EAA agonists of plant origin**

| Agonist at: | Compound | Origin |
| --- | --- | --- |
| NMDA receptor | Ibotenate | Fungus (*Amanita muscaria*) |
| | β-*N*-Methylamino-L-alanine | Cycad fruit |
| | | (*Cycas circanalis*) |
| AMPA receptor | Quiqualate | |
| | Willardiine | Acacia seeds |
| | | (*Aracia willardinia*) |
| | β-*N*-Oxalyl-L-α,β-diaminopropionate | Chick pea (*Lathyrus sativus*) |
| Kainate receptor | α-Kainate | Seaweed (*Digenea simplex*) |
| | Domoate | Seaweed (*Chondria armata*) |
| | | Diatom (*Nitschia pungens*) |
| | | Blue mussel (*Mytilus edulis*) |
| | Acromelate A | Fungus |
| | | (*Clitocybe acromelalga*) |

*Modified from Meldrum [213]*

Another example of dietary excitotoxins emerged from a link between the upper motor neuron disease neurolathyrism and β-*N*-oxalyl-L-a-b-diaminopropionate, an AMPA receptor agonist found in chick-peas. Symptoms (spastic paraplegia) are irreversible and non-progressive (once consumption of the toxin has ceased), and are probably attributable to marked degeneration of the corticospinal neurons. The loss of pyramidal cell perikarya from the motor cortex does suggest that this brain area is selectively vulnerable to the excitotoxic action of this compound [145,146]. Similarly, the NMDA receptor agonist β-*N*-methylamino-L-alanine, found in the cycad fruit, has been linked to the Western Pacific (Guam) form of motor neuron disease [146]. However, it now appears that the β-*N*-methylamino-L-alanine present in the cycad fruit is almost entirely eliminated during the preparation to form a flour. The heavy metal zinc has now emerged as the most likely candidate for the toxic component of cycad flour [147].

## Chronic neurodegeneration

### Acquired immune deficiency syndrome (AIDS)

AIDS is a world-wide epidemic on a very large scale. More than 60% of patients with AIDS eventually develop neurological symptoms, such as impairments in cognition, movement or sensation. Loss of synapses and large neurons is a feature of the disease [148], which is caused by human immunodeficiency virus type 1 (HIV-1). This retrovirus contains a core of RNA surrounded by an envelope composed of protein, lipid and glycoprotein; a prominent coat glycoprotein is gp120. This molecule or a fragment of it was shown to be toxic in primary rodent neurone [149, 150]. Moreover, expression of gp120 in the brain of transgenic mice produced a pathology very similar to that of CNS-HIV and this damage correlated with gp120 expression [151]. Studies *in vitro* showed that toxicity occurred only in the presence of glial cells and was blocked by NMDA receptor antagonists [152]. In addition, HIV-infected monocytes or macrophages secrete a factor (that has not yet been identified) that is also neurotoxic by an NMDA-receptor-mediated mechanism [153]. Persons infected by HIV-1 also have elevated CSF concentrations of quinolinic acid, an 'excitotoxic' metabolite and an agonist of NMDA receptors [154]. Although there is no direct evidence for excitotoxic cell death in the brains of people with AIDS, it remains possible that EAA receptor antagonists might slow or halt the course of neurodegeneration. Surprisingly, such treatment may have already been tried, albeit inadvertently. Pentamidine, a drug often given prophylactically to patients with AIDS in order to protect them against pneumonia, has been shown to be a potent NMDA receptor antagonist [155].

## Motor neuron disease

Motor neuron disease (or amyotrophic lateral sclerosis) is a disorder characterized by a selective and progressive degeneration of the lower motor neurons in the spinal cord and the upper motor neurons in the cerebral cortex. The case for excitotoxicity was initially made on the basis of data indicating increased plasma and CSF concentrations of aspartate and glutamate in patients with motor neuron disease [156,157]; however, an increased CSF concentration of EAAs is not a universal finding [155]. Further support derives from the findings of reduced concentrations of EAAs in the spinal cord and motor cortex [159,160]; reduced EAA concentrations have also been reported in the spinal cord (but not motor cortex) in a hereditary canine motor neuron disease, a disorder with clinical and neuropathological similarities to motor neuron disease [161]. In addition, reduced uptake of [$^3$H]glutamate into synaptosomes prepared from the spinal cord, motor cortex and somatosensory cortex has been demonstrated in subjects with motor neuron disease [162]. Although the cause of the defect in EAA transport is not known, it is clear that a faulty transporter has the potential to contribute to excitotoxic cell death. However, the finding that, in a mouse model of motor neuron disease, pronounced pathological changes occur before the loss of EAA uptake suggests that the decline in EAA uptake is a consequence rather than a cause of motor neuron disease [163]. Oxidative stress appears to play a role, since antioxidants were neuroprotective in an *in vitro* model of motor neuron toxicity that utilized cell death induced by EAA uptake inhibitors in organotypic cultures of the rat spinal cord [164].

## Huntington's disease

Huntington's disease is an autosomal dominant disorder with full penetrance (linked to the short arm of chromosome 4), which is characterized by disturbances in movement (chorea), psychiatric symptoms and progressive dementia. The brain area most affected is the neostriatum, a structure consisting of two main types of neuron: spiny and non-spiny. It is the more abundant spiny neurons that are lost in the disease. By contrast the non-spiny neurons are relatively spared [165]. A role for EAAs in the pathogenesis of this disease was first suggested in 1976, based on the observation that kainic acid reproduced many of the neurological, neurochemical and histological features of Huntington's disease [166]. However, unlike in Huntington's disease, medium aspiny neurons (identified by localization of NADPH diaphorase) and large aspinay neurons were not spared. Subsequent studies showed that quinolinic acid [167] and other NMDA receptor agonists [168] more accurately reproduced the pathology of the disease, since spiny neurons were lost and neurons staining positive for NADPH diaphorase were relatively spared. In addition, Beal and colleagues [169] demonstrated that the intrastriatal injection of amino-oxyacetic acid produced a lesion that was blocked by the NMDA receptor antagonist dizocilpine and was associated with more pronounced sparing of NADPH-diaphorase-containing neurons than was seen with NMDA receptor agonists. However, this form of excitotoxicity is mediated by an indirect mechanism, since amino-oxyacetic acid has no direct depolarizing action on striatal neurons. Amino-oxyacetic acid is an inhibitor of the malate–aspartate shunt (which is essential in moving reducing equivalents from the cytoplasm to the mitochondria), and its toxicity appears to be caused by the induction of an intracellular equivalent of tissue ischaemia. Similar results have been obtained with the mitrochondrial toxin 3-nitropropionic acid [170]. This ability of metabolic poisons to accurately reproduce the pathology of Huntington's disease suggests that an impairment of energy metabolism may underlie EAA-mediated cell death associated with this disease. Support for this is derived from NMR spectroscopic studies which have revealed increased lactic acid concentrations in the brains of living Huntington patients [171]. The delayed onset of Huntington's disease can then be explained (at least in part) by the age-dependent vulnerability of the neostriatum to metabolic inhibitors [172].

## Parkinson's disease

Another movement disorder that has been linked to excitotoxic injury is Parkinson's disease, a progressive disorder of late life characterized by unintentional tremor, rigidity and slowness of movement, coupled with pronounced loss of neurons from the substantia nigra. Both the clinical and pathological features of the disease have been reproduced by intravenous administration of 1-methyl-4-phenyl-1,2,5,6-tetrahydropyridine in both humans and primates, by accident in the former and by intent in the latter. This compound is oxidized to 1-methyl-4-phenyl pyridine ($MPP^+$) by monoamine oxidase B, an enzyme largely located in astrocytes. $MPP^+$ is selectively taken up into dopamine nerve terminals, where it is concentrated in mitochondria. It then selectively inhibits NADP:ubiquinone oxidoreductase (complex I) and thus blocks the electron transport chain, leading to the formation of ROS [173]. There are several experimental models of Parkinson's disease, including rodents or primates treated with amphetamine or metamphetamine. These compound both increase the efflux of cytosolic dopamine and cause toxicity to nigrostriatal dopaminergic neurons. This toxicity is blocked by NMDA receptor antagonists [174]. Similar protection by NMDA receptor antagonists has been reported with $MPP^+$-induced neurotoxicity [175]; however, other studies have failed to observe any such protective effect [174,176]. Because of this controversy, as well as the uncertainty of the relationship between these models and Parkinson's disease itself, definite conclusions regarding the role of excitotoxicity in the pathophysiology of Parkinson's disease are not yet warranted. Nonetheless, post-mortem evidence indicating that the substantia nigra is subject to oxidative stress [177,178] is at least consistent with the hypothesis that oxidative stress induces excitotoxic injury in the substantia nigra.

Traditional therapy for Parkinson's disease has relied on supplying the immediate precursor of dopamine, i.e. L-dihydroxyphenylalanine (L-DOPA). However, this treatment does not provide consistent clinical benefit, particularly after prolonged treatment, when 'on–off' symptoms develop. This may be related to toxicity caused by L-DOPA itself [179], which can spontaneously form 2,4,5-trihydroxyphenylalanine, a potent AMPA receptor antagonist [180,181]. Although it is not clear if this toxicity contributes to the pathogenesis of Parkinson's disease, it may well be involved in some of the unwanted side effects associated with L-DOPA therapy.

## Alzheimer's disease

The hypothesis that excitotoxicity contributes to the pathophysiology of Alzheimer's disease was proposed by Young and colleagues in 1987 [180]. However, in the absence of an animal model of Alzheimer's disease, the hypothesis remains speculative. There is evidence for the loss of EAA uptake sites, but this probably reflects loss of EAA nerve terminals [183,184]. If excitotoxic injury occurs in the disease, then the regional specificity of neuropathological changes [185] should be expected to have some correspondence with the density of EAA receptors. This, however, has not been borne out by experimental data [186,187]. Nonetheless, it is possible that secondary excitotoxicity occurs in response to oxidative stress, which not only appears to be a feature of the disease but also occurs early in the disease process. Data from cerebral biopsy tissue indicate increased glucose utilization, increased oxygen uptake, elevated concentrations of aspartate and decreased concentrations of glutamate [181,185,186]. This change in the aspartate/glutamate ratio is consistent with glutamate being used as an energy source, which occurs when the glucose concentration becomes limiting (see Fig. 1). Increased lipid peroxidation is also a feature of the disease, since elevations of both basal and free-radical-stimulated concentrations of malondialdehyde (a product of lipid peroxidation) have been demonstrated in the neocortex of Alzheimer subjects [190–192].

## Aging

There is evidence of slowing or impairment in many functional capacities later in life, including decreased reaction times and mild impairments in memory. The aging human brain is associated with neuronal loss [193], but the role (if any) of excitotoxicity in this process is not yet clear. There is some evidence for an age-related loss of EAA nerve terminals in human brain, based on assessment of the sodium-dependent binding of D-[³H]aspartate [184]. However, assessment of two functional markers of EAA neurons (high-affinity D-[³H]aspartate uptake and the K⁺-evoked superfused efflux of endogenous aspartate and glutamate) in rat brain suggests that EAA neurotransmission is, in at least a functional sense, preserved with normal aging. In addition, the sensitivity of both EAA release and uptake to oxidative stress was also unaffected by age [194]. The potency of cyanide in inhibiting D-[³H]aspartate uptake displayed marked regional variability. Cyanide was most potent in the hippocampus and least potent in the neostriatum [194]. Thus it is possible that the pronounced sensitivity to oxidative stress of EAA nerve terminals in the hippocampus renders this region vulnerable to subsequent excitotoxin injury and contributes to the age- and Alzheimer's disease-related neurodegeneration associated with this region.

Further evidence for a role of excitotoxicity in age-related neurodegeneration derives from data demonstrating an age-related increase in the vulnerability of striatal neurons to 3-nitropropionic acid-induced toxicity [172]. Excitotoxicity may play a role in this process, since 3-nitropropionic acid-induced toxicity can be blocked by NMDA receptor antagonists. In addition, an age-related loss of NMDA receptors has been reported in a number of studies of post-mortem human brain [187,195–197]. This is consistent with a role of excitotoxicity in age-associated neurodegeneration, possibly occurring as a consequence of increased vulnerability to oxidative stress.

## Anti-excitotoxic therapies

### Preclinical studies

There are a number possible sites of therapeutic intervention in the cascade of excitotoxic injury including the agonist, co agonist and modulatary sites of the NMDA receptor, AMPA/kinate receptors, EAA metabotropic receptors, voltage-sensitive Ca²⁺ channels, EAA release, Ca²⁺-dependent release, Ca²⁺-dependent enzymes, immediate early genes and pro-oxidant molecules. Demonstrating a neuroprotective action of anti-excitotoxic therapy in models of brain injury not only provides important confirmation of the critical role of excitotoxicity but also raises hope that clinical conditions associated with excitotoxic injury may soon have a therapy that attenuates neurodegenerative changes and thus improves the behavioural outcome of injured persons. It is EAA receptor antagonists, NMDA receptor antagonists in particular, that have formed the major focus of preclinical drug development in the pharmaceutical industry. The neuroprotective efficacy of EAA receptor antagonists has been demonstrated *in vitro* in cortical and hippocampal neurons exposed to EAA receptor agonists, or to hypoglycaemia (no glucose in the perfusing medium), anoxia (no oxygen in the medium) or a combination of both [198]. Neuroprotective efficacy has also been demonstrated in experimental models of cerebral ischaemia [114], hypoglycaemia [121,122], epilepsy [129] and TBI (Table 2), with most studies describing the effects of NMDA receptor antagonists.

## Clinical studies

To be of practical use, the impressive neuroprotective efficacy of EAA receptor antagonists needs to be translated into the successful treatment of persons with excitotoxin-mediated brain injury. A number of anti-excitotoxic agents are currently in clinical trials for stroke and brain trauma (Table 4). Most therapies have targetted the NMDA receptor and include antagonists that compete for the agonist (EAA) and co-agonist (glycine) site and non-competitive antagonists. Development of AMPA/kainate receptor antagonists is less well advanced. Other anti-excitotoxic therapies under development include inhibitors of EAA release, such as $Na^+$ channel blockers and inhibitors of N-type $Ca^{2+}$ channels (SNX-111, Neurex). Inhibitors of voltage-activated $Ca^{2+}$ channels have also been pursued, but such molecules are less efficacious than EAA receptor antagonists in preclinical models. Although clinical trials of acute brain injury are dominated by NMDA receptor antagonists, this approach is severely limited by the emergence of side effects such as hypertension, tachycardia, agitation, hallucinations, confusion, paranoia and delirium. In addition, dizocilpine and phencyclidine cause reversible morphological changes (vacuolization and neuronal swelling) in the posterior cingulate cortex and the strosplenial cortex [199]. The key issue for clinical trials is the concentration at which behavioural changes are manifest relative to the therapeutic dose. These values are similar for non-competitive antagonists such as dizocilpine, but for competitive antagonists such as CGS-19755, and particularly antagonists at the polyamine site, behavioural alterations occur at concentrations greater than that required for anti-excitotoxic action [200, 201].

Major anti-excitotoxic therapies have been directed towards ischaemic injury. A number of factors make it difficult to demonstrate the efficacy of neuroprotective drugs in cerebral ischaemia. First, stroke patients are often elderly, so their outcome is often determined by other factors, such as likelihood of hospitalization (stroke in elderly subjects is not treated as an emergency in many countries) and cardiovascular health (myocardial ischaemia may be a greater determinant of mortality). Secondly, the chances of success of neuroprotective therapy decline with time following the initial injury. Early treatment of stroke is difficult because occlusive stroke is frequently difficult to diagnose, and strokes are not managed as emergencies unless patients are in a coma. Finally, because of the heterogeneity of factors that determine outcome, extremely large numbers of

**Table 4    Anti-excitotoxic neuroprotective agents in clinical trials**

| NMDA antagonists | Code | Name | Company | Status | Indication |
|---|---|---|---|---|---|
| Competitive Antagonists | SDZ-EAA-494 | d-CPP-ene | Norvatis | Phase III | Stroke, TBI |
|  | CGS 19755 | Selfotel | Norvatis | Phase II† | Stroke, TBI |
| Noncompetitive Antagonists | CNS 1102 | Aptiganel | Cambridge Neurosciences | Phase III | Stroke, TBI |
|  | RPL 12924AA | Remacemide | Astra | Phase II | Stroke, epilepsy |
|  | HU 211 | Dexanabinol* | Pharmos | Phase II | Stroke |
|  | Memantine |  | Neurobiological Technologies | Phase II | AIDS dementia |
| Glycine site antagonists | Acea 1021 |  | CoCensys (Norvatis) | Phase I | Stroke, TBI |
|  | ZD 9379 |  | Zeneca | Phase I | Stroke, TBI |
|  | GV 150526 |  | GlaxoWellcome | Phase II | Stroke |
| Subtype selective antagonists | SL 82.0715 | Eliprodil | Synthelabo | Phase III | TBI |
| **AMPA/kainate receptor antagonists** |  |  |  |  |  |
|  | CP 101,606 |  | Pfizer | Phase I | PD |
|  | YM-90K |  | Yamanouchi | Phase I | Stroke |
|  | LY 293558 |  | Eli Lilly | Phase I | Stroke |
| **EAA release inhibitors** |  |  |  |  |  |
| Sodium channel modulators | BW 619C89 |  | GlaxoWellcome | Phase III† | Stroke, TBI |
|  | Fosphenytoin | Cerebyx | Parke-Davis | Phase II | Stroke |
|  |  | Luteluzole | Janssen | Phase III | Stroke |
|  | Riluzole | Rilutek | Rhone-Poulenc | Marketed | ALS |
|  | Lamotrigine | Lamictal | GlaxoWellcome | Marketed | Epilepsy |

*Dexanibinol binds with low affinity to the NMDA receptor and has antioxidant and anti-inflammatory properties. †Study halted.

Abbreviations used: TBI, traumatic brain injury; PD, Parkinson's disease; ALS, amyotrophic lateral sclerosis patients are required in order to demonstrate a significant effect of treatment. These limitations are less severe in trials of patients with severe head injury, since victims are usually young, are treated as emergency cases and are available for treatment soon after the time of injury. The numbers necessary to demonstrate significant effects of treatment are therefore much smaller than for a clinical trial of stroke victims.

In addition to the latency of treatment, the duration of therapy is also a critical factor in the success of any treatment regime. The observation that elevations in the concentrations of aspartate and glutamate in the ventricular CSF of severely head-injured patients are maintained over several days (Fig. 7) suggests that sustained application of EAA receptor antagonists may be necessary in order to achieve maximum therapeutic benefit.

There is evidence that moderate post-injury hypothermia is neuroprotective in models of both cerebral ischaemia and TBI [142,202,203]. Initial studies of head-injured patients cooled to 32°C suggest that this approach has great potential in the treatment of excitotoxic brain injuries such as stroke and severe head injury [204]. The neuroprotection conferred by hypothermia is probably mediated by a slowing of metabolism, thus reducing cellular energy demand. This will, in turn, decrease the vulnerability of the cell to the massive excitation caused by overactivation of EAA receptors. Therefore additional cytoprotection can be expected if hypothermic treatment is supplemented by pharmacological blockade of EAA receptors. Another potential combination therapy that could provide improved neuroprotection is an anti-excitotoxic drug with antioxidant properties, e.g. an NMDA receptor antagonist that is also a potent antioxidant [205].

## Summary and conclusions

The toxic action of EAAs was first described in 1957. Since then, much has been learned about the anatomy, neurochemistry, physiology and pharmacology of EAA synaptic transmission and plasticity. Similarly great strides have been made in understanding the mechanism by which EAAs kill cells. It is clear that the massive influx of $Ca^{2+}$ ions through the NMDA receptor gated channel is a critical factor, together with activation of AMPA/kainate receptors, which not only activate NMDA receptors but, in some cases, also permit a toxic influx of $Ca^{2+}$ ions into the cell; the role of EAA metabotropic receptors in excitotoxicity is not yet clear. In addition, it is now becoming apparent that excitotoxic injury is tightly linked to the toxicity associated with oxidative stress and the inflammatory response. The precise mechanism by which ROS, cytokines and trophic factors interact with EAAs in the complex cascade of neurochemical changes that accompanies brain injury has only just begun to be unravelled.

Excitotoxicity clearly plays an important role in mediating neurodegenerative changes in stroke, severe hypoglycaemia, severe epilepsy and severe head injury. Although the involvement of excitotoxic mechanisms in chronic brain injury is less clear-cut, evidence is emerging for a critical role in mediating the neurodegenerative changes associated with AIDS. Excitotoxicity coupled with oxidative stress may also be central to the neuropathology associated with motor neuron disease and with Huntington's, Parkinson's and Alzheimer's diseases. Since all of these diseases are age-related, disease-induced changes are probably potentiated by the neurodegeneration associated with normal aging, which itself may be caused by excitotoxic injury, probably in conjunction with oxidative stress. Thus there is great potential for drugs that block the cascade of destruction associated with excitotoxic injury. Preclinical studies have now established the utility of NMDA and AMPA/kainate receptor antagonists in the treatment of acute brain injury, and a number of NMDA receptor antagonists are now in clinical trial for stroke and severe head injury. It therefore seems only a matter of time before an EAA receptor antagonist with an acceptable side-effect profile is launched on to the market, and hence the prospects of successful recovery from acute brain injury are markedly increased. The prospects for the long-term treatment of chronic neurodegenerative diseases are less certain, but will be greatly improved with the emergence of an anti-excitotoxic agent with a good side-effect profile from clinical trials of acute brain injury.

## References


1. Collingridge, G.L. and Lester, R.A.J. (1989) Pharmacol. Rev. **40**, 143–210
2. Fonnum, F. (1984) J. Neurochem. **41**, 1–11
3. Nicholls, D.G. (1989) J. Neurochem. **52**, 331–341
4. Fleck, M.W., Henze, D.A., Barrionuevo, G. and Palmer, A.M. (1993) J. Neurosci. **13**, 3944–3955
5. Okamoto, S. (1951) J. Physiol. Soc. Jpn. **13**, 555–562
6. Hayashi, T. (1952) Jpn. J. Physiol. **3**, 46–64
7. Curtis, D.R., Phillis, J.W. and Watkins, J.C. (1959) Nature (London) **183**, 611–612
8. Krnjevic, K. and Phillis, J.W. (1963) J. Physiol. (London) **165**, 274–304
9. Salt, T. and Herrling, P. (1991) in Excitatory Amino Acids and Synaptic Transmission (Wheal, H. and Thomson, A., eds.), pp. 155–170, Academic Press, London
10. Deutsch, S.I., Mastropaolo, J., Schwartz, B.L., Rosse, R.B. and Morihisa, J.M. (1989) Clin. Neuropharmacol. **12**, 1–13
11. Palmer, A.M. and Gershon, S. (1990) FASEB J. **4**, 1–9
12. Lucas, D.R. and Newhouse, J.P. (1957) AMA Arch. Ophthalmol. **58**, 193–201
13. Olney, J.W. (1969) J. Neuropathol. Exp. Neurol. **28**, 455–474
14. Olney, J.W. (1969) Science **164**, 719–721
15. Olney, J.W. and Ho, O.L. (1970) Nature (London) **227**, 609–610
16. Olney, J.W., Sharpe, L.G. and Feigin, R.D. (1972) J. Neuropathol. Exp. Neurol. **31**, 464–488
17. Burde, R.M., Schainker, B. and Kayes, J. (1971) Nature (London) **233**, 58–63
18. Olney, J.W., Oi Lan, H. and Rhee, V. (1971) Exp. Brain Res. **14**, 61–76
19. Olney, J.W. (1990) Annu. Rev. Pharmacol. Toxicol. **30**, 47–71
20. Curtis, D.R., Phillis, J.W. and Watkins, J.C. (1960) J. Physiol. (London) **150**, 656–682
21. Winfield, D.A., Gatter, K.C. and Powell, T.P.S. (1980) Brain **103**, 245–258
22. Streit, P. (1980) J. Comp. Neurol. **191**, 429–463
23. Spencer, H.J. (1976) Brain Res. **102**, 91–101
24. Herrling, P.L. (1985) Neuroscience **14**, 417–426
25. McGeer, P.L., McGeer, E.G., Scherer, U. and Singh, K. (1977) Brain Res. **128**, 369–373
26. Palmer, A.M., Hutson, P.H., Lowe, S.L. and Bowen, D.M. (1989) Exp. Brain Res. **340**, 1–5

27.  Storm-Mathisen, J. and Ottersen, O.P. (1988) in Neurotransmitters and Cortical Function: From Molecules to Mind (Avoli, M., Reader, T.A., Dykes, R.W. and Gloor, P., eds.), pp. 39–70, Plenum Press, New York

28.  Giuffrida, R. and Rustioni, A. (1989) Exp. Brain Res **78**, 41–46

29.  Hicks, T., Ruwe, W., Veale, W. and Veenhuizen, J. (1985) Exp. Brain Res. **58**, 421–425

30.  Erecinska, M. and Silver, I.A. (1990) Prog. Neurobiol. **35**, 245–296

31.  Derouiche, A. and Frotscher, M. (1991) Brain Res. **552**, 346–350

32.  Pellerin, L. and Magistretti, P.J. (1994) Proc. Natl. Acad. Sci. U.S.A. **91**, 10625–10629

33.  Benveniste, H., Hansen, A.J. and Ottosen, N.S. (1989) J. Neurochem. **52**, 1741–1750

34.  Naito, S. and Ueda, T. (1985) J. Neurochem. **44**, 99–109

35.  Palmer, A.M. and Reiter, C.T. (1994) Neurochem. Int. **25**, 441–450

36.  Bouvier, M., Szatkowski, M., Amato, A. and Attwell, D. (1992) Nature (London) **360**, 471–474

37.  Opitz, T. and Reymann, K.G. (1991) NeuroReport **2**, 455–457

38.  Monaghan, D.T., Bridges, R.J. and Cotman, C.W. (1989) Annu. Rev. Pharmacol. Toxicol. **29**, 365–402

39.  Bekkers, J.M. and Stevens, C.F. (1989) Nature (London) **341**, 230–233

40.  Jones, K.A. and Baughman, R.W. (1991) Neuron **7**, 593–603

41.  Hestrin, S., Nicoll, R.A., Perkel, D.J. and Sah, P. (1990) J. Physiol. (London) **422**, 203–225

42.  Lester, R.A.J., Clements, J.D., Westbrook, G.L. and Jahr, C.E. (1990) Nature (London) **346**, 565–566

43.  Iino, M., Ozawa, S. and Tsuzuki, K. (1990) J. Physiol. (London) **424**, 151–165

44.  Brorson, J., Bleakman, D., Chard, P. and Miller, R. (1992) Mol. Pharmacol. **41**, 603–608

45.  Hollmann, M., Hartley, M. and Heinemann, S. (1991) Science **252**, 851–853

46.  Verdoorn, T., Burnashev, N., Monyer, H., Seeburg, P.H., and Sakmann, B. (1991) Science **252**, 1715–1718

47.  Brorson, J., Manzolillo, P. and Miller, R. (1994) J. Neurosci. **14**, 187–197

48.  Turetsky, D., Canzoniero, L., Sensi, S., Weiss, J., Goldberg, M. and Choi, D. (1994) Neurobiol. Dis. **1**, 101–110

49.  Reynolds, I.J. (1990) Life Sci. **47**, 1785–1792

50.  Traynelis, S.F. and Cull-Candy, S.G. (1991) J. Physiol. (London) **433**, 727–763

51.  Aizenman, E., Lipton, S.A. and Loring, R.H. (1989) Neuron **2**, 1257–1263

52.  Choi, D.W., Maulucci-Gedde, M. and Kriegstein, A.R. (1987) J. Neurosci. **7**, 357–368

53.  Garthwaite, G. and Garthwaite, J. (1987) Neurosci. Lett. **83**, 241–246

54.  Rajdev, S. and Reynolds, I.J. (1993) Neurosci. Lett. **162**, 149–152

55.  Lei, S.Z., Zhang, D., Abele, A.E. and Lipton, S.A. (1992) Brain Res. **598**, 196–202

56.  Manev, H., Costa, E., Wroblewski, J.T. and Guidotti, A. (1990) FASEB J. **4**, 2789–2797

57.  Farooqui, A.A. and Horrocks, L.A. (1991) Brain Res. Rev. **16**, 171–191

58.  Wang, K. and Yuen, P.-W. (1994) Trends Neurosci. **15**, 412–419

59.  Siman, R. and Card, J.P. (1988) Neuroscience **26**, 433–447

60.  Martin, D.P., Ito, A., Horigome, K., Lampe, P.A. and Johnson, Jr., E.M. (1992) J. Neurobiol. **23**, 1205–1220

61.  Franklin, J.L. and Johnson, E.N. (1992) Trends Neurosci. **15**, 501–508

62.  Kure, S., Tominga, T., Yoshimoto, T., Tada, K. and Narisawa, K. (1991) Biochem. Biophys. Res. Commun. **179**, 39–45

63.  Deshpande, J., Bergstedt, K., Linden, T., Kalimo, H. and Wieloch, T. (1992) Exp. Brain Res. **88**, 91–105

64.  Dessi, F., Charriaut-Marlangue, C., Khrestchatisky, M. and Ben-Ari, Y. (1993) J. Neurochem. **60**, 1953–1955

65.  Samples, S.D. and Dubinsky, J.M. (1993) J. Neurochem. **61**, 382–385

66.  Roberts-Lewis, J.M., Marcy, V.R., Zhao, Y., Vaught, J.L., Siman, R. and Lewis, M.E. (1993) J. Neurochem. **61**, 378–381

67.  Lerea, L.S., Butler, L.S. and McNamara, J.O. (1992) J. Neurosci. **12**, 2973–2981

68.  Smeyne, R.J., Vendrell, M., Hayward, M., Baker, S.J., Miao, G.G., Schilling, K., Robertson, L.M., Curran, T. and Morgan, J.I. (1993) Nature (London) **363**, 166–169

69.  Novelli, A., Reilly, J.A., Lysko, P.G. and Henneberry, R.C. (1988) Brain Res. **451**, 205–212

70.  Zeevalk, G.D. and Nicklas, W.J. (1991) J. Pharmacol. Exp. Ther. **257**, 870–878

71.  Sapolsky, R.M., Krey, L.C. and McEwen, B.S. (1985) J. Neurosci. **5**, 1222–1227

72. Sapolsky, R.M., Uno, H., Rebert, C.S. and Finch, C.E. (1990) J. Neurosci. **10**, 2897–2902
73. Sapolsky, R.M. and Pulsinelli, W.A. (1985) Science **229**, 1397–1400
74. Stein-Behrens, B.A., Chang, I., Mattson, M.P. and Sapolsky, R.M. (1992) J. Neurochem. **58**, 1730–1736
75. Armanini, M.P., Hutchins, C., Stein, B.A. and Sapolsky, R.M. (1990) Brain Res. **532**, 7–12
76. Sapolsky, R.M. (1986) J. Neurosci. **6**, 2240–2244
77. Garvey, W., Huecksteadt, T., Lima, F. and Birnbaum, M. (1989) Mol. Endocrinol. **3**, 1122–1138
78. Sies, H. (1985) Oxidative Stress, Academic Press, London
79. Dumuis, A., Sebben, M., Haynes, L., Pin, J.-P. and Bockaert, J. (1988) Nature (London) **336**, 68–70
80. Pellerin, L. and Wolfe, L.S. (1991) Neurochem. Res. **16**, 983–989
81. McCord, J. (1985) N. Engl. J. Med. **312**, 159–163
82. Kiedrowski, L., Costa, E. and Wroblewski, J.T. (1992) J. Neurochem. **58**, 335–341
83. Beckman, J.S., Beckman, T.W., Chen, J., Marshall, P.A. and Freeman, B.A. (1990) Proc. Natl. Acad. Sci. U.S.A. **87**, 1620–1624
84. Murphy, T.H., Miyamoto, M., Saster, A., Schnaar, R.L. and Coyle, J.T. (1989) Neuron **2**, 1547–1558
85. Bridges, R.J., Koh, J.-Y., Hatalski, C.G. and Cotman, C.W. (1991) Eur. J. Pharmacol. **192**, 199–200
86. Bondy, S.C. and Lee, D.K. (1993) Brain Res. **610**, 229–233
87. Sun, A.Y., Cheng, Y. and Sun, G.Y. (1992) Prog. Brain Res. **94**, 271–280
88. Puttfarcken, P.S., Getz, R.L. and Coyle, J.T. (1993) Brain Res. **624**, 223–232
89. Murphy, T., Parikh, R., Schnaar, R. and Coyle, J. (1989) Ann. N.Y. Acad. Sci **349**, 474–481
90. Lafon-Cazal, M., Pletri, S., Culcasi, M. and Bockaert, J. (1993) Nature (London) **364**, 535–537
91. Monyer, H., Hartley, D.M. and Choi, D.W. (1990) Neuron **5**, 121–126
92. Dykens, J.A., Stern, A. and Trenkner, E. (1987) J. Neurochem. **49**, 1222–1228
93. Miyamoto, M., Murphy, T.H., Schnaar, R.L. and Coyle, J.T. (1989) J. Pharmacol. Exp. Ther. **250**, 1132–1140
94. Chan, P., Chu, L., Chen, S., Carlson, E. and Epstein, C. (1990) Stroke **21**, 80–82
95. Pellegrini-Giampietro, D.E., Cherici, G., Alesiani, M., Carla, V. and Moroni, F. (1990) J. Neurosci. **10**, 1035–1041
96. Schor, N.F. (1988) Brain Res. **456**, 17–21
97. Darbour, B., Szatkowski, M., Ingledew, N. and Atwell, D. (1989) Nature (London) **342**, 918–920
98. Moon, S. and Betz, A.L. (1991) Stroke **22**, 915–921
99. Giulian, M. (1988) Brain Behav. Immun. **2**, 352–358
100. Rothwell, N. and Hopkins, S. (1995) Trends Neurosci. **18**, 130–136
101. Relton, J. and Rothwell, N. (1992) Brain Res. Bull. **29**, 243–246
102. DeKosky, S., Styren, S., O'Malley, M., Goss, J., Kochanek, P., Marion, D., Evans, C. and Robbins, P. (1996) Ann. Neurol., **39**, 123–127
103. Pan, Z. and Perez-Polo, R. (1993) J. Neurochem. **61**, 1713–1721
104. Mattson, M.P. and Scheff, S.W. (1994) J. Neurotrauma **11**, 3–33
105. Kirino, T. (1982) Brain Res. **239**, 57–69
106. Petito, C., Feldmann, E., Pulsinelli, W. and Plum, F. (1987) Neurology **37**, 1281–1286
107. Benveniste, H., Drejer, J., Schousboe, A. and Diemer, N.H. (1984) J. Neurochem. **43**, 1369–1374
108. Rothman, S. (1984) J. Neurosci. **7**, 1884–1891
109. Graham, D.I. (1992) in Greenfield's Neuropathology (Adams, J.H. and Ducher, L.W., eds.), pp. 153–168, Edward Arnold, London
110. Choi, D.W. (1985) Neurosci. Lett. **58**, 293–297
111. Simon, R.P., Griffiths, T., Evans, M.C., Swan, J.H. and Meldrum, B.S. (1984) J. Cereb. Blood Flow Metab. **4**, 350–361,
112. Ito, U., Spatz, M., Walker, Jr., J.T. and Klatzo, I. (1975) Acta Neuropathol. **32**, 209–223
113. Wieloch, T., Lindvall, O., Bloomqvist, P. and Gage, F.H. (1985) Neurochem. Res. **7**, 24–26
114. Buchan, A.M. (1990) Cerebrovasc. Brain Metab. Rev. **2**, 1–26
115. Sheardown, M.J., Nielsen, E.O., Hansen, A.J., Jacobsen, P. and Honoré, T. (1990) Science **247**, 571–574

116. Wahl, F., Obrenovitch, T.P., Hardy, A.M., Plotkine, M., Boulu, R. and Symon, L. (1994) J. Neurochem. **63**, 1003–1011
117. Mitani, A., Andou, Y., Matsuda, S., Arai, T., Sakanaka, M. and Kataoka, K. (1994) J. Neurochem. **63**, 2152–2164
118. Auer, R.N., Olsson, Y. and Siesjö, B.K. (1984) Diabetes **33**, 1090–1098
119. Sandberg, M., Butcher, S.P. and Hagberg, M. (1986) J. Neurochem. **47**, 178–184
120. Auer, R., Kalimo, H., Olsson, Y. and Wieloch, T. (1985) Acta Neuropathol. **67**, 279–288
121. Wieloch, T. (1985) Science **230**, 681–683
122. Nellgard, B. and Wieloch, T. (1992) Exp. Brain Res. **92**, 259–266
123. Dingledine, R., McBain, C.J. and McNamara, J.O. (1990) Trends Pharmacol. Sci. **11**, 334–338
124. Meldrum, B.S. (1993) Brain Pathol. **3**, 405–412
125. Weitz, M.D., Merlob, P., Amir, J. and Reisner, S.H. (1981) Arch Neurol **38**, 258–259
126. Ronne-Engström, E., Hillered, L., Flink, R., Spännare, B., Ungerstedt, U. and Carlson, H. (1992) J. Cereb. Blood Flow Metab. **12**, 873–876
127. During, M. and Spencer, D. (1993) Lancet **341**, 1607–1610
128. Masukawa, L.M., Higashima, M., Hart, G.J., Spencer, D.D. and O'Conner, M.J. (1991) Brain Res. **562**, 179–180
129. Meldrum, B.S. (1992) Epilepsy Res. **12**, 189–196
130. Cohadon, F., Richer, E. and Castel, J.P. (1991) Neurol. Sci. **103**, S23–S27
131. Faden, A.I., Demediuk, P., Panter, S.S. and Vink, R. (1989) Science **244**, 798–800
132. Katayama, Y., Becker, D.P., Tamura, T. and Hovda, D.A. (1990) J. Neurosurg. **73**, 889–900
133. Nilsson, P., Hillered, L., Pontén, U. and Ungerstedt, U. (1990) J. Cereb. Blood Flow Metab. **10**, 631–637
134. Palmer, A.M., Marion, D.W., Botscheller, M.L., Swedlow, P.E., Styren, S.D. and DeKosky, S.T. (1993) J. Neurochem. **61**, 2015–2024
135. Bernert, H., Loschmann, P.A. and Turski, L. (1992) Soc. Neurosci. Abstr. **18**, 4572
136. Palmer, A., Marion, D., Botscheller, M., Bowen, D. and DeKosky, S. (1994) NeuroReport **6**, 153–156
137. Baker, A.J., Moulton, R.J., MacMillan, V.H. and Shedden, P.M. (1993) J. Neurosurg. **79**, 369–372
138. Zeevalk, G.D., Hyndman, A.G. and Nicklas, W.J. (1989) J. Neurochem. **53**, 1610–1619
139. Zeevalk, G.D. and Nicklas, W.J. (1993) J. Neurochem. **61**, 1445–1453
140. Busto, R., Globus, M.Y.-T., Dietrich, D., Martinez, E., Valdés, I. and Ginsberg, M.D. (1989) Stroke **20**, 904–910
141. Mitani, A. and Kataoka, K. (1991) Neuroscience **42**, 661–670
142. Palmer, A.M., Marion, D.W., Botscheller, M.L. and Redd, E. (1994) J. Neurotrauma **4**, 363–372
143. Perl, T.M., Bédard, L., Kosatsky, T., Hockin, J.C., Todd, E.C.D. and Remis, R.S. (1990) N. Engl. J. Med. **322**, 1775–1780
144. Teitelbaum, J.S., Zatorre, R.J., Carpenter, S., Gendron, D., Evans, A.C., Gjedde, A. and Cashman, N.R. (1990) N. Engl. J. Med. **322**, 1781–1787
145. Ludolph, A.C., Hugeon, J., Dwievedi, M.P., Schaumburg, H.H. and Spencer, P.S. (1987) Brain **110**, 149–165
146. Spencer, P.S., Ludolph, A.S. and Kisby, G.E. (1992) Ann. N.Y. Acad. Sci. **648**, 154–160
147. Duncan, M.W., Marini, A.M., Watters, R., Kopin, I.J. and Markey, S.P. (1992) J. Neurosci. **12**, 1523–1537
148. Wiley, C.A., Masliah, E., Morey, M., Lemere, C., DeTeresa, R., Grafe, M., Hansen, L. and Terry, R. (1991) Ann. Neurol. **29**, 651–657
149. Dryer, E.B., Kaiser, P.K., Offerman, J.T. and Lipton, S.A. (1990) Science **248**, 364–367
150. Brenneman, D.F., Westbrook. G.L., Fitzgerald, S.P., Ennist, D.L., Elkins, K.L., Ruff, M.R. and Pert, C.B. (1988) Science **335**, 639–642
151. Tuggas, S.M., Masliah, E. Ruckenstein, E.M., Rall, G.F., Abraham, C.R.and Macke, L. (1994) Nature (London) **367**, 188–193
152. Lipton, S.A., Sucher, N.J., Kaiser, P.K. and Dreyer, E.B. (1991) Neuron **7**, 111–118
153. Giulian, D., Vaca, K. and Noonan, C.A. (1990) Science **250**, 1593–1596
154. Heyes, M.P., Brew, B.J., Martin, A., et al. (1991) Ann. Neurol. **29**, 202–209
155. Reynolds, I.J. and Aizenman, E. (1992) J. Neurosci. **12**, 970–975
156. Plaitakis, A. and Caroscio, J. (1987) Ann. Neurol. **22**, 275–279

157. Rothstein, J.D., Tsai, G., Kuncl, R.W., Clawson, L., Cornblath, D.R., Drachman, D.B., Pestronk, A., Stauch, B.L. and Coyle, J.T. (1990) Ann. Neurol. **28**, 18–25

158. Perry, T.L., Krieger, C., Hansen, S. and Eisen, A. (1990) Ann. Neurol. **28**, 12–17

159. Plaitakis, A., Constantakakis, E. and Smith, J. (1988) Ann. Neurol. **24**, 446–449

160. Tsai, G., Stauch-Slusher, B., Sim, L., Hedreen, J., Rothstein, J., Kuncl, R. and Coyle, J. (1991) Brain Res. **556**, 151–156

161. Tsai, G., Cork, L.C., Slusher, B.S., Price, D. and Coyle, J.T. (1993) Brain Res. **629**, 305–309

162. Rothstein, J.D., Martin, L.J. and Kuncl, R.W. (1992) N. Engl. J. Med. **326**, 1464–1468

163. Battaglioli, G., Martin, D.L., Plummer, J. and Messer, A. (1993) J. Neurochem **60**, 1567–1569

164. Rothstein, J. and Kuncl, R. (1995) J. Neurochem. **65**, 643–651

165. DiFiglia, M. (1990) Trends Neurosci. **13**, 286–289

166. Coyle, J. and Schwartz, R. (1976) Nature (London) **263**, 244–246

167. Schwartz, R., Whetsell, W. and Mangano, R. (1983) Science **219**, 316–318

168. Beal, M., Kowall, N., Swartz, K. and Ferrante, R. (1990) Neurosci. Lett. **108**, 36–42

169. Beal, M.F., Swartz, K.J., Hyman, B.T., Storey, E., Finn, S.F. and Koroshetz, W. (1991) J. Neurochem. **57**, 1068–1073

170. Beal, M.F., Brouillet, E., Jenkins, B.G., et al. (1993) J. Neurosci. **13**, 4181–4192

171. Jenkins, B.G., Koroshetz, W.J., Beal, M.F. and Rosen, B.R. (1993) Neurology **43**, 2689–2695

172. Brouillet, E., Jenkins, B.G., Hyman, B.T., Ferrante, R.J., Kowall, N.W., Srivastava, R., Roy, D.S., Rosen, B.R. and Beal, M.F. (1993) J. Neurochem. **60**, 356–359

173. Tipton, K. and Singer, T. (1993) J. Neurochem. **61**, 1191–1205

174. Sonsalla, P.K., Nicklas, W.J. and Heikkila, R.E. (1989) Science **243**, 398–400

175. Turski, L., Bressler, K., Rettig, K.-J., Löschmann, P.-A. and Wachtel, H. (1991) Nature (London) **349**, 414–418

176. Sonsalla, P.K., Zeevalk, G.D., Manzino, L., Giovanni, A. and Nicklas, W.J. (1992) J. Neurochem. **58**, 1979–1982

177. Jenner, P., Dexter, D.T., Sian, J., Schapira, A.H.V. and Marsden, C.D. (1992) Ann. Neurol. **32**, S82–S87

178. Fahn, S. and Cohen, G. (1992) Ann. Neurol. **32**, 804–812

179. Mena, M.A., Pardo, B., Paino, C.L. and Garcia De Yebenes, J. (1993) NeuroReport **4**, 438–440

180. Aizenman, E., White, W.F., Loring, R.H. and Rosenberg, P.A. (1990) Neurosci. Lett. **116**, 168–171

181. Cha, J. H.J., Dure, L.S., Sakurai, S.Y., Penney, J.B. and Young, A.B. (1991) Neurosci. Lett. **132**, 55–58

182. Maragos, W.F., Greenamyre, J.T., Penney, J.B. and Young, A.B. (1987) Trends Neurosci. **10**, 65–68

183. Palmer, A.M., Procter, A.W., Stratmann, G.C. and Bowen, D.M. (1986) Neurosci. Lett. **66**, 199–204

184. Procter, A.W., Palmer, A.M., Francis, P.T., Lowe, S.L., Neary, D., Murphy, E., Doshi, R. and Bowen, D.M. (1988) J. Neurochem. **50**, 790–802

185. Pearson, R.C.A. and Powell, T.P.S. (1989) Rev. Neurosci. **2**, 101–122

186. Penney, J.B., Maragos, W.F., Greenamyre, J.T., Debowey, D.L., Hollingsworth, Z. and Young, A.B. (1990) J. Neurol. Neurosurg. Psychiatry **53**, 314–320

187. Palmer, A.M. and Burns, M.A. (1994) J. Neurochem. **62**, 187–196

188. Sims, N.R. Firegan, J.M., Blass, J.P., Bowen, D.M. and Neary D. (1987) Brain Res. **436**, 30–38

189. Sims, N.R., Marek, K.I., Bowen, D.M. and Davison, A.N. (1982) J. Neurochem. **38**, 488–492

190. Palmer, A.M. and Burns, M.A. (1994) Brain Res. **645**, 338–342

191. Subbarao, K.V., Richardson, J.S. and Ang, L.C. (1990) J. Neurochem. **55**, 342–345

192. Hajimohammadreza, I. and Brammer, M. (1990) Neurosci. Lett. **112**, 333–337

193. DeKosky, S. and Palmer, A. (1994) in Clinical Neurology of Aging (Albert, M. and Knoefel, J., eds.), pp. 79–101, Oxford University Press, New York

194. Palmer, A.M., Robichaud, P.J. and Reiter, C.T. (1994) Neurobiol. Aging **15**, 1–9

195. Piggott, M.A., Perry, E.K., Perry, R.H. and Court, J.A. (1992) Brain Res. **588**, 277–286

196. Procter, A.W., Wong, E.H.F., Stratmann, G.C., Lowe, S.L. and Bowen, D.M. (1989) J. Neurochem. **53**, 698–704

197. Mouradian, M.M., Contreras, P.C., Monahan, J.B. and Chase, T.N. (1988) Neurosci. Lett. **93**, 225–230

198. Choi, D.W. and Rothman, S.M. (1990) Annu. Rev. Neurosci. **13**, 171–182
199. Olney, J.W., Labruyere, J. and Price, M.T. (1989) Science **244**, 1360–1362
200. Gotti, B., Benavides, J., MacKenzie, E. and Scatton, B. (1990) Brain Res. **522**, 290–307
201. Koek, W. and Colpaert, F. (1989) J. Pharmacol. Exp. Ther. **252**, 349–357
202. Ginsberg, M.D., Sternau, L.L., Globus, M.Y.-T., Dietrich, W.D. and Busto, R. (1992) Cerebrovasc. Brain Metab. Rev. **4**, 189–225
203. Clifton, G.L., Jiang, J.Y., Lyeth, B.G., Jenkins, L.W., Hamm, R.J. and Hayes, R.L. (1991) J. Cereb. Blood Flow Metab. **11**, 114–121
204. Marion, D.W., Penrod, L.E., Kelsey, S.F., Obvist, W.D., Kochavok, P.M., Palmer, A.M., Wisliewski, S.R. and DeKosky, S.T. (1997) New Eng. J. Med **336**, 540–546
205. Moroni, F., Alesiani, M., Galli, A., Mori, F., Pecorari, R., Carla, V., Cherici, G. and Pellicciari, R. (1991) Eur. J. Pharmacol. **199**, 227–232
206. Henneberry, R.C., Novelli, A., Cox, J.A. and Lysko, P.G. (1989) Ann. N.Y. Acad. Sci. **568**, 225–233
207. Rosenberg, P.A. and Aizenman, E. (1989) Neurosci. Lett. **103**, 162–168
208. Garthwaite, G., Williams, G.D. and Garthwaite, J. (1992) Eur. J. Neurosci. **4**, 353–360
209. Hayes, R.L., Jenkins, L.W., Lyeth, B.G., Balster, R.L., Robinson, S.E., Clifton, G.L., Stubbins, J.F. and Young, H.F. (1988) J. Neurotrauma **5**, 259–274
210. McIntosh, T., Vink, R., Soares, H., Hayes, R. and Simon, R. (1990) J. Neurochem. **55**, 1170–1179
211. Shapira, Y., Yadid, G., Niska, A. and Shohami, E. (1990) J. Neurotrauma **7**, 131–139
212. Smith, D., Okiyama, K. and McIntosh, T. (1991) Soc. Neurosci. Abstr. **17**, 66.12
213. Meldrum, B. (1993) Brain Res. Rev. **18**, 293–314

# 11

# Long-term potentiation in the hippocampus

**Zafar I. Bashir**

Department of Anatomy, School of Medical Sciences, University of Bristol,
University Walk, Bristol BS8 1TD, U.K.

## Introduction

Understanding the synaptic and molecular basis of learning and memory is one of the fundamental questions facing neuroscientists today. Long-term potentiation (LTP) of synaptic transmission is one of the best understood and most widely used models of synaptic plasticity, and may provide clues as to the mechanisms underlying learning and memory in higher vertebrates, including humans. LTP has been most intensely studied in the hippocampus, but has also been shown to occur in many other regions of the central nervous system. In this chapter I will concentrate on the role of amino acid neurotransmitters in the induction and the expression of LTP, reviewing some of the well established features and considering some recent developments in this field.

LTP was first described in the dentate region of the hippocampus as a sustained increase in the size of post-synaptic field potentials following a period of high-frequency (tetanic) stimulation of the perforant path [1]. Following the induction of LTP, potentiation can last for many hours *in vitro* and for days *in vivo*. In the ensuing years there has been a plethora of information describing LTP in different regions of the central nervous system, both *in vivo* and *in vitro* [2–4]. Different forms of LTP are characterized by some, or all three, of the following basic properties: input specificity, co-operativity and associativity. Input specificity relates to the effect whereby, when separate inputs on to a cell or group of cells are stimulated, only the afferent input which receives tetanic stimulation is potentiated [1,5–7]. Co-operativity is the property whereby the strength of the afferent input has to be above a threshold level for tetanic stimulation to induce LTP [8–10]. Associativity is the property whereby an afferent input that is below the co-operativity threshold can undergo LTP if a second converging input is stimulated sufficiently strongly in temporal association with stimulation of the first input [9,11–13]. It was postulated by Hebb [14] that synaptic plasticity should occur by repeated presynaptic activation of a post-synaptic neuron. Of the conditions outlined above, associativity and input specificity especially provide means of fulfilling the conditions of 'Hebbian' synapses. Associative learning, such as classical conditioning, could be subserved by Hebbian synapses whereby pre- and post-synaptic activity are temporally correlated by inputs from two separate afferent pathways.

Two main forms of LTP will be considered: (1) the associative form that occurs at a wide variety of pathways, including the Schaffer collateral commissural (SCC)–CA1 synapses, medial perforant path–granule cell synapses and within many other areas of the cerebral cortex, and (2) the form of LTP that is found to occur at mossy fibre–CA3 synapses and is generally considered to be non-associative. Since most of the pertinent details concerning the underlying mechanisms of associative and non-associative LTP have come through studies on SCC–CA1 synapses and mossy fibre–CA3 synapses respectively, it is these two pathways that this chapter will concentrate on. A schematic representation of these hippocampal pathways is shown in Fig. 1.

## Fig. 1.  LTP in the hippocampus

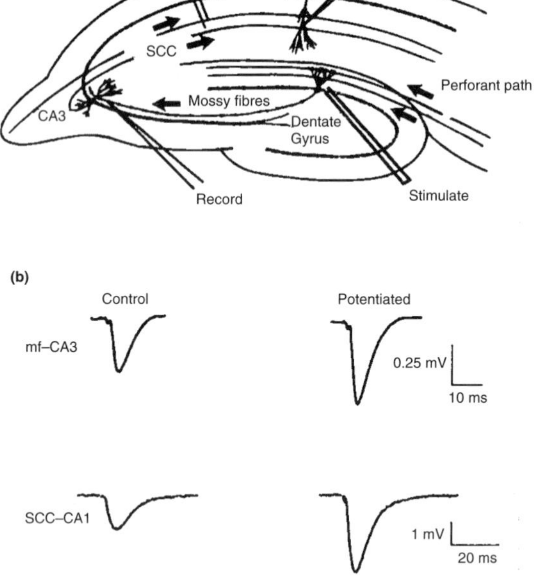


*(a) Schematic representation illustrating the major pathways in the hippocampal slice and the placing of recording and stimulating electrodes. Extracellular field EPSPs are recorded from the CA1 and CA3 regions in response to electrical stimulation of the SCC pathway and mossy fibres respectively. (b) Synaptic responses evoked by single-shock stimulation of the SCC pathway before (control) and following the induction of LTP (potentiated), induced by tetanic (100 Hz, 1 s) stimulation. mf, mossy fibre.*

# Synaptic transmission

Excitatory synaptic transmission at pathways that undergo LTP is mediated by the acidic amino acid L-glutamate [2], which acts on a variety of different pre- and post-synaptic receptors. Thus glutamate can mediate its actions via kainate receptors, $\alpha$-amino-3-hydroxy-5-methyl-4-isoxazolepropionic acid (AMPA) receptors, $N$-methyl-D-aspartate (NMDA) receptors and metabotropic glutamate receptors (mGluRs). Over the past few years, selective antagonists for AMPA receptors have been reported [15–17]. These 2,3-benzodiazepine derivatives (e.g. GYKI 52466 and 53655) have allowed the isolation of a post-synaptic kainate-receptor-mediated inward current in hippocampal [17] and cerebellar [18] cultures. In addition, pre- and post-synaptic effects of kainate have been decribed in area CA1 of the hippocampus [19]. Little is still known regarding the function of kainate receptors in the hippocampus and these, as is generally the case, will be grouped with AMPA receptors in the remainder of this chapter.

## AMPA receptors

AMPA receptors are linked directly to an ion channel permeable to $Na^+$ and $K^+$, although a variant lacking the GluR2 subunit is also permeable to $Ca^{2+}$ [20]. In bathing medium containing 'physiological' levels of $Mg^{2+}$ and $Ca^{2+}$ (1 mM $Mg^{2+}$, 2 mM $Ca^{2+}$), AMPA/kainate receptor antagonists, such as 6-cyano-7-nitro-quinoxaline-2,3-dione, virtually abolish synaptic transmission evoked by single-shock, low-frequency (0.033 Hz) stimulation in both the SCC [21] and mossy fibre [22] pathways of the hippocampus. These results suggest that AMPA/kainate receptor activation essentially underlies excitatory synaptic transmission in the CA1 and CA3 regions of the hippocampus. The contributions of different amino acid receptors to synaptic transmission are depicted in Fig. 2.

## NMDA receptors

The NMDA receptor is coupled directly to a channel permeable to $Na^+$, $K^+$ and $Ca^{2+}$. The selective NMDA receptor antagonist 2-amino-5-phosphonopentanoate (AP5) [23] has virtually no effect on evoked synaptic transmission under the same conditions as described above [24]. The reasons for this lack of involvement, in the CA1 region, are due to the biophysical properties of the NMDA-receptor-associated channel (hereafter called the NMDA receptor channel). Firstly, the NMDA receptor channel has a region of negative slope conductance [25] which is due to a voltage-dependent [26] blockade of the NMDA receptor channel by physiological concentrations of extracellular $Mg^{2+}$ [27]. Secondly, the rise and decay times of NMDA-receptor-mediated responses are slower than those of AMPA-receptor-mediated responses [28]. Following single-shock stimulation of the SCC pathway, AMPA-receptor-mediated depolarization is rapidly curtailed by $\gamma$-aminobutyric acid (GABA)$_A$- and GABA$_B$-receptor-mediated feed-forward inhibition, which returns the post-synaptic membrane to a hyperpolarized (relative

**Fig. 2.**        **Synaptic transmission at SCC–CA1 synapses**

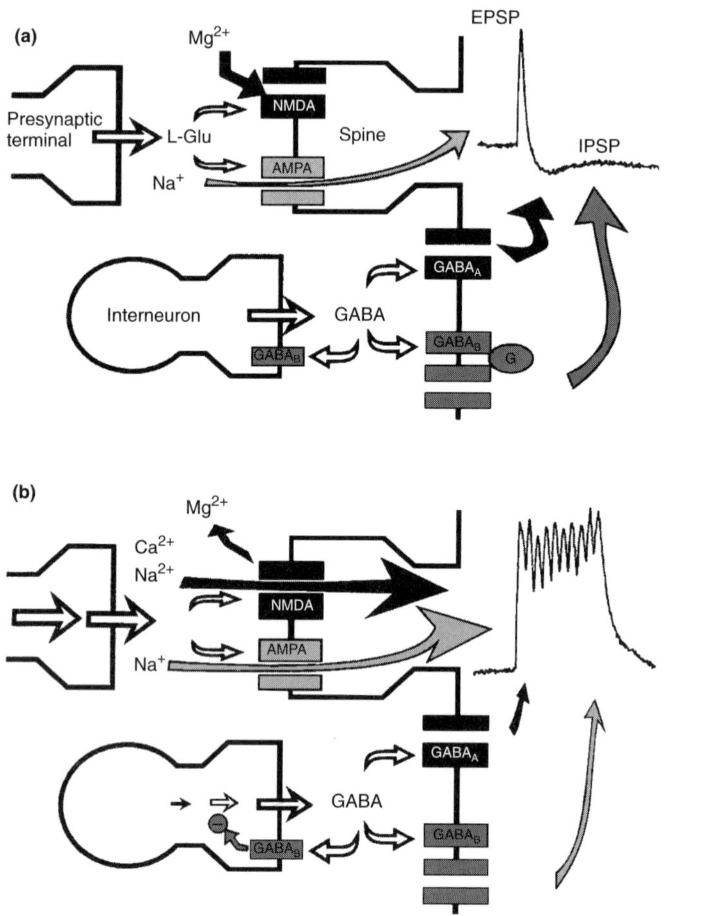



*(a) Under conditions of single-shock stimulation, AMPA receptor and GABA receptor (GABA$_A$ and GABA$_B$) activation mediates the EPSP and inhibitory post-synaptic potential respectively. The NMDA receptor channel is blocked by extracellular Mg$^{2+}$. G, G-protein. (b) During high-frequency stimulation, NMDA receptors contribute to synaptic transmission. A combination of depression of GABAergic inhibition and summation of AMPA EPSPs results in sufficient depolarization to alleviate the blockade of the NMDA receptor channel by Mg$^{2+}$. The resultant Ca$^{2+}$ entry is believed to provide the trigger for the induction of LTP. Modified from [4].*

to rest) state (Fig. 2a). Therefore the voltage-dependence and slow kinetics of the NMDA receptor channel preclude it from contributing significantly to synaptic transmission under these conditions in the CA1 region. However, under conditions of, for example, high-frequency stimulation [29], post-synaptic depolarization [30], blockade of inhibition [31] or removal of extracellular $Mg^{2+}$ [32], NMDA receptor antagonists block a greater degree of synaptic transmission in the CA1 region than under 'normal' conditions. Thus activation of NMDA receptors can be made to contribute to synaptic transmission under certain conditions (Fig. 2b). In the mossy fibre–CA3 field there is a relative paucity of NMDA receptors compared with the SCC–CA3 terminal field [33]. NMDA-receptor-mediated synaptic responses can be evoked by stimulation of the mossy fibres [34,35], but these responses are relatively small in amplitude and, compared with NMDA receptor responses in other regions, appear to have a very shallow $I/V$ curve at holding potentials negative to zero [34].

## Metabotropic receptors

L-Glutamate also activates metabotropic receptors, a class of G-protein-coupled receptors which are not linked directly to ion channels (see Chapter 6 of this volume, by Henley et al.). Molecular biology cloning techniques have demonstrated that at least eight distinct subtypes (mGluR1–mGluR8) exist. These have been divided into three groups; group 1 constitutes mGluR1 and mGluR5, group 2 contains mGluR2 and mGluR3, and group 3 contains mGluR4, mGluR6 and mGluR7. mGluR8 has been classified into both groups 2 [36] and 3 [37]. Group 1 mGluRs stimulate phospholipase C, which catalyses the hydrolysis of phosphatidylinositol bisphosphate to inositol 1,4,5-trisphosphate, which in turn mobilizes $Ca^{2+}$ from intracellular stores. Additionally, stimulation of phospholipase C results in the formation of diacylglycerol, which activates protein kinase C. Groups 2 and 3 are negatively coupled to cyclic AMP (cAMP) turnover, but can also potentiate cAMP formation induced by other G-protein-coupled receptors.

Exogenous activation of mGluRs by $1S,3R$-aminocyclopentane dicarboxylic acid ($1S,3R$-ACPD), a broad-spectrum mGluR agonist, can activate both pre-synaptic (depression of synaptic transmission [38]) and post-synaptic (e.g. blockade of spike accommodation [39,40]) effects, presumably by acting at different mGluR subtypes. There are currently several different mGluR agonists which can preferentially activate different mGluR subtypes. In order to elucidate the role of mGluR subtypes in synaptic transmission, potent and specific antagonists of these receptors are required [41]. α-Methyl-4-carboxyphenylglycine (MCPG), a broad-spectrum mGluR antagonist, can reverse the above-described effects of exogenously applied $1S,3R$-ACPD [40,41] and is proving useful for probing the role of mGluRs. Under conditions of low-frequency stimulation, MCPG has no effect on synaptic transmission [40]; thus MCPG-sensitive mGluRs do not play a regulatory role in synaptic transmission under these conditions. A synaptically evoked MCPG-sensitive inward current can be observed [42] at mossy fibre–CA3 synapses and was demonstrated in organotypic cultures during blockade of ionotropic excitatory transmission.

# Mechanisms of induction of LTP in the CA1 region

The mechanism of associative LTP hinges on the fact that NMDA receptor channels are subject to a voltage-dependent blockade by $Mg^{2+}$, a blockade which is alleviated under conditions of depolarization. Thus the NMDA receptor can act as a coincidence detector by sensing the post-synaptic level of depolarization and the presynaptic release of glutamate. This makes the NMDA receptor an ideal candidate for underpinning Hebbian forms of synaptic plasticity.

## NMDA receptor activation

The requirement for the activation of NMDA receptors in the induction of LTP was first demonstrated in the CA1 region [24] and is now well established in this and many other pathways. The conditions that can result in the activation of NMDA receptors have been outlined above. High-frequency stimulation, of the sort that induces LTP, has been shown to increase the contribution of NMDA receptors to synaptic transmission [29]. The myriad factors [4] that can enable NMDA receptor activation to contribute to high-frequency synaptic transmission are beyond the scope of this chapter. However, one important factor that will be considered is the frequency-dependent depression of inhibitory synaptic transmission [43]. It has been shown in the CA1 region that it is possible to study GABAergic inhibition directly by placing a stimulatory electrode close to the post-synaptic cell in the presence of excitatory amino acid receptor antagonists [44]. It was demonstrated that the frequency-dependent depression of GABAergic inhibition is mediated by synaptically released GABA acting on presynaptic $GABA_B$ autoreceptors to depress the subsequent release of GABA [44]. The depression was maximal at approximately the theta rhythm, a frequency of stimulation very effective in inducing LTP [45,46]. It was subsequently demonstrated that antagonism of $GABA_B$ receptors prevented the induction of theta rhythm-induced (but not tetanus-induced) LTP [47], illustrating that frequency-dependent depression of GABAergic inhibition is an important factor in allowing sufficient depolarization, and hence NMDA receptor activation, for the induction of LTP under certain circumstances (represented schematically in Fig. 2).

## Role of $Ca^{2+}$

Following the activation of NMDA receptors, a post-synaptic increase in the $Ca^{2+}$ concentration is thought to provide the trigger for the subsequent mechanisms that underlie synaptic plasticity. Thus loading cells with high concentrations of $Ca^{2+}$ chelators has been shown to block the induction of LTP [48,49]. Since, under most conditions, pharmacological antagonism of NMDA receptor activity blocks the induction of LTP, but blockade of voltage-gated calcium channels does not [4], it is assumed that the increase in the post-synaptic $Ca^{2+}$ concentration is a result of $Ca^{2+}$ entry through NMDA receptor channels. In addition, the finding that thapsigargin, which depletes intracellular $Ca^{2+}$ stores, can also block the induction of LTP [50] suggests that $Ca^{2+}$ entry via NMDA receptor channels may lead to release of $Ca^{2+}$ from intracellular stores, which may also play a role in the induction of LTP. Indeed, it has been demonstrated that, following high-frequency stimulation of the SCC pathway, the rise in post-synaptic dendritic $Ca^{2+}$ results from activation of both NMDA receptors and mGluRs and $Ca^{2+}$ release from intracellular stores [51,52].

## Role of mGluRs

It has long been suspected that NMDA receptor activation alone may not be sufficient for the induction of LTP. For example, the application of NMDA to slices of hippocampus generally does not produce LTP, but results in short-term potentiation (STP) which decreases over the course of 30–60 min. In addition, since tetanic stimulation in the presence of blockers of protein kinases results in the induction not of LTP but of STP, it has been considered that some mechanism that is linked directly to the activation of protein kinases may also be required for the induction of LTP. Therefore, since mGluRs are activated by synaptically released glutamate and provide a direct link to the intracellular biochemistry of the cell, the role of these receptors in synaptic plasticity has come under close scrutiny. It has been demonstrated that $1S,3R$-ACPD can augment the magnitude of LTP [53] and convert STP [54] to LTP. In addition, $1S,3R$-ACPD can augment NMDA-receptor-mediated responses [54,55]. The mechanisms of this action have been shown to be $Ca^{2+}$-, but not protein kinase C-, dependent ([55], but see [54]). The modulation by mGluRs of NMDA receptor function may therefore be different from the protein kinase C-induced alleviation of the $Mg^{2+}$ blockade of the NMDA receptor channel [56]. Thus it is possible that activation of mGluRs can augment LTP or convert STP to LTP by increasing NMDA receptor function during tetanic stimulation. In addition, it has also been shown that the application of $1S,3R$-ACPD to brain slices can result in a slowly developing synaptic potentiation which, in hippocampal slices, is AP5-insensitive and occludes with tetanus-induced LTP [57]. The time course of this potentiation is complementary to STP and thus may reflect the later kinase-dependent phase of LTP [58]. Early studies using the then available antagonists (L-2-amino-3- and 4-phosphonobutyrate) of mGluRs reported that these agents shortened the time course of LTP [59,60]. However, the specificity and potency of these drugs is questionable. More recently, the role of mGluRs has been explored further by the use of more selective pharmacological antagonists [41]. It has been shown that MCPG blocks $1S,3R$-ACPD-induced LTP [40] and tetanus-induced LTP *in vitro* [40,61] and *in vivo* [62], such that only STP remains. In addition, it has been shown that MCPG blocks $1S,3R$-ACPD- and tetanus-induced LTP of NMDA-receptor-mediated synaptic transmission [63] in the dentate gyrus *in vitro*. However, these findings are not universal; one study [64] has reported that MCPG does not block, but reduces the magnitude of, LTP, while two reports suggested that LTP is not affected by MCPG [65,66]. The reasons for some of these discrepancies are still unclear. However, the findings of one of the above groups [66] can be explained because, in contrast with all other groups, they reported that MCPG is not an mGluR antagonist.

A complementary approach to the use of pharmacological methods in addressing the role of specific receptor types is to use gene knock-out techniques to remove the receptor or receptor subtype of interest. It has been demonstrated that there is no difference in LTP induced in either area CA1 or the lateral or medial perforant paths in slices from mice lacking mGluR1 (mGluR1$^{-/-}$) compared with LTP in slices from normal mice [67]. Taken at face value, these results suggest that there is no role for mGluR1 in the induction of LTP. However, it is possible that any loss of mGluR1 function may be compensated for by up-regulation of another subtype (e.g. mGluR5), which may take over the normal role of mGluR1. In contrast with the above results, it has been reported that LTP in the CA1 region of mGluR1$^{-/-}$ mice is depressed by about 50% compared with that in normal animals; in addition, in normal animals MCPG reduced LTP by approx. 50% [64]. It was concluded, therefore, that mGluR1 activation, rather than being involved in the induction of LTP, is responsible for regulating the magnitude of LTP.

It has been demonstrated that the role of mGluRs in the induction of LTP is more complex than first thought, and it is possible that this may go some way to explaining some of the discrepancies described above. It was found that MCPG did not block the induction of LTP providing that LTP had previously been induced in the same pathway [68]. It was postulated that the initial induction of LTP sets up a long-lasting MCPG-sensitive mGluR-dependent mechanism (possibly through kinase activation) which negates the requirement for further mGluR activation for the induction of subsequent LTP [68]. It has been shown that prolonged stimulation (900 stimuli) at 1–2 Hz can result in an MCPG-sensitive [69,70] reversal (depoten-tiaton) of LTP [71,72]. The conditioning effect described above can be reversed by low-frequency stimulation, such that MCPG blocks the induction of subsequent LTP [68]. Therefore both MCPG-sensitive mGluR and NMDA receptor activation are required for the induction of LTP. NMDA receptor activation generates a transient induction signal, whereas mGluR activation sets a lasting 'conditioning' mechanism (Fig. 3). Once this conditioning has been set, the induction of sub-sequent LTP requires only the activation of NMDA receptors. It is also possible that the induction of subsequent LTP requires the activation of MCPG-insensitive mGluRs and of NMDA receptors.

## Expression of associative LTP

The expression of LTP could result from an increase in synaptic transmission via the glutamate receptors (AMPA receptors) responsible for mediating non-potentiated synaptic responses, or could result from other glutamate receptor types contributing to the potentiated response. Following the induction of LTP, neither AP5 [24] nor MCPG [40] has any effect on synaptic transmission, whereas

**Fig. 3.**              **Three-state model for the induction of LTP**

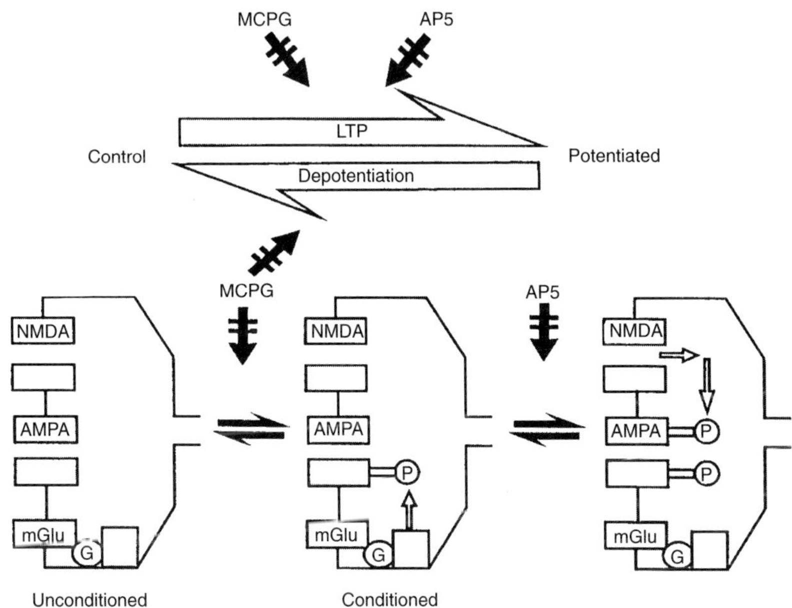


*It is proposed that there is one potentiated and two (an unconditioned and a conditioned) non-potentiated states. Under normal circumstances, tetanic stimulation results in the induction of LTP by activation of both NMDA receptors and mGluRs; the induction of LTP is MCPG- and AP5-sensitive. However, if LTP has already been induced, subsequent LTP is MCPG-insensitive [68]. Therefore it is proposed that there is an intermediate MCPG-sensitive mGluR-dependent conditioned state. Once this state has been reached, only NMDA receptors need to be activated to induce LTP. MCPG-sensitive depotentiation can reverse both the potentiation and the conditioning. Because both LTP and conditioning are protein kinase-dependent, it is assumed that both of these stages require phosphorylation of some intracellular substrate. Modified from [41].*

potentiated and non-potentiated responses are blocked by AMPA receptor antagonists [73,74]. Thus AMPA receptor activation mediates normal and potentiated synaptic responses. The increase in synaptic transmission could result from a myriad of different possibilities. Of these, there has been a debate raging as to whether an increase in presynaptic transmitter release or a change in the post-synaptic receptor–channel complex mediates the expression of LTP. The locus of expression of LTP has been investigated in various ways. Earlier reports suggested that, under conditions in which both AMPA and NMDA receptor activation contribute to synaptic transmission, there was a selective or far greater increase in the AMPA-receptor-mediated component relative to the NMDA-receptor-mediated component of the response. However, treatments traditionally considered to increase transmitter release increased both the AMPA- and NMDA-receptor-mediated components of synaptic responses [73,75]. In addition, it was reported that 'pure' NMDA-receptor-mediated excitatory post-synaptic potentials (EPSPs) did not undergo LTP [76]. It was therefore suggested that the locus of expression of LTP is a post-synaptic phenomenon, since a presynaptic locus should result in potentiation of both AMPA- and NMDA-receptor-mediated components of mixed EPSPs. However, contrary to the earlier reports, it has now been shown that both AMPA- and NMDA-receptor-mediated components of dual-component responses can undergo LTPs of roughly equal magnitude [77,78] and that LTP of pure NMDA receptor EPSPs/EPSCs (where EPSC is the current underlying an EPSP), comparable in magnitude to LTP of AMPA receptor EPSPs, can also be induced [79–82]. Although these results are contrary to the earlier reports, they do not necessarily promote a presynaptic over a post-synaptic locus for LTP expression. They could also be explained by postulating an approximately equal post-synaptic up-regulation of both AMPA receptors and NMDA receptors. Ionophoretic ejection of AMPA into the dendritic region of the CA1 area demonstrated that there was increased sensitivity to AMPA following the induction of LTP [74]. This effect was delayed and developed slowly after the induction of tetanus-induced LTP. These results suggest, therefore, that the expression of LTP is partly post-synaptic, as reflected by the increased sensitivity to AMPA, but also, prior to this, partly presynaptic [74].

The application of quantal analysis to the issue of pre- versus post-synaptic expression of LTP has not resulted in any consensus of views. Evidence has been found for an increase in quantal size and an increase in amplitude of miniature EPSCs (mEPSCs), both of which are usually thought to reflect a post-synaptic change [83]. Other studies have reported an increase in quantal content, a decrease in failures and an increase in the frequency of mEPSCs, all of which are usually considered to reflect a presynaptic locus of LTP expression [84–87]. However, in keeping with the view expressed above, some reports have suggested changes in both pre- and post-synaptic parameters [88,89].

Another method employed to investigate the locus of LTP has utilized the use-dependent blockade of NMDA receptor channels by MK-801. The rate of blockade of NMDA-receptor-mediated responses by MK-801 is dependent on the probability of release ($P_r$) of glutamate [90,91]; using this method, it has been suggested that two populations of synapses with a 6-fold difference in $P_r$ exist within the SCC–CA1 pathway [90]. Those synapses having a low $P_r$ would, in theory, be able to increase $P_r$ to increase overall transmitter release. It was subsequently reported that, whereas manipulations involving an increase in transmitter release, such as paired pulse facilitation, did alter $P_r$, there was no change in $P_r$ following the induction of LTP [92]. However, these results relied on MK-801 blockade of NMDA-receptor-mediated responses, whereas LTP in these experiments was expressed by AMPA receptors. In addition, the effect of MK-801 was tested some 80 min after the induction of LTP, while it has been shown that presynaptic effects may be more prominent earlier in the expression of LTP [74].

Using stimuli sufficient to activate a single afferent fibre, the probability of evoking mEPSCs before and after the induction of LTP has been studied [93]. These results showed that there was an increase in the probability of evoking mEPSCs but no increase in the size of these synaptic responses following the induction of LTP. Therefore it was suggested that LTP is expressed purely by an increase in the probability of synaptic transmission, the usual interpretation of which is an increase in transmitter release.

More recently, the expression of LTP has been postulated to result from the uncovering or addition of post-synaptic 'silent' AMPA receptors [94]. It was subsequently demonstrated that there are indeed synapses at which only functional NMDA receptors exist. Following the induction of LTP, these same synapses subsequently expressed transmission by both NMDA and AMPA receptors [95,96]. Such a mechanism may explain how a post-synaptic change underlying the expression of LTP could be interpreted by some types of analysis (e.g. change in failures, coefficient of variance) as being presynaptic.

It is not entirely clear why there is such a discrepancy between the various reports concerning the locus of expression of LTP. It is possible that subtle differences in methodology may produce significant differences in the expression of LTP. If the relative contributions of pre- and post-synaptic mechanisms to the expression of LTP are temporally regulated, then this may be one factor that could go some way to explaining some of the discrepancies that are reported in the literature.

## Mechanisms of induction of non-associative LTP

The CA3 region of the hippocampus has two major synaptic inputs. Of these, the mossy fibre–CA3 synapses exhibit LTP which is widely believed to be non-associative [97] (but see below), whereas LTP at commissural–CA3 synapses is associative [97]. In all respects so far tested, LTP at commissural–CA3 synapses appears to be identical to that at SCC–CA1 synapses. Mossy fibre LTP is not dependent on the activation of NMDA receptors [98], even though NMDA-receptor-mediated synaptic responses can be evoked in this pathway [34] and can undergo LTP [35], despite the relatively low density of such receptors in the mossy fibre terminal field [33].

The mechanisms underlying the induction of mossy fibre LTP are still far from clear. Indeed, the widely held view of the non-associative nature of mossy fibre LTP has been challenged by the demonstration that the induction of LTP may be dependent on the activation of the post-synaptic cell. Firstly, it has been shown that, as in the CA1 region, hyperpolarizing the post-synaptic cell blocks the induction of LTP, and post-synaptic depolarization, in conjunction with high-frequency mossy fibre stimulation, can result in the enhancement of LTP [99]. Secondly, loading the post-synaptic cell with Ca²⁺ chelators has been shown to block the induction of LTP [100]. Furthermore, it has now been reported that mossy fibre LTP is associative. Pairing weak tetanic stimulation of mossy fibres with strong stimulation of the commissural fibres has been reported to result in LTP in the mossy fibre pathway [101]. The reasons for the discrepancies regarding the role of the post-synaptic cell and the associative or non-associative nature of LTP in the mossy fibre pathway are still unclear.

## Receptor types involved in the induction of mossy fibre LTP

It has been shown that MCPG blocks the induction of LTP at the mossy fibre–CA3 synapse [40], although this finding has been disputed [65]. Using hippocampal slices from mice lacking the gene for mGluR1, it was demonstrated that there is a significant deficit of mossy fibre LTP compared with that in normal animals. These results suggest that glutamate, by acting at mGluR1, may be involved in the induction of LTP at mossy fibre synapses [67].

In addition, other transmitter systems have been postulated to be involved in LTP in this pathway. Opioid peptides are contained in, and released from, mossy fibre terminals [102]. Opioid receptor antagonists, while having no effect on low-frequency-evoked synaptic responses, prevent the induction of tetanus-induced LTP ([103–105], but see [106]). The mechanism of the role of opioid receptor activation in the induction of LTP is not known, but may be due to the demonstrated opioid-induced depression of GABAergic inhibition [107,108] or to a direct increase in CA3 excitability [107].

## Role of Ca²⁺

Blockade of ionotropic glutamate receptors with kynurenate and AP5, while blocking synaptic transmission, has been shown not to block the induction of mossy fibre LTP [109]. Since synaptic transmission was blocked in this study, it was assumed that post-synaptic voltage-dependent calcium channels could not be activated, and hence that there could be no role for post-synaptic Ca²⁺ in LTP. Furthermore, since LTP was not induced in the absence of extracellular calcium, it was suggested that this demonstrated a role for presynaptic Ca²⁺ in the induction of LTP. However, some other source of post-synaptic calcium, e.g. through activation of mGluRs (which were not blocked), could be involved, since calcium chelators within the post-synaptic cell [100], MCPG [40] and mGluR1 gene deletion [67] can all block the induction of LTP.

## Expression of mossy fibre LTP

As with LTP in the CA1 region, the locus of expression of mossy fibre LTP is also controversial. It has been shown that paired pulse facilitation, a process thought to be presynaptically mediated, is affected by the induction of LTP [97,110]. In addition, results from studies using quantal analysis [111] and the use-dependent blockade of NMDA receptor activity [35] support the proposal that this form of LTP may be expressed presynaptically. In contrast, however, Yamamoto et al. [112] have demonstrated an increase in post-synaptic sensitivity to AMPA following the induction of LTP.

Recent studies have indicated a role for cAMP in the expression of LTP in this pathway [113,114]. It has been postulated that, following high-frequency stimulation, presynaptic calcium entry can activate adenylate cyclase, through the action of $Ca^{2+}$/calmodulin, with the subsequent increase in cAMP activating protein kinase A. This in turn could promote the persistent increase in glutamate release which may underlie the expression of LTP.

## Conclusions

The role of NMDA receptors in the induction of associative LTP has been well established. The characteristics of the NMDA receptor provide the molecular basis whereby pre- and post-synaptic activity can be temporally correlated, and can thus provide the underlying basis of associativity. The role of mGluRs in LTP has only recently begun to be investigated, but already it would appear that these receptors play a subtle but complex role in synaptic plasticity. Finally, the locus of expression of LTP in both the CA1 and CA3 regions remains controversial, as do the mechanisms of induction of LTP in the mossy fibre–CA3 pathway.

## References


1. Bliss, T.V.P. and Lomo, T. (1973) J. Physiol. (London) **232**, 331–356
2. Collingridge, G.L. and Lester, R.A. (1989) J. Pharmacol. Rev. **41**, 143–210
3. Artola, A. and Singer, W. (1993) Trends Neurosci. **16**, 480–487
4. Bliss, T.V.P. and Collingridge, G.L. (1993) Nature (London) **361**, 31–39
5. Lynch, G., Dunwiddie, T. and Gribkoff, V. (1977) Nature (London) **266**, 737–739
6. Andersen, P., Sundberg, S.H., Sveen, O. and Wigström, H. (1977) Nature (London) **266**, 736–737
7. Dunwiddie, T. and Lynch, G. (1978) J. Physiol. (London) **276**, 353–367
8. Bliss, T.V.P. and Gardner-Medwin, A.R. (1973) J. Physiol. (London) **232**, 357–374
9. Lee, K.S. (1983) J. Neurosci. **3**, 1369–1372
10. Schwartzkroin, P. and Wester, K. (1975) Brain Res. **89**, 107–119
11. Levy, W.B. and Steward, O. (1979) Brain Res. **175**, 233–245
12. McNaughton, B.L., Douglas, R.M. and Goddard, G.V. (1978) Brain Res. **157**, 277–294
13. Wigström, H. and Gustafsson, B. (1983) Acta Physiol. Scand. **119**, 455–458
14. Hebb, D.O. (1949) The Organization of Behaviour, John Wiley, New York
15. Donevan, S.D. and Rogawski, M.A. (1993) Neuron **10**, 51–59
16. Zorumski, C.F., Yamada, K.A., Price, M.D. and Olney, J.W. (1993) Neuron **10**, 61–67
17. Patermain, A.V., Morales, M. and Lerma, J. (1995) Neuron **14**, 185–189
18. Renard, A., Crepel, F. and Audinat, E. (1995) Neuropharmacology **34**, 335–347
19. Chittajulu, R., Vignes, M., Dev, K.K., Barnes, J.M., Collingridge, G.L. and Henley, J.M. (1996) Nature (London) **379**, 78–81
20. Gasic, G.P. and Hollmann, M. (1992) Annu. Rev. Physiol. **54**, 507–536

21.  Blake, J.F., Brown, M.W. and Collingridge, G.L. (1988) Neurosci. Lett. **89**, 182–186
22.  Neuman, R.S., Ben-Ari, Y., Gho, M. and Cherubini, E. (1988) Neurosci. Lett. **92**, 64–68
23.  Evans, R.H., Francis, A.A., Jones, A.W., Smith, D.A.S. and Watkins, J.C. (1982) Br. J. Pharmacol. **75**, 65–75
24.  Collingridge, G.L., Kehl, S.J. and McLennan, H. (1983) J. Physiol. (London) **334**, 33–46
25.  MacDonald, J.F., Porietis, A.V. and Wojtowicz, J.M. (1982) Brain Res. **237**, 248–253
26.  Nowak, L., Bregestovski, P., Ascher, P., Herbet, A. and Prochiantz, A. (1984) Nature (London) **307**, 462–465
27.  Ault, B., Evans, R.H., Francis, A.A., Oakes, D.J. and Watkins, J.C. (1980) J. Physiol. (London) **307**, 413–428
28.  Dale, N. and Roberts, A. (1985) J. Physiol. (London) **363**, 35–59
29.  Herron, C.E., Lester, R.A.J., Coan, E.J. and Collingridge, G.L. (1986) Nature (London) **322**, 265–268
30.  Collingridge, G.L., Herron, C.E. and Lester, R.A.J. (1988) J. Physiol. (London) **399**, 283–300
31.  Herron, C.E., Williamson, R. and Collingridge, G.L. (1985) Neurosci. Lett. **61**, 255–260
32.  Herron, C.E., Lester, R.A.J., Coan, E.J. and Collingridge, G.L. (1985) Neurosci. Lett. **60**, 19–23
33.  Monaghan, D.T. and Cotman, C.W. (1985) J. Neurosci. **5**, 2909–2919
34.  Jonas, P., Major, G. and Sakmann, B. (1993) J. Physiol. (London) **472**, 615–663
35.  Weisskopf, M.G. and Nicoll, R.A. (1995) Nature (London) **376**, 256–259
36.  Nakanishi, S. (1994) Neuron **13**, 1031–1037
37.  Pin, J.-P. and Duvoisin, R. (1995) Neuropharmacology **34**, 1–26
38.  Baskys, A. and Malenka, R.C. (1991) J. Physiol. (London) **444**, 687–701
39.  Stratton, K.R., Worley, P.F. and Baraban, J.M. (1989) Eur. J. Pharmacol. **173**, 235–237
40.  Bashir, Z.I., Bortolotto, Z.A., Davies, C.H., et al. (1993) Nature (London) **363**, 347–350
41.  Watkins, J. and Collingridge, G. (1994) Trends Pharmacol. Sci. **15**, 333–342
42.  Charpak, S. and Gähwiler, B.H. (1991) Proc. R. Soc. London B **243**, 221–226
43.  Ben-Ari, Y., Krnjevic, K. and Reinhardt, W. (1979) Can. J. Physiol. Pharmacol. **57**, 1462–1466
44.  Davies, C.H., Davies, S.N. and Collingridge, G.L. (1990) J. Physiol. (London) **424**, 513–531
45.  Larson, J. and Lynch, G. (1986) Science **232**, 985–990
46.  Staubli, U. and Lynch, G. (1987) Brain Res. **435**, 227–234
47.  Davies, C.H., Starkey, S.J., Pozza, M.F. and Collingridge, G.L. (1991) Nature (London) **349**, 609–611
48.  Lynch, G., Larson, J., Kelso, S., Barrionuevo, G. and Schottler, F. (1983) Nature (London) **305**, 719–721
49.  Malenka, R.C., Kauer, J.A., Zucker, R.S. and Nicoll, R.A. (1988) Science **242**, 81–84
50.  Harvey, J. and Collingridge, G.L. (1992) Neurosci. Lett. **139**, 197–200
51.  Alford, S., Frenguelli, B.G., Schofield, J.G. and Collingridge, G.L. (1993) J. Physiol. (London) **469**, 693–716
52.  Frenguelli, B.G., Potier, B., Slater, N.T., Alford, S. and Collingridge, G.L. (1993) Neuropharmacology **32**, 1229–1237
53.  McGuinness, N., Anwyl, R. and Rowan, M. (1991) Eur. J. Pharmacol. **197**, 231–232
54.  Aniksztejn, L., Otani, S. and Ben-Ari, Y. (1992) Eur. J. Neurosci. **4**, 500–505
55.  Harvey, J. and Collingridge, G.L. (1993) Br. J. Pharmacol **109**, 1085–1090
56.  Chen, L. and Huang, L.-Y.M. (1992) Nature (London) **356**, 521–523
57.  Bortolotto, Z.A. and Collingridge, G.L. (1993) Neuropharmacology **32**, 1–9
58.  Lovinger, D.M., Wong, K.L., Murakami, K. and Routtenberg, A. (1987) Brain Res. **436**, 177–183
59.  Reymann, K.G. and Matthies, H. (1989) Neurosci. Lett. **98**, 166–171
60.  Izumi, Y., Clifford, D.B. and Zorumski, C.F. (1991) Neurosci. Lett. **122**, 187–190
61.  Sergueeva, O.A., Fedorov, N.B. and Reymann, K.G. (1993) Neuropharmacology **32**, 933–935
62.  Riedel, G. and Reymann, K. (1993) Neuropharmacology **32**, 929–931
63.  O'Connor, J., Rowan, M. and Anwyl, R. (1994) Nature (London) **367**, 557–559
64.  Alba, A., Chen, C., Herrup, K., Rosenmund, C., Stevens, C.F. and Tonegawa, S. (1994) Cell **79**, 365–375
65.  Manzoni, O.J., Weisskopf, M.G. and Nicoll, R.A. (1994) Eur. J. Neurosci. **6**, 1050–1054

66. Chinestra, P., Aniksztejn, L., Diabira, D. and Ben-Ari, Y. (1993) J. Neurophysiol. **70**, 2684–2689

67. Conquet, F., Bashir, Z.I., Davies, C.H., et al. (1994) Nature (London) **372**, 237–243

68. Bortolotto, Z.A., Bashir, Z.I., Davies, C.H. and Collingridge, G.L. (1994) Nature (London) **368**, 740–743

69. Bashir, Z.I., Jane, D.E., Sunter, D.C., Watkins, J.C. and Collingridge, G.L. (1993) Eur. J. Pharmacol. **239**, 265–266

70. Bashir, Z.I. and Collingridge, G.L. (1994) Exp. Brain Res. **100**, 437–443

71. Staubli, U. and Lynch, G. (1990) Brain Res. **513**, 113–118

72. Fujii, S., Saito, K., Miyakawa, H., Ito, K. and Kato, H. (1991) Brain Res. **555**, 112–122

73. Muller, D., Joly, M. and Lynch, G. (1988) Science **242**, 1694–1697

74. Davies, S.N., Lester, R.A.J., Reymann, K.G. and Collingridge, G.L. (1989) Nature (London) **338**, 500–503

75. Kauer, J.A., Malenka, R.C. and Nicoll, R.A. (1988) Neuron **1**, 911–917

76. Muller, D., Larson, J. and Lynch, G. (1989) Brain Res. **477**, 396–399

77. Asztely, F., Wigström, H. and Gustafsson, B. (1992) Eur. J. Neurosci. **4**, 681–690

78. Clark, K.A. and Collingridge, G.L. (1995) J. Physiol. (London) **482**, 39–52

79. Bashir, Z.I., Alford, S., Davies, S.N., Randall, A.D. and Collingridge, G.L. (1991) Nature **349**, 156–158

80. Berretta, N., Berton, F., Bianchi, R., Brunelli, M., Capogna, M. and Francesconi, W. (1991) Eur. J. Neurosci. **3**, 850–854

81. Tsien, R.W. and Malinow, R. (1990) Cold Spring Harbour Symp. Quant. Biol. **55**, 147–159

82. Xie, X., Berger, T.W. and Barrionuevo, G. (1992) J. Neurophysiol. **67**, 1009–1013

83. Foster, T.C. and McNaughton, B.L. (1991) Hippocampus **1**, 79–91

84. Malinow, R. and Tsien, R.W. (1990) Nature (London) **346**, 177–180

85. Bekkers, J.M. and Stevens, C.F.(1990) Nature (London) **346**, 724–729

86. Voronin, L.L., Kuhnt, U., Hess, G., Gusev, A.G. and Roschin, V. (1992) Exp. Brain Res. **89**, 248–264

87. Malinow, R. (1991) Science **252**, 722–724

88. Kullmann, D.M. and Nicoll, R.A. (1992) Nature (London) **357**, 240–244

89. Larkman, A., Hannay, T., Stratford, K. and Jack, J. (1992) Nature (London) **360**, 70–73

90. Hessler, N.A., Shirke, A.M. and Malinow, R. (1993) Nature (London) **366**, 569–572

91. Rosenmund, C., Clements, J.D. and Westbrook, G.L. (1993) Science **262**, 754–757

92. Manabe, T. and Nicoll, R.A. (1994) Science **265**, 1888–1892

93. Stevens, C.F. and Wang, Y. (1994) Nature (London) **371**, 704–707

94. Kullman, D.M. (1994) Neuron **12**, 1111–1120

95. Liao, D., Hessler, N.A. and Malinow, R. (1995) Nature (London) **375**, 400–404

96. Issac, J.T.R., Nicoll, R.A. and Malenka, R.C. (1995) Neuron **15**, 427–434

97. Zalutsky, R.A. and Nicoll, R.A. (1990) Science **249**, 1619–1624

98. Harris, E.W. and Cotman, C.W. (1986) Neurosci. Lett. **70**, 132–137

99. Jaffe, D. and Johnston, D. (1990) J. Neurophysiol. **64**, 948–960

100. Williams, S. and Johnston, D.(1989) Neuron **3**, 583–588

101. Derrick, B.E. and Martinez, Jr., J.L. (1994) Proc. Natl. Acad. Sci. U.S.A. **91**, 10290–10294

102. Gall, C.M., Brecha, H.G., Chang, K. and Karten, H (1981) J. Comp. Neurol. **198**, 335–350

103. Martin, M.R. (1983) Neuropeptides **4**, 45–50

104. Yamamoto, C., Sawada, S. and Ohno-Shosaku, T. (1993) Hippocampus **3**, 87–92

105. Breindl, A., Derrick, B.E., Rodriguez, S.B. and Martinez, Jr., J.L. (1994) Brain Res. Bull. **33**, 17–24

106. Weisskopf, M.G., Zalutsky, R.A. and Nicoll, R.A. (1993) Nature (London) **362**, 423–427

107. Henriksen, S.J., Wiesner, J.B. and Chouvet, G. (1988) NIDA Res. Monogr. **52**, 67–93

108. Derrick, B.E., Weinberger, S.B. and Martinez, Jr., J.L. (1991) Brain Res. Bull. **27**, 219–223

109. Castillo, P.E., Weisskopf, M.G. and Nicoll, R.A. (1994) Neuron **12**, 261–269

110. Staubli, U., Larson, J. and Lynch, G. (1990) Synapse **5**, 333–335

111. Xiang, Z., Greenwood, A.C., Kairiss, E.W. and Brown, T.H. (1994) J. Neurophysiol. **71**, 2552–2556

112. Yamamoto, C., Sawada, S. and Kamiya, H. (1992) Neurosci. Lett. **138**, 111–114

113. Weisskopf, M.G., Castillo, P.E., Zalutsky, R.A. and Nicoll, R.A. (1994) Science **265**, 1878–1882

114. Huang, Y.-H., Li, X.-C. and Kandel, E.R. (1994) Cell **79**, 69–79

# Index